Molecular Biology for Oncologists

SECOND EDITION

Edited by

J. R. Yarnold
Royal Marsden NHS Trust, UK

M. R. Stratton
Institute of Cancer Research, UK

and

T. J. McMillan
Lancaster University, UK

CHAPMAN & HALL

London · Weinheim · New York · Tokyo · Melbourne · Madras

Published by Chapman & Hall, 2–6 Boundary Row, London SE1 8HN, UK

Chapman & Hall, 2–6 Boundary Row, London SE1 8HN, UK

Chapman & Hall GmbH, Pappelallee 3, 69469 Weinheim, Germany

Chapman & Hall USA, 115 Fifth Avenue, New York, NY 10003, USA

Chapman & Hall Japan, ITP-Japan, Kyowa Building, 3F, 2-2-1 Hirakawacho, Chiyoda-ku, Tokyo 102, Japan

Chapman & Hall Australia, 102 Dodds Street, South Melbourne, Victoria 3205, Australia

Chapman & Hall India, R. Seshadri, 32 Second Main Road, CIT East, Madras 600 035, India

First edition 1993
Second edition 1996

Typeset in Times Roman 10/12pt by Columns Design Ltd, Reading
Printed in Great Britain at the University Press, Cambridge

ISBN 0 412 71270 9

A catalogue record for this book is available from the British Library

Library of Congress Catalog Card Number: 96-84903

Contents

Part Three

GENE STRUCTURE AND FUNCTION

Contributors

Elizabeth Anderson, Tumour Biochemistry Laboratory, Christie Hospital NHS Trust, Wilmslow Road, Manchester M20 4BX, UK

P. Rani Anné, Department of Radiation Oncology, Massachusetts General Hospital, Havard Medical School, Boston, MA 02114, USA

Robert Brown, CRC Department of Medical Oncology, CRC Beaston Laboratories, Garscube Estate, Glasgow, UK

Antony M. Carr, MCR Cell Mutation Unit, Sussex University, Falmer, Brighton BN1 9RR, UK

Kerry A. Chester, Department of Clinical Oncology, Royal Free Hospital School of Medicine, Rowland Hill Street, London NW3 2PF, UK

Mark Crompton, Section of Cell Biology and Experimental Pathology, Institute of Cancer Research, 15 Cotswold Road, Sutton, Surrey SM2 5NG, UK

Rachel C. Davies, MRC Human Genetics unit, Western General Hospital, Crewe Road, Edinburgh EH4 2XU, UK

Sally L. Davies, Imperial Cancer Research Fund Laboratories, Institute of Molecular Medicine, John Radcliffe Hospital, Oxford OX3 9DU, UK

Julian Downward, Imperial Cancer Research Fund, 44 Lincoln's Inn Fields, London WC2A 3PX, UK

Martin J. S. Dyer, Senior Lecturer/Honorary Consultant Physician, Academic Haematology and Cytogenetics, Haddow Laboratories, Institute of Cancer Research/Royal Marsden Hospital, Sutton, Surrey SM2 5NG, UK

Rosalind A. Eeles, Clinical Senior Lecturer and Honorary Consultant, Institute of Cancer Research and Royal Marsden NHS Trust, Downs Road, Sutton, Surrey SM2 5PT, UK

Michael J. Fry, Section of Cell Biology and Experimental Pathology, Institute of Cancer Research, Haddow Laboratories, 15 Cotswold Road, Sutton, Surrey SM2 5NG, UK

Anthony T. Gordon, Radiotherapy Research Unit, Institute of Cancer Research, Cotswold Road, Sutton, Surrey SM2 5NG, UK

Adrian L. Harris, Molecular Angiogenesis Group, Imperial Cancer Research Fund, Institute of Molecular Medicine, John Radcliffe Hospital, Headington, Oxford OX3 9DU, UK

Jonathan D. Harris, Gene Therapy and Neurogenetics Laboratory, Department of Surgery and Medicine, Yale University School of Medicine, TMP 522, 330 Cedar Street, PO Box 208039, New Haven, CT 06520-8039, USA

Ian R. Hart, Richard Dimbleby Department of Cancer Research, ICRF Laboratory, St Thomas' Hospital, Lambeth Palace Road, London SE1 7EH

Robert E. Hawkins, Centre for Protein Engineering and Department of Oncology, MRC Centre and Addenbrooke's Hospital, Hills Road, Cambridge CB2 2HQ, UK

Emma S. Hickman, ABL Basic Research Program, NCI-FCRDC, West 7th Street, Frederick, MD 2170, USA

John A. Hickman, CRC Molecular and Cellular Pharmacology Group, School of Biological Sciences, University of Manchester, Manchester, M13 9PT, UK

Ian D. Hickson, Imperial Cancer Research Fund Laboratories, Institute of Molecular Medicine, John Radcliffe Hospital, Headington, Oxford OX3 9DU, UK

Liza Ho, Section of Cell Biology and Experimental Pathology, Institute of Cancer Research, Haddow Laboratories, 15 Cotswold Road, Sutton, Surrey SM2 5NG, UK

Anthony Howell, CRC Department of Medical Oncology, Christie Hospital NHS Trust, Wilmslow Road, Manchester M20 4BX, UK

Dean A. Jackson, Sir William Dunn School of Pathology, University of Oxford, South Parks Road, Oxford OX1 3RE, UK

Rhys T. Jaggar, Molecular Angiogenesis Group, Imperial Cancer Research Fund, Institute of Molecular Medicine, John Radcliffe Hospital, Headington, Oxford OX3 9DU, UK

Lisa A. Kachnic, Department of Radiation Oncology, Havard Medical School, Massachusetts General Hospital, Boston, MA 02114, USA

Val Macaulay, MRC Clinical Scientist, Imperial Cancer Research Fund Laboratories, Institute of Molecular Medicine, John Radcliffe Hospital, Headington, Oxford OX3 9DU, UK

Trevor J. McMillan, Division of Biological Sciences, Institute of Environmental and Biological Sciences, Lancaster University, Lancaster LA1 4YQ, UK

Malcolm D. Mason, Department of Clinical Oncolgy, Velindre Hospital, Whitchurch, Cardiff CF4 7XL, UK

Philip Mitchell, Section of Cell Biology and Experimental Pathology, Institute of Cancer Research, 15 Cotswold Road, Belmont, Sutton, Surrey SM2 5NG, UK

Victoria Murday, Department of Clinical Genetics, St George's Hospital Medical School, Cranmer Terrace, London SW17 ORE, UK

Simon N. Powell, Department of Radiation Oncology, Cox 302, 100 Blossom Street, Massachusetts General Hospital, Boston, MA 02114, USA

Philippe J. Rocques, Section of Molecular Carcinogenesis, Institute of Cancer Research, 15 Cotswold Road, Belmont, Sutton, Surrey SM2 SNG, UK

Karol Sikora, Department of Clinical Oncology, Hammersmith Hospital, Du Cane Road, London W12 ONN, UK

Caroline J. Springer, CRC Centre for Cancer Therapeutics, Institute of Cancer Research, 15 Cotswold Road, Sutton, Surrey SM2 5NG, UK

Alasdair Stamps, Section of Pathology, Institute of Cancer Research, 15 Cotswold Road, Sutton, Surrey SM2 5NG, UK

Michael R. Stratton, Section of Molecular Carcinogenesis, Institute of Cancer Research, 15 Cotswold Road, Sutton, Surrey SM2 5NG, UK

Stanley Venitt, Section of Molecular Carcinogenesis, Institute of Cancer Research, 15 Cotswold Road, Sutton, Surrey SM2 5NG, UK

Karen H. Vousden, ABL Basic Research Program, NCI-FCRDC, West 7th Street, Frederick, MD 2170, USA

William Warren, Section of Molecular Carcinogenesis, Institute of Cancer Research, 15 Cotswold Road, Sutton, Surrey SM2 5NG, UK

Richard Wooster, The Sanger Centre, Hinxton Hall, Hinxton, Cambridge CB10 1RQ, UK

Andrew H. Wyllie, CRC Laboratories, Edinburgh University Medical School, Teviot Place, Edinburgh EH8 9AG, UK

John R. Yarnold, Academic Radiotherapy Unit, Royal Marsden NHS Trust, Downs Road, Sutton, Surrey, SM2 5PT, UK

Foreword

The pathophysiology, cell proliferation kinetics and metastasis of cancers and their responses to the currently employed therapies are increasingly being understood in terms of their genetic control mechanisms. Parallel genetic investigations are being aggressively pursued for normal tissue systems. The resultant greatly enhanced understanding is leading to clinically useful predictors of prognosis and the beginnings of new therapeutic strategies so as to augment the differential effect on tumour and normal tissues. This includes not only manoeuvres which deploy the currently available treatment methods with greater effect, but also the introduction into the clinic of conceptually different treatment strategies, with the potential of great positive impact.

Development in the field of basic genetics of cancer biology, cell cycle control, angiogenesis, immunology, etc. have been extraordinarily rapid with exceptional importance to medicine. This area has attracted some of the keenest intellects and greatest enthusiasm in biology and medicine. To illustrate the vigour and the growth rate of the research activity in this area, note that the numbers of journal articles published with the words gene and cancer (carcinoma, sarcoma, neoplasm, tumour) in the title or abstract for 1975 was 14 and for 1995 was 14 593. This number has increased dramatically over the past two decades. The entire field of oncology (clinical and laboratory) has, likewise, been growing at a remarkable pace. Namely, the numbers of articles published with the word cancer (carcinoma, sarcoma, neoplasm, tumour) in the title or abstract in 1975 and 1995 were 7245 and 20 623, respectively. There is hardly an issue of a clinical journal in the field of oncology which does not have at least one article devoted in part, as a minimum, to some aspect of the genetic factors of cancer. Further, our newspapers and major national magazines rarely have less than one major story per week on some aspect of molecular biology, largely related to oncology. The DNA molecule is appreciated in some detail by schoolchildren.

This volume provides the clinical oncologist with a clearly needed and comprehensive statement of the current status of research into the genetic basis of cancer and relevant normal tissues. Without knowledge of this subject, so well presented here, there is little prospect for the clinician to understand the developments in diagnostics, therapeutics, family counselling, epidemiology, etc. which are impinging on the clinical practice of oncology. Indeed, the standards of practice are evolving at a readily perceived rate due to these researchers and, with a virtual certainty, the style, substance and efficacy of practice will be changed even more dramatically over the next decade.

Herman D. Suit
Massachusetts General Hospital
Boston

Preface

Molecular Biology for Oncologists introduces clinicians to the extraordinary power, elegance and importance of molecular biology in the research and treatment of cancer. The volume explains how new techniques involving the manipulation of genetic material in cells and organisms are providing dramatic insights into how cell behaviour is subverted in common human cancers

The book also shows how techniques of cell and molecular biology are improving the understanding and effecfiveness of radiation and drug actions, as well as leading to the development of gene therapy. Ultimately, techniques of molecular biology will help to identify the specific genetic and environmental factors predisposing to cancer, offering the means of cancer prevention.

Molecular Biology for Oncologists is based on an annual three-day teaching course in London that shows us how quickly and keenly clinicians with no previous knowledge of the subject grasp its scope and opportunities. The book does not attempt a comprehensive coverage of its subject, but chooses topics of a particular interest and importance. These illustrate the applications of molecular biology to an understanding of conventional therapies as well as to the development of new strategies of prevention and cure.

J. R. Yarnold
M. R. Stratton
T. J. McMillan

PART ONE

Cancer Genetics

What are cancer genes and how do they upset cell behaviour?

1

John R. Yarnold

1.1 MUTATIONS IN CELLULAR GENES AND INFECTION BY ONCOGENIC VIRUSES UPSET CELL BEHAVIOUR IN CANCER

Normal homeostatic mechanisms that regulate many aspects of cell behaviour such as proliferation, motility and death are upset in cancer cells. The deregulation is caused by sporadic or inherited mutations that increase or decrease the expression of cellular genes, or by foreign genes introduced by viruses. The abnormal gene expression causes interference with a wide range of biochemical pathways that determine cellular responses. Because the fundamental defects arise in the genome, cancer is described as a genetic disease.

1.2 CANCER GENES WERE FIRST IDENTIFIED THROUGH THE ACTIONS OF ONCOGENIC VIRUSES IN ANIMALS

Oncogenic (cancer-causing) gene sequences can be carried into animal cells by RNA or DNA viruses. Each type of oncogenic virus introduces a gene sequence capable of transforming normal cells into malignant cells. Many RNA tumour viruses (retroviruses) have been identified, each associated with a distinct nucleotide sequence responsible for its oncogenic properties. In every case, the viral nucleotide sequence is a mutated fragment of a normal cellular gene accidentally removed from its natural animal host in the course of viral evolution. Each

Molecular Biology for Oncologists. Edited by J. R. Yarnold, M. R. Stratton and T. J. McMillan. Published in 1996 by Chapman & Hall, London. ISBN 0 412 71270 9

oncogenic retrovirus is associated with a specific animal host and a specific DNA sequence responsible for its carcinogenic effects. For example, the Rous sarcoma (src) virus infects only chickens. The cellular *src* gene (cellular-*src* or c-*src*) sequence in the chicken encodes a plasma membrane protein playing a regulatory role in cell adhesion and migration. The virus carries a mutated fragment (viral-*src* or v-*src*) of the chicken gene which is constitutively active (expressed regardless of its genetic environment) and responsible for its oncogenic effects.

Retroviruses are historically important but they are rare causes of cancer in humans. Oncogenic DNA viruses on the other hand are important causative agents in human cancer (see Table 1.1).

Table 1.1 Viruses associated with human cancers

Oncogenic viruses	Associated human cancers
RNA viruses	
Human T cell leukaemia virus type I (HTLV-I)	Adult T cell leukaemia/lymphoma
DNA viruses	
Human papilloma virus (HPV)	Cervix uteri
Hepatitus B virus (HBV)	Hepatocellular carcinoma
Epstein–Barr virus (EBV)	Burkitt's lymphoma

Epstein–Barr, hepatitis-B and papilloma viruses contribute to Burkitt's lymphoma, hepatocellular carcinoma and cancer of the cervix uteri respectively. These viruses introduce cellular genes that so far have few homologies (similar amino acid sequences) with known mammalian proteins.

1.3 MOST HUMAN CANCER GENES ARE GENERATED BY MUTATIONS IN CELLULAR GENES (ONCOGENES AND TUMOUR SUPPRESSOR GENES)

A cancer gene is any gene sequence contributing directly to neoplastic change. Most human cancer genes are generated by mutations in normal (wild-type) cellular genes. Mutations develop as a result of uncorrected errors in DNA replication, for example, or after exposure to physical and chemical carcinogens. Gene function is either activated or inactivated by these mutations. Normal genes whose cellular functions are activated or enhanced by mutation are called **proto-oncogenes** (Figure 1.1).

The mutant form of a proto-oncogene is called an **oncogene**. The protein product of an oncogene is called an **oncoprotein**. Some mutations that activate or enhance gene expression cause overproduction of an oncoprotein which is structurally identical to the wild-type protein. One example is a chromosome translocation in Burkitt's lymphoma which stimulates the overexpression of wild-type c-Myc protein in malignant B cells. Alternatively, activating mutations can lead to

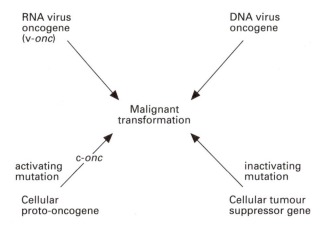

Figure 1.1 Schema describing classification of cancer genes. Cancer genes can be introduced into the cell by oncogenic viruses but most human cancers involve mutations in the cells' own genes. Genes whose cellular functions are activated by mutation are called proto-oncogenes. Genes whose cellular functions are inactivated by mutation are called tumour suppressor genes.

structurally altered proteins. For example, mutant Ras proteins have a single amino acid substitution that alters their biochemical properties. Mutations in proto-oncogenes are referred to as activating or gain-in-function mutations because they enhance, or confer new properties on, the cellular functions of the encoded oncoproteins.

Genes whose cellular functions are inactivated by mutation are called **tumour suppressor genes**. Examples include inactivation of the retinoblastoma and *p53* genes. These mutations are referred to as inactivating or loss-of-function mutations because they inactivate the cellular functions of the wild-type proteins. This removes a natural constraint on cell proliferation, cell adhesion or some other cellular function that controls cell behaviour. Several dozen human tumour suppressor genes have been isolated and partially characterized in terms of biochemical activities and cellular effects (Table 1.2).

1.4 ACTIVATED CANCER GENES ARE DOMINANT AT THE CELLULAR LEVEL

Cancer genes that are activated by mutation are referred to as **dominant** oncogenes. Borrowing a term from Mendelian genetics, dominant oncogenes exert their cellular effects despite the presence of normal gene product from the homologous allele. For example, mutation of one c-H-*ras* allele on chromosome 11 results in a single amino acid substitution which alters its biochemical properties. It exerts its effects despite the presence of wild-type (normal) Ras protein in the cell from

Table 1.2 Cloned human tumour suppressor genes

Gene	Main cancer type	Chromosome	Subcellular location	Cellular functions
RB1	Retinoblastoma	13q	Nucleus	Cell cycle control
NF1	Neurofibromatosis (von Recklinghausen)	17q	Cytoplasm	Growth signal transduction
P53	Many types	17p	Nucleus	Genome stability, apoptosis and cell cycle
NF2	Meningomas	22q	Cytoskeleton	Cytoskelton–matrix interactions
WT1	Wilms	11p	Nucleus	Cell proliferation and differentiation
DCC	Colon	18q	Plasma membrane	Cell–cell adhesion
APC	Colon (polyposis coli)	5q	Cytoplasm	Cell–matrix adhesion
VHL	Multiple primaries (von Hippel–Lindau)	3p	Unknown	Unknown
MTS1 (p16)	Melanoma	9p	Nucleus	Cell cycle control
MSH2	Colon	2p	Nucleus	DNA repair
MLH1	Colon	3p	Nucleus	DNA repair
BRCA1	Breast and ovary	17q	Unknown	Unknown

the homologous *ras* allele. The cancer genes carried into cells by RNA and DNA viruses also function as gain-in-function or dominant oncogenes at the cellular level.

1.5 INACTIVATED CANCER GENES (TUMOUR SUPPRESSOR GENES) ARE RECESSIVE AT THE CELLULAR LEVEL

Normal cellular genes whose protein products and cellular functions are inactivated by mutation during carcinogenesis are called tumour suppressor genes. Tumour suppressor genes are described as recessive because they exert their cellular effects only in the absence of normal protein. This usually means that both alleles must be inactivated by mutation for cell behaviour to be neoplastic. This contrasts with proto-oncogenes, where only one allele needs to be activated by mutation in order to exert its cellular effects (Table 1.3).

The retinoblastoma (*Rb*) gene was the first tumour suppressor gene to be isolated. As predicted from its recessive phenotype, insertion of the wild-type *Rb*

Table 1.3 Characteristics of proto-oncogenes and tumour suppressor genes

	Proto-oncogenes	Tumour suppressor genes
Number of alleles in normal somatic cells	2	2
Effect of mutations on cellular function of gene products	Enhanced	Reduced
Adjectives used to describe mutations	Activating Gain-in-function Dominant	Inactivating Loss-of-function Recessive

gene sequence into retinoblastoma cells (which lack functional Rb protein) suppresses malignant behaviour in culture and abolishes the ability of transplanted retinoblastoma cells to form tumours in animals.

1.6 MUTATIONS IN TUMOUR SUPPRESSOR GENES TRANSMITTED VIA THE GERMLINE CAUSE INHERITED CANCER PREDISPOSITIONS

Mutations occur in germ cells as well as somatic cells. Mutations in cancer genes can be transmitted via the germline from either parent to cause an inherited cancer predisposition. The transmission of cancer predisposition genes is nearly always confined to tumour suppressor genes. This is presumably because the mutant allele is functionally silent in the heterozygous state and does not disrupt embryogenesis. Every cell in the developing embryo is heterozygous for the inactivating mutation, and sooner or later a random mutation inactivates the other (wild-type) allele in at least one somatic cell. The complete loss of functional protein confers a selective growth advantage on the cell and its descendants.

Linkage analysis of genetic loci in large family pedigrees with multiple cancers has been exploited to localize and isolate genes responsible for hereditary cancer predispositions. Germline transmission of a mutation in one allele of the retinoblastoma (*Rb*) suppressor gene on chromosome 13 is a paradigm for hereditary predisposition. The disease is inherited as a Mendelian autosomal dominant trait which usually presents within the first few months of life. The dominant inheritance of the trait at the clinical level is nevertheless consistent with a recessive disorder at the cellular level (Figure 1.2).

In the hereditary form of retinoblastoma every fetal cell inherits one inactivated *Rb* allele from one or other parent. The defect is compatible with normal embryonic development because sufficient Rb protein is expressed from the homologous allele. The germline mutation is recessive at the cellular level because tumours form only when the other *Rb* allele is also inactivated. This produces a tumour only if it happens in a retinal cell at a very restricted stage during eye develop-

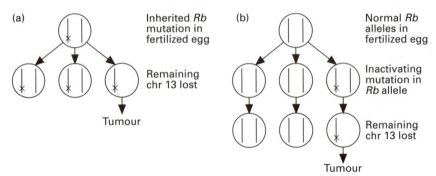

Figure 1.2 Hereditary and sporadic retinoblastoma (Rb). In the hereditary form **(a)**, every cell in the body, including the retinal cells, carries an inactivating mutation in one *Rb* allele. There is a very high probability of a somatic mutation inactivating the healthy *Rb* allele in at least one retinal cell during embryogenesis, leading to sustained proliferation and tumour formation. Several tumours may arise in the same patient and involvement of both eyes is common. In sporadic retinoblastoma **(b)**, both mutations are somatic. The probability of independent somatic mutations inactivating both *Rb* loci in more than one retinal cell in the same patient is remote, hence tumours are nearly always unilateral and single.

ment. Retinoblastoma is transmitted as an autosomal dominant trait because of the very high probability of a somatic mutation inactivating the homologous *Rb* locus in at least one of the 10^6 retinoblasts in each retina. Independent inactivating mutations of the homologous wild type *Rb* allele in multiple retinal cells cause multiple tumours. These are randomly distributed between the two retinas and explain the bilateral clinical presentation. By contrast, sporadic retinoblastoma arises from two independent somatic mutations which inactivate both *Rb* alleles in the same retinoblast. This occurs very rarely and explains why these tumours are unilateral and single.

It is unlikely that mutations inactivating both retinoblastoma alleles are sufficient to explain all manifestations of the disease. Although mutations at other genetic loci are almost certainly involved, mutations at the *Rb* locus are rate-limiting and determine the age at which the disease presents. The cancer predisposition is not confined to the retina. Rb protein is expressed in most cells throughout life and patients with hereditary retinoblastoma are susceptible to multiple tumour types, especially osteosarcoma. These usually occur later in life, possibly because several other genes have to undergo mutation as well.

The inheritance of a dominantly acting oncogene is usually lethal to the developing embryo. The only dominant oncogene known to be transmitted via the germline is the *ret* oncogene that is mutated in patients with multiple endocrine neoplasia. Ret protein has a tissue distribution confined to selected endocrine glands and is presumably not needed for normal embryogenesis.

1.7 MULTISTAGE CARCINOGENESIS HAS A GENETIC BASIS

Changes in epithelia ranging from atypical hyperplasia to carcinoma-*in-situ* are observed for many years before the appearance of some carcinomas, e.g. cervix uteri and colon (Chapter 12). Precancerous changes are consistent with the concept of multistage carcinogenesis, according to which multiple genetic events contribute to loss of normal regulatory controls and changes in cell behaviour. In the majority of patients, there is no inherited defect and somatic cell mutations are solely responsible for the development of disease. It takes years, possibly decades, for the required somatic genetic events to accumulate. The genetic abnormalities underlying multistage carcinogenesis are best understood so far in colon cancer (Table 1.4).

Table 1.4 Genetic abnormalities reported in human colon cancer (*FCC* = familial colon cancer (with polyposis coli); *DCC* = deleted in colon cancer; HNPCC = human non-polyposis colon cancer, in which several genes can be affected)

Gene	Chromosome	Tumours with mutations (%)	Category of mutation
c-K-*ras*	12	~ 50	Activating
Cyclins	Various	4	Activating
c-*erbB2*	17	2	Activating
c-*myc*	8	2	Activating
FCC	2	~ 15	Inactivating
APC	5	> 70	Inactivating
DCC	18	> 70	Inactivating
p53	17	> 70	Inactivating
HNPCC	Various	< 10	Inactivating

1.8 THE BIOLOGICAL FUNCTIONS OF DIFFERENT CANCER GENES ARE DIVERSE, AFFECTING GROWTH REGULATION, CELL DEATH, ADHESION AND MIGRATION

Many oncogenes and tumour suppressor genes work by interfering with regulatory processes responsible for maintaining an appropriate balance between cell production and cell loss (Table 1.5).

The cellular effects of activating mutations in c-*myc*, c-H-*ras* and c-*erbB2* involve the undermining of normal controls of cell proliferation. The cellular effects of the *bcl-2* gene overexpression in nodular B cell lymphoma is to inhibit programmed cell death (**apoptosis**). This accounts for the accumulation of long-lived postmitotic malignant B cells responsible for the polyadenopathy typical of this condition. The deregulation of cell differentiation is another process upset by

Table 1.5 Cell growth regulation disrupted in cancer

- Increased positive growth signals cause increased cell proliferation
- Reduced negative growth signals lift constraints on cell proliferation
- Proliferative capacity retained due to blocks in differentiation pathways
- Inhibition of programmed cell death (apoptosis) maintains cell viability

cancer genes. For example, inappropriate expression of c-Myc protein is thought to block normal lymphoid differentiation pathways in Burkitt's lymphoma and permits continued proliferation instead of the normal transition to a postmitotic cell phenotype.

Compared to deregulation of cell growth, less is understood about the cellular basis of invasion and metastasis. Cell–cell contacts and cell–matrix adhesion are important constraints on cell migration and circulation. The *DCC* gene (**d**eleted in **c**olon **c**ancer) is a tumour suppressor gene whose protein product helps to anchor colonic epithelial cells to the extracellular matrix. Loss of both *DCC* alleles predisposes to cell detachment from the matrix, a first step towards invasion and metastasis. Genes controlling cell–cell contacts, cell motility, access to the circulation and tumour angiogenesis are also potential candidates for deregulation by mutation.

1.9 THE PROTEIN PRODUCTS OF CANCER GENES ARE INVOLVED IN A WIDE RANGE OF BIOCHEMICAL PATHWAYS

The cellular processes upset in cancer involve deregulation of a wide range of biochemical pathways, many of which are still poorly characterized. The genes encoding proteins responsible for transmitting biochemical signals from plasma membrane to cytoplasm and nucleus are important targets of mutation in cancer (Chapters 5 and 6). For example, normal plasma membrane receptors transfer (transduce) extracellular growth signals across the cell membrane only when activated by specific extracellular growth factors. Mutations in growth factor receptor genes can result in receptor activation despite the absence of appropriate ligand (growth factor). This causes inappropriate stimulation of downstream proteins, called **signal transduction molecules**, that transduce growth signals from the cell surface to the nucleus. Ras proteins are examples of signal transduction molecules that are targets of mutations in cancer (Figure 1.3).

Within the nucleus, the expression of transcription factors is regulated by signal transduction pathways. Transcription factors are proteins that bind to DNA in a sequence-specific fashion and stimulate or repress gene expression. Transcription factors such as c-Fos, c-Jun and p53 can be targets for mutations, as well as the downstream genes they control, especially genes regulating DNA replication and cell division. An example is the *bcl-1* gene which is overexpressed in some non-

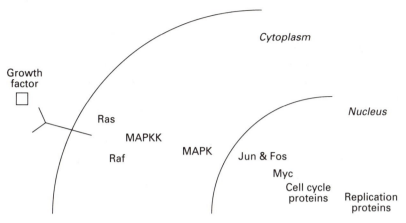

Figure 1.3 Schema of a growth signal transduction pathway. Growth signals are transferred or transduced from outside the cell to the nucleus by transient activation of biochemical pathways. Peptide growth factors transfer the signal inside the cell by activation of a plasma membrane receptor. Inside the cell membrane, the signal is passed as a wave of phosphorylation through a chain of second messenger molecules to transcription factors in the nucleus. Activated transcription factors (Jun, Fos and Myc) alter the transcription of genes controlling passage through the cell cycle which in turn regulate the expression of genes needed for DNA replication and cell division. MAPK stands for a protein called mitosis-activated protein kinase; it is phosphorylated by an upstream enzyme called MAPK kinase (MAPKK). Many such pathways across the cell exist and there is extensive crosstalk between them.

Hodgkin's lymphomas. It encodes cylin D1, one of the proteins regulating passage from G1 into S phase of the cell cycle.

The biochemistry of invasion and metastasis is less understood. However, the protein encoded by the *DCC* tumour suppressor gene has a similar amino acid sequence to neural cell adhesion molecule (N-CAM), a cell surface protein known to anchor cells to the extracellular matrix. Inactivation of both alleles in colonic cancer, with complete loss of functional protein at the cell surface, is thought to contribute to acquisition of invasive and metastatic properties in a proportion of tumours.

Phosphorylation is a recurring biochemical theme in cell metabolism, including the biochemical pathways disrupted in cancer. The transfer of a high-energy phosphate group from ATP or GTP to a protein is catalysed by enzymes called **kinases**. Phosphorylation acts as an efficient 'molecular switch', either activating or inactivating a specialized biochemical function. Phosphorylation of the hydroxyl group on tyrosine, serine and threonine is performed by kinases that are highly specific for their protein substrates. Phosphorylation regulates the biochemical activity of a molecule by altering its shape, or conformation, in a region critical for a specialized function, often an enzymic function. Many cancer genes encode proteins with enzymic activities involved in the phosphorylation of tyro-

sine residues. For example, the specific binding of an extracellular growth factor
to the plasma membrane receptor c-ErbB2, activates a tyrosine kinase domain on
the intracellular portion of the receptor molecule. The activated kinase domain
phosphorylates second messenger molecules acting 'downstream' in the signal
pathway. The messenger molecules acquire kinase activity in their own right as a
result of this phosphorylation. One activated molecule can typically phosphorylate several thousand substrate molecules before it is inactivated (within seconds).
Most kinases are associated with specific phosphatases, enzymes that regulate the
rate of dephosphorylation. A typical example of phosphorylation and dephosphorylation affecting biochemical activity is seen in the Rb protein (Figure 1.4).

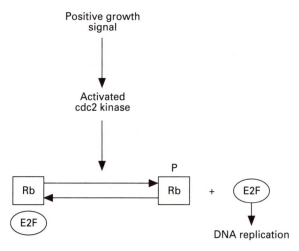

Figure 1.4 Phosphorylation of normal retinoblastoma (Rb) protein causes it to release a
molecule required for entry of the cell into S phase. Growth signal transduction pathways
converge on molecules that regulate passage through the cell cycle, including a protein
called cdc2 kinase (cdc stands for cell division cycle). Rb protein is one of the most
important substrates of cdc2 kinase activity. Phosphorylation causes Rb protein to release
E2F, a transcription factor needed for passage into S phase and the transcription of genes
required for DNA replication.

1.10 DIFFERENT CANCER GENES ARE FOUND IN THE SAME
   BIOCHEMICAL PATHWAY

The protein products of different cancer genes can act in the same biochemical
pathway, although not necessarily in the same individual. For example, the product of the c-H-*ras* gene on chromosome 11 has a molecular weight of 21 000 and
is called p21*ras* (Chapters 5 and 19). It is a small GTP-binding protein located on
the inner surface of the plasma membrane which responds to activation of plasma
membrane receptors, including the T cell receptor and epidermal growth factor

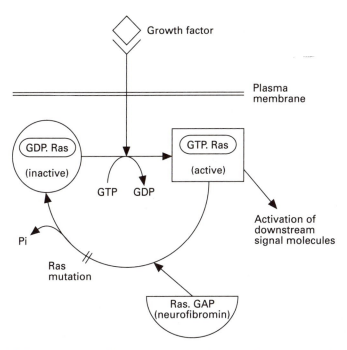

Figure 1.5 Schema of normal Ras function and its disruption by mutations in Ras and GAP. Ras is a small GDP/GTP binding protein that acts as a molecular switch in many biochemical pathways, including those transmitting growth signals from the plasma membrane to the nucleus. A growth signal arriving at the cell membrane activates specific plasma membrane receptors which indirectly result in Ras binding GTP. GTP-bound Ras activates specific downstream proteins in the signal pathways. To turn off the stimulation, GTP-bound Ras is inactivated by an intrinsic GTPase activity which catalyses the conversion of GTP to GDP. A GAP protein (**GTP**ase **a**ctivating **p**rotein) is an important cofactor for Ras GTPase activity. Selected point mutations in Ras are oncogenic because they inactivate the GTPase function and lock the molecule in the active Ras configuration. Inactivation of GAP function by homozygous mutations in both alleles has the same effect by a different means. Neurofibromin is a GAP protein that acts as a tumour suppressor gene in von Recklinghausen's disease.

receptor. In many cancers, p21^{ras} protein is activated by a point mutation that keeps it in the 'switched on' position. In this activated state, it transmits a continuous growth signal to the nucleus in the absence of growth factor stimulation (Figure 1.5).

The tumour suppressor genes responsible for von Recklinghausen's disease codes for a protein called neurofibromin located close to the p21^{ras} on the inner plasma membrane. Neurofibromin is a member of a family of proteins called **GTPase activating protein (GAP)**. Its normal biochemical activity is to activate a p21^{ras} phosphatase activity shortly after p21^{ras} activation by phosphorylation. In the Schwann cells of patients with von Recklinghausen's disease, an inherited deficiency of neurofibromin permits wild-type p21^{ras} to remain in the activated phosphorylated state and transmit a continuous growth signal to the nucleus (Figure 1.5). This contributes to the development of neurofibromas and other benign and malignant tumours in affected individuals.

An equally spectacular example of the close biochemical associations between different oncogenes includes the retinoblastoma (Rb) protein, p53 and human papilloma virus (Chapters 7, 8 and 9). Wild-type Rb and p53 are important nuclear proteins that control the passage of cells through the cell cycle. The wild-type forms of Rb and p53 bind and inactivate specific proteins needed for cell cycle progression. The oncoproteins produced by human papilloma viruses overcome this constraint on cell cycle progression by binding Rb and p53. The binding of viral oncoproteins to Rb and p53 proteins causes the latter to release the specific proteins needed to promote cell cycle progression in infected cervical squamous epithelium (Figure 1.6).

1.11 THE BIOCHEMICAL EFFECT OF SOME CANCER GENES IS TO INCREASE THE MUTATION RATE IN CELLULAR DNA

Tumour suppressor genes have been identified that predispose to four types of familial human non-polyposis colon cancer (HNPCC). The four genes encode distinct proteins that normally act cooperatively in recognizing base-pairing errors

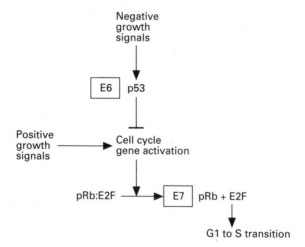

Figure 1.6 Human papilloma virus oncogenes bind and inactivate normal cellular Rb and p53 proteins. p53 and Rb act as constraints on normal cell cycle progression and inactivation of either protein by homozygous gene mutations contributes to carcinogenesis. In cancers of the uterine cervix, the *p53* and *Rb* genes usually remain unaffected but the normal protein is functionally inactivated by the binding of viral oncoprotein. E6 and E7 are the two commonest human papilloma virus oncoproteins responsible for these effects in human cancer of the uterine cervix.

(mismatches) that arise during DNA replication (Chapter 13). In affected family members, random base-pair mismatches accumulate in somatic cells during successive rounds of DNA replication and cell division. Most have no consequences for the cell but some activate or inactivate cancer genes in colonic mucosa. The cellular phenotype conferred by this genetic defect is referred to as a 'mutator phenotype'. One of the proteins particularly susceptible to inactivation as a result of this defect appears to be the TGF-ß receptor, a plasma membrane protein that normally transmits growth inhibitory signals in colonic epithelial cells.

1.12 MUTATIONS IN CANCER GENES CAUSE MULTIPLE INDIRECT (EPIGENETIC) EFFECTS VIA DOWNSTREAM BIOCHEMICAL PATHWAYS

It cannot be assumed that over- or underexpression of a protein in a neoplasm is the direct result of mutation in the gene encoding that protein. Activation of a proto-oncogene or inactivation of a tumour supressor gene indirectly alters the expression of multiple proteins in downstream biochemical pathways. Observed changes in protein expression are therefore usually remote consequences of mutation. A classic example is the variable oestrogen receptor expression found in breast cancers. Oestrogen receptor negative tumours are not caused by inactivating mutations in both alleles of the oestrogen receptor gene, nor are high levels caused by gene amplification or some other activating mutation. Mutations have been reported in the ligand-binding domain which result in constitutive activation of the receptor in the absence of oestrogen, but these appear to be uncommon.

1.13 FURTHER READING

Bishop, J. M. (1987) The molecular genetics of cancer. *Science*, **235**, 305–311.

Kerr, R. A. (1994) Oncogenes reach a milestone. *Science*, **266**, 1942–1943.

Knudson, A. G. (1985) Hereditary cancer, oncogenes, and antioncogenes. *Cancer Res.*, **45**, 1437–1443.

Nowell, P. C. (1986) Mechanisms of tumour progression. *Cancer Res.*, **46**, 2203–2207.

Skuse, G. R. and Ludlow, J. W. (1995) Tumour suppressor genes in disease and therapy. *Lancet*, **345**, 902–906.

2 | **Mechanisms of activation and inactivation of dominant oncogenes and tumour suppressor genes**

Michael R. Stratton

2.1 INTRODUCTION

In the course of a lifetime a variety of mutations accumulate in cells throughout the body. These arise through exposure to chemicals, radiation, viruses and as a consequence of mistakes made by the DNA replicative machinery. Most mutations make no difference to the functioning of the individual cell in which they occur. A few may alter its viability so that the cell dies. In either case there are no noticeable ill effects on the well-being of the body as a whole.

Occasionally, however, a single cell suffers a series of mutations which, when converted into abnormal cellular machinery, cause the cell and its descendants to proliferate in the unrestrained and uncoordinated fashion characteristic of cancer cells. The subject of this chapter is the nature of the mutations, the patterns in which they occur and some of the functional consequences that they entail.

2.2 CATEGORIES OF MUTATION INVOLVED IN HUMAN TUMOURS

The following types of mutation are involved in the activation or inactivation of oncogenes in human tumours:

- point mutation;

Molecular Biology for Oncologists. Edited by J. R. Yarnold, M. R. Stratton and T. J. McMillan. Published in 1996 by Chapman & Hall, London. ISBN 0 412 71270 9

- translocation (gene rearrangement);
- gene amplification;
- deletion.

Point mutation

The term 'point mutation' describes the substitution of one base pair of a DNA sequence by another, for example, substitution of a G:C base pair by an A:T. In the protein this single base pair substitution may have a number of effects depending upon its position:

- The mutation may result in the substitution of one amino acid for another (a **missense** mutation), resulting in a protein identical to the wild type except for the single amino acid change. The functional consequences of this change will depend on the nature of the amino acid substitution and the biological activity of the altered part of the protein.
- If the mutated sequence encodes a signal for the termination of translation (a **stop codon**) then the protein will be prematurely terminated at this position. As a consequence, a substantial part of the protein may be omitted, rendering it non-functional.
- During the production of mRNA two exons of coding sequence are joined together and the intron between them is removed. The process is known as **splicing** and takes place at certain consensus sequences near intron–exon junctions. Should a mutation arise in one of these sequences the correct splicing events may not take place and a whole exon may be missed out of the mRNA. When translated into protein the amino acids encoded by this exon will be absent and a frame shift may be introduced, altering substantially the predicted protein sequence downstream of the mutation.

Translocation

In a translocation part of one chromosome is joined to another. The outcome is a hybrid chromosome that may be detectable by karyotypic (cytogenetic) analysis of tumour cells. At the molecular level, a gene located at or near the breakpoint in one of the two chromosomes is fused to sequences from the other chromosome. This rearrangement of DNA sequences may generate a structurally altered version of the gene and its protein or may place it under new transcriptional control.

Gene amplification

The diploid genome of each eukaryotic human cell normally carries two copies of each gene, one originating from each parent (genes on X and Y chromosomes are exceptions). Under certain circumstances one copy may be multiplied up to several thousand-fold, a phenomenon known as gene amplification. Amplified genes may form two types of microscopically abnormal chromosomal configuration known as **double minutes (DM)** and **homogeneously staining regions (HSRs)**. DMs are tiny, paired, extrachromosomal chromatin bodies that segregate randomly during mitosis and are not linked to a centromere. HSRs are expanded chromosomal regions which do not exhibit the usual banding pattern of normal chromosomes but which, being linked to a centromere, segregate in the normal way. Gene amplification is believed to contribute to oncogenesis by increasing the levels of mRNA that are transcribed from the gene, and as a consequence increasing the levels of protein that it encodes.

Increased or decreased expression of certain proteins without amplification or any other form of mutation of the DNA encoding them is a common feature of many tumours. In some cases this may have a role in the generation of the neoplastic phenotype. In others, however, changes in expression are more likely to be a consequence of neoplastic change rather than a cause.

Deletion

A wide range of DNA deletions occurs in tumour cells. At one extreme, a single base pair may be removed. Larger deletions may encompass part or all of a gene. Finally, a deletion may be large enough to be visible under the microscope by karyotype analysis or may remove a whole chromosome.

While a large deletion will obviously remove from a cell and hence inactivate all the genes that are included within it, smaller, intragenic deletions may have a number of effects. If the DNA fragment that is deleted results in removal from the mRNA of a segment that is a multiple of three base pairs, then only the amino acid sequence encoded by that segment will be absent from the protein. The consequences of this structural abnormality will depend upon the functional properties of this segment in the normal protein.

If however, the deleted segment in the mRNA is not a multiple of three base pairs, then a frame shift will be introduced. The likely consequence is that the amino acid sequence of the protein downstream from the mutation will bear little resemblance to the normal and will often be prematurely terminated because of the presence of a stop codon. The effects of insertions into DNA are subject to the same considerations as deletions but tend to be less common as mutational events in cancer genes.

2.3 CATEGORIES OF MUTATION IN DOMINANTLY ACTING ONCOGENES AND TUMOUR SUPPRESSOR GENES

Much of the repertoire of genetic abnormalities reviewed above is used in both the activation of dominantly acting oncogenes and in the inactivation of tumour suppressor genes. However, because of the different outcomes as far as gene function is concerned, the spectrum and pattern of changes differs markedly between the two classes of gene. These differences are emphasized in the account below.

2.4 PATTERNS OF MUTATION IN DOMINANTLY ACTING ONCOGENES AND THEIR FUNCTIONAL CONSEQUENCES

The proteins encoded by genes of the *ras* family are activated by missense point mutations at only three amino acids

Genes of the *ras* family, H-, K- and N-*ras*, provide the most celebrated examples of oncogene activation by point mutation in human tumours (Chapter 6). Ras proteins are composed of 188 or 189 amino acid residues. Normally, they undergo a cycle of activation and inactivation associated with the hydrolysis of GTP to GDP. Mutated Ras proteins in human tumours, however, are fixed in the activated conformation. The alterations in DNA that result in this change in biochemical activity are almost exclusively point mutations. All are missense and hence allow translation of a full length Ras protein. Of particular importance, however, is the restriction of the mutations to certain sites within the gene, namely codons 12, 13 and 61 (that encode amino acids 12,13 and 61 of the Ras protein). Mutations at other sites are not found in human tumours, nor do most of them result in oncogenic activation in experimental models. Thus it appears that the biological characteristics of Ras proteins place considerable constraints upon the type and location of activating point mutations. Moreover, these structural and biochemical alterations are apparently sufficient for an activated Ras protein to contribute to oncogenesis. Changes in the level of expression of the protein are unnecessary (although occasionally present) and other mutational mechanisms such as translocation or gene amplification are rare.

Activation by translocation

Cytogenetic studies have indicated that chromosomal abnormalities are common in human tumour cells. Closer inspection has revealed that some of these abnormalities are consistently associated with a particular tumour type. For example, the majority of cases of chronic myelogenous leukaemia (CML) carry a translocation between particular regions of chromosomes 9 and 22 termed the Philadelphia translocation (Chapter 4). Similarly, translocations involving chromosomes 8 and 14 are found in Burkitt's lymphoma and between chromosomes 14 and 18 in other B-cell lymphomas. Chromosomal translocations have also been reported in solid tumours, for example, the t(11:22) in Ewing's sarcoma and the t(X:18) in synovial sarcoma. The role of the *bcl-2* gene in the t(14:18) of B-cell lymphoma is described in another chapter. Here the t(9:22) of CML and the t(8:14) of Burkitt's lymphoma will be used to illustrate the ways in which chromosomal translocations activate genes in oncogenesis.

The translocation between chromosomes 9 and 22 in chronic myelogenous leukaemia generates a hybrid protein encoded by sequences from both chromosomes

Between 90% and 95% of CML carries the translocation between chromosomes 9 and 22 termed the Philadelphia translocation. A clue to the important molecular events reflected by this translocation was provided by the localization of the human homologue of the v-*abl* gene to precisely the region in which the break-point was located on chromosome 9 (v-*abl* is the oncogene carried by the Abelson murine leukaemia virus). It was subsequently demonstrated that in the t(9:22) translocation one allele of the c-*abl* gene is split and joined to a gene on chromo-some 22 known as *bcr* (for **b**reakpoint **c**luster **r**egion). The consequence of the rearrangement is that the *abl* gene loses its most 5' exon and is joined to an allele of the *bcr* gene which has lost its 3' region (Figure 2.1).

When the region is transcribed, the resulting mRNA is a hybrid, with the 5' region composed of *bcr* and the 3' region of *abl* sequences. Similarly, the protein is a hybrid of Bcr amino acid sequence at the amino terminus and Abl amino acid sequence at the carboxy terminus. A biochemical feature of the Abl protein believed to be of importance both in its normal functions and in oncogenesis is its ability to phosphorylate other proteins on tyrosine residues. The Bcr–Abl fusion protein has an elevated tyrosine kinase activity and this may contribute to its onco-genic action.

The breakpoints underlying the Philadelphia translocation are usually within introns. In the *abl* gene the breakpoint separates the 3' part of the gene carrying exons 2–11 from the 5' part carrying exons 1a or 1b. However, in the *bcr* gene there is more than one option, with the breakpoint located in different introns in different tumours. Consequently several *bcr–abl* fusion mRNAs and proteins have been reported in which the 3' region, composed of *abl* sequences, is constant and the 5' region is constituted by different *bcr* sequences. Notably, one set of break-

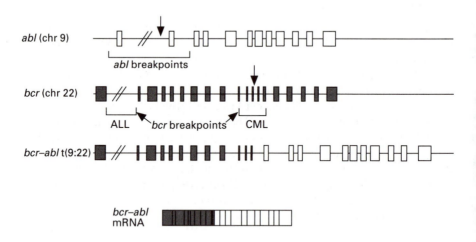

Figure 2.1 Translocation of the *abl* gene on chromosome 9 to the *bcr* gene on chromo-some 22. This results in the formation of a fusion gene that encodes a hybrid mRNA and protein. The *abl* exons are represented as pale boxes and the bcr exons as dark boxes. The usual sites of the breakpoints in *abl* and in *bcr* (both in CML and ALL) are indicated as are the locations of the breakpoints in the particular case illustrated (vertical arrows).

points is associated predominantly with the t(9;22) in CML and generates a 210 kDa protein, while the more breakpoints in a more 5' intron are often associated with the t(9;22) in acute lymphoblastic leukaemia.

The translocation between chromosomes 8 and 14 in Burkitt's lymphoma results in deregulated expression of a structurally normal c-*myc* gene

In 75% of cases of Burkitt's lymphoma the distal end of chromosome 8 is translocated to the long arm of chromosome 14q32. In the remaining 25% a cytogenetically indistinguishable segment of chromosome 8 translocates either to the long arms of chromosomes 2 or 22. Thus translocation of the segment 8q24-ter is a feature common to most and probably all Burkitt's lymphomas. The loci to which this segment is transferred are the positions of the immunoglobulin heavy (chromosome 14), lambda (chromosome 22) and kappa (chromosome 2) chains.

The c-*myc* gene is located at the common Burkitt's lymphoma breakpoint at chromosome 8q24 (c-*myc* is the human homologue of an oncogene carried by an acutely transforming retrovirus known as Avian **my**elo**c**ytomatosis virus). Although some of the Burkitt's lymphoma breakpoints interrupt the c-*myc* gene itself, this is usually in untranslated regions and rarely in the coding sequence. Indeed many of the breakpoints are not within the c-*myc* gene at all and some are many thousands of kilobases away from it. Irrespective of these differences, the regulation of expression of the translocated c-*myc* allele is abnormal and is influenced by factors controlling the expression of the immunoglobulin gene to which it has been joined. Thus in contrast to the situation in chronic myelogenous leukaemia with Bcr–Abl, a fusion protein is not produced by the Burkitt's lymphoma translocations, but instead the deregulation of a structurally normal c-Myc protein occurs.

Activation by gene amplification

Gene amplification contributes to the transformed phenotype by increasing the level of mRNA and hence protein that is expressed from a proto-oncogene. However, expression of the amplified allele is sometimes increased out of proportion to the increase in gene copy number, indicating that abnormalities of transcriptional regulation may also be operative in amplified genes. Moreover, the amplicon, the segment of DNA that is amplified, is often large, extending over several hundred kilobases or even megabases. The amplified fragment therefore usually stretches well beyond the boundaries of the proto-oncogene and encompasses several other genes.

Many types of oncogene are activated by gene amplification. These include the genes for growth factor receptors (for example epidermal growth factor receptor (*EGFR*) in malignant gliomas and *erbB2* in breast cancer), and nuclear oncogenes such as N-*myc* in neuroblastoma. Gene amplification is usually a late and hence progressive step in oncogenesis rather than an initiating or early event. Moreover, amplification can sometimes be correlated with poor outcome and late stage, for example, amplification of the N-*myc* gene in neuroblastoma is now recognized as an independent clinical indicator of outcome.

Activation by intragenic deletions

While there can be no transcript or protein from a gene that has been completely deleted, small intragenic deletions may result in activation of a proto-oncogene protein. In principle this could occur through removal of a domain of the protein that normally mediates inhibition of the protein's activity, or by inducing conformational changes that mimic those normally brought about by stimulatory signals.

The clearest example of this phenomenon in human tumours that has been reported is in the epidermal growth factor receptor gene (*EGFR*) in brain tumours. Amplification of *EGFR* is a common late event in the development of glioblastoma multiforme. It transpires however, that in a proportion of cases, not only is the gene amplified, but an in-frame deletion is present within the amplified copy. The deletion results in the removal of exons 2–7, which encode part of the extracellular domain. As a result a truncated EGFR protein is produced. Studies of the functional consequences of this change are in progress, but it is presumed that it contributes further to the constitutive activation of the protein.

2.5 INACTIVATION OF TUMOUR SUPPRESSOR GENES (RECESSIVE ONCOGENES)

The mutations described above that result in activation of dominant oncogenes are usually found in only one of the two alleles of the gene that are present in diploid human cells. By contrast, inactivation of a tumour suppressor protein usually requires mutation of both alleles of the gene that encodes it. Therefore two independent mutations or 'hits' are required.

There are differences in the patterns of mutation between dominantly acting genes and recessive (tumour suppressor) genes. In addition, there are differences between the first and the second mutations that result in the inactivation of tumour suppressor genes. These are discussed below.

The first inactivating event in a tumour suppressor gene is usually a 'small mutation'

The mutation or hit that inactivates the first allele of a tumour suppressor gene is usually confined to the gene itself or involves the gene and chromosomal DNA that is immediately adjacent to it. In contrast to the dominantly acting genes that are mutated in a consistent manner either by point mutation (*ras*) or by translocation (*abl*) or by gene amplification (N-*myc*), the types of alterations that inactivate a particular tumour suppressor such as the retinoblastoma gene are diverse and include point mutations, deletions and rearrangements. Moreover, in contrast to the point mutations, deletions and translocations of dominantly acting oncogenes, the mutations in tumour suppressor genes are much less constrained both in their type and position within the gene. For example, the point mutations that inactivate the retinoblastoma or *p53* genes are located at a large number of positions within the gene in contrast to the three codons that are point mutated in activated *ras* genes. Moreover, although they may be missense, they can also be nonsense (encoding translational stop codons) or affect splice sites. Similarly, deletions may affect part or all of the gene and may be in or out of frame. Translocations may occur at virtually any site that will interrupt the production of an intact mRNA.

The reasons for this difference in pattern are not difficult to appreciate but may be elucidated by considering the parallel of a car travelling along a road. 'Activation' of the car (as in activation of a dominant oncogene) so that it fails to respond when instructed to stop, requires tampering with one particular part of the mechanism, the brake system. On the other hand 'inactivation' of the car (as in inactivation of a tumour suppressor gene) so that it is stopped dead can be the result of a variety of malfunctions from subtle electrical defects to glue in the ignition, a puncture or an axe being driven into the engine.

The pattern of inactivating mutations differs in different tumour suppressor genes

While the overall pattern described above is applicable to most tumour suppressor genes, there are differences between tumour suppressor genes in the predominance of types of mutation. For example, the *APC* (**a**denomatous **p**olyposis **c**oli) gene, which is the responsible for the syndrome of familial polyposis coli and is mutated in a large proportion of colonic tumours, is subject to a preponderance of nonsense mutations such as stop codons, or small deletions and insertions which introduce frameshifts. Translation of the APC protein is usually prematurely terminated and hence the protein is inactivated.

By contrast, in the *p53* gene there is a much higher proportion of missense mutations that result in a full length protein with just a single amino acid substitution. The reasons for this contrasting pattern are probably complicated. It seems likely, however, that some p53 proteins with missense mutations can interact with and functionally inactivate normal p53 protein in the cell without the need for a mutation at the DNA level in the remaining allele (Chapter 9). In order to do this, however, the carboxy terminus of the protein needs to remain intact, which will not be the case if there is a stop codon upstream. Conversely, most missense mutations in *APC* may not be sufficient to inactivate the function of the protein encoded by the mutated allele and hence there is selection for the major disruptions of protein structure introduced by stop codons and frameshifts.

The second inactivating event in a tumour suppressor gene is usually a 'large' mutation

The mutation that inactivates the second allele of a tumour suppressor gene is in only a minority of cases similar in character to the first mutation, i.e. confined to the gene and neighbouring chromosomal DNA. More commonly it involves loss of a large part and even all of the chromosome upon which the second allele of the tumour suppressor gene is situated. The reason for this difference is probably that the second hit often arises due to errors of mitosis. Such errors, which include non–disjunction (incorrect separation of chromatids during mitosis) or mitotic recombination (exchange of segments of DNA by homologous chromosomes) usually involve such large stretches of DNA up to and including a whole chromosome.

This phenomenon has been of considerable importance in the detection and localization of tumour suppressor genes. By using polymorphic probes (Chapter 3) it is possible to visualize on Southern blots both parental copies of a particular region independently. If one copy has been lost during oncogenesis this will manifest itself as loss of heterozygosity (reduction from two bands to one) in the tumour compared to the germline DNA. Because the second hit in a tumour suppressor gene is large, the probe can be located several millions of base pairs away from the gene itself and still detect its presence by loss of heterozygosity. By performing studies in which a set of tumours are examined using at least one polymorphic probe on each chromosomal arm (termed an **allelotype**) it is possible to obtain an approximate picture of the number and genetic locations of the tumour suppressor genes involved in the development of a tumour class.

2.6 MUTATIONS IN ONCOGENES MAY ALSO REFLECT THE MUTAGEN THAT INDUCED THEM

The emphasis so far has been on the way constraints introduced by the modes of function of oncogenic proteins determine the type and pattern of mutations that result in oncogenic change. However, superimposed upon these influences are the effects of involvement of certain types of mutagen. Evidence for this notion in human tumours has emerged through study of mutations in the *p53* gene in many types of cancer. Together these investigations suggest that in many tumours (for example colon cancer) the predominant mutation type is attributable to errors introduced by the DNA replicative machinery rather than to exogenous carcinogenic influences. This type of mutation is a G:C to A:T transition mutation that characteristically targets CpG dinucleotides (where a C is just 5' (or upstream) to a G). The proposed reason for this change is as follows. The C in CpG dinucleotides can become methylated. If it is subsequently deaminated, DNA polymerases may encounter problems distinguishing between the deaminated methylated C and a T residue. The polymerase may then insert an A opposite the C rather than the correct G resulting in a G:C to A:T transition mutation. (A **transition mutation** is one in which a purine is replaced by a purine or a pyrimidine by a pyrimidine; a **transversion mutation** is one in which a purine is replaced by a pyrimidine or a pyrimidine by a purine.)

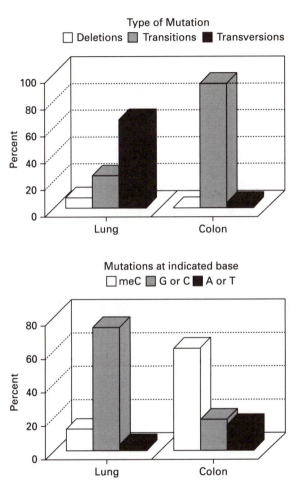

Figure 2.2 Exposure to certain types of carcinogen can affect the type and location of a point mutation. In the upper panel the types of mutation in lung and colon cancer are compared. In colon cancer most of the changes are transition mutations while, by contrast, in lung cancer most are transversions. In the lower panel the sites of the mutations in these tumours are compared. In colon cancer most mutations are at CpG dinucleotides at which the C may become methylated while in lung cancer most are not. The most likely interpretation of these differences is that the mutations in colon cancer are endogenous and caused by errors made during DNA replication due to deamination of methyl cytosine. In contrast the mutations in lung cancer may be due to the particular pattern of DNA binding by mutagenic carcinogens in tobacco smoke.

By contrast, a number of tumour types are found to have a different spectrum of mutations, which in some cases can be related to a known mutagen. For example, a proportion of squamous carcinomas of the skin show a particularly characteristic CpC to TpT double transition mutation which is very infrequent in visceral neoplasms. This is highly reminiscent of the type of mutation induced by UV light, a known risk factor for skin cancer. Similarly, in smoking induced lung cancers the predominant mutation type is a G:C to T:A transversion, consistent with the type of mutation induced by mutagens in tobacco smoke (Figure 2.2). Moreover, in the case of hepatoma, it is believed that exposure to aflatoxin results not just in a particular type of base change, but also in clustering at a particular codon of p53 (codon 249) around which the DNA sequence may preferentially encourage carcinogen binding.

2.7 FURTHER READING

Bos, J. L. (1988) The *ras* gene family and human carcinogenesis. *Mutation Res.*, **195**, 255–271.

Hollstein, M., Sidransky, D., Vogelstein, B. and Harris, C. C. (1991) p53 mutations in human cancers. *Science*, **253**, 49–53.

Korsmeyer, S. (1992) Chromosomal translocations in lymphoid malignancies reveal novel protooncogenes. *Ann. Rev. Immunol.*, **10**, 785–807.

Powell, S., Zilz, N., Beazer-Barclay, Y. *et al.* (1992) APC mutations occur early during colorectal tumorigenesis. *Nature*, **359**, 235–237.

Sugawa, N., Ekstrand, A. J., James, C. D. and Collins, V. P. (1990) Identical splicing of aberrant epidermal growth factor receptor transcripts from amplified rearranged genes in human glioblastomas. *Proc. Nat. Acad. Sci. USA*, **87**, 8602–8606.

Hereditary predisposition and gene discovery | 3

Rosalind A. Eeles

3.1 WHAT IS HEREDITARY PREDISPOSITION TO CANCER?

Cancer is described as a genetic disease at the cellular level because alterations in genes result in cancer formation. In some individuals, the first genetic alteration is inherited and accounts for an inherited predisposition to cancer. This predisposition does not necessarily make cancer inevitable, but it does greatly increase the risk. The risk does not rise to 100% in individuals carrying a predisposition gene because several other independent genetic changes also have to occur. The chance that the disease will develop as the result of inheriting a cancer predisposition is called the **penetrance**. The fact that penetrance is not 100% increases the difficulties of searching for cancer-predisposition genes.

In general, inherited predisposition to cancer is likely to be present when cases cluster together in genetically related individuals more often than one would expect by chance. There are two instances when this occurs, the first being when rare cancers cluster in families. If each cancer is a rare event, the chance of two or more cancers representing a random cluster is very low. An example of this is the multiple endocrine neoplasia syndrome, in which medullary thyroid cancer clusters with phaeochromocytoma (both are rare tumours in the general population). The second instance is when common cancers cluster in families at a much younger age than in the general population, or where many cases occur in one family. For example, four cases of breast cancer, even at older age of onset, are unlikely to be due to chance alone in the same genetic lineage. Multiple instances of cancer in the same patient are also more common in cancer due to a genetic predisposition (e.g. bilateral breast cancer).

Molecular Biology for Oncologists. Edited by J. R. Yarnold, M. R. Stratton and T. J. McMillan. Published in 1996 by Chapman & Hall, London. ISBN 0 412 71270 9

3.2 WHEN IS A FAMILY LIKELY TO HAVE AN INHERITED CANCER-PREDISPOSITION?

From the information above, it can be deduced that a family is more likely to have a cancer-predisposition gene if:

- there are multiple young cases of a common cancer site on one side of the family;
- there are clusters of rare-site cancers;
- there are clusters of cancers at different sites known to be associated with particular cancer-predisposition genes.

Examples of these points are illustrated in Figures 3.1 and 3.2.

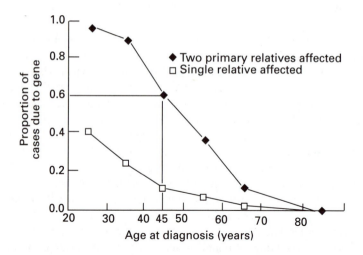

Figure 3.1 Probability that a breast cancer case is due to the inheritance of a cancer-predisposition gene. The figure shows the overall proportion of cases due to a breast-cancer-predisposition gene (of which there are several) by age. The lower curve is for singleton breast cancer cases in a family. The upper curve is for pairs of affected sisters. This illustrates the marked increase in the chance of the presence of a breast-cancer-predisposition gene if two cases are present in the family, particularly if they are affected at a young age. The vertical line indicates that two sisters diagnosed with breast cancer at 45 years have a 60% chance of carrying a breast-cancer-predisposition gene; a single case at that age would have a 10% chance of carrying a breast-cancer-predisposition gene. The data are from the Cancer and Steroid Hormone study (USA), conducted to assess the effects of hormones on breast cancer risk, in which family history data were collected.

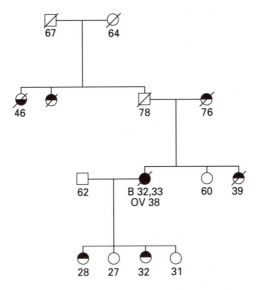

Figure 3.2 Family pedigree transmitting a mutation in the breast/ovarian-cancer-predisposition gene on 17q (*BRCA*1). This gene predisposes to both breast and ovarian cancer and so when these cancer sites cluster together in a family, the chances of the family pattern of cancers being due to the presence of this gene increases. Pedigree (family tree) notation is as follows: males are denoted by squares, females by circles, deceased members by a diagonal line through their symbol, married individuals by a horizontal line joining their symbols, and children by a vertical line from the horizontal line joining their parents. Affected individuals have filled-in (solid) symbols. Pedigree notation follows the same format in the literature, with the exception of the degree to which symbols are filled in to denote an affected member. There are efforts to standardize this notation but, until these are accepted, the notation for affected individuals has to be explained for each published pedigree. Filled-in symbols denote individuals affected with both breast and ovarian cancer, lower-half-filled symbols are those with ovarian cancer and upper-half-filled symbols are those with breast cancer. The numbers refer to age at diagnosis or death or current age. (Source: reproduced by permission of D. Easton, Breast Cancer Linkage Consortium.)

3.3 SEGREGATION ANALYSIS OF FAMILIES HELPS DEFINE THE GENETIC MODEL

Segregation analysis of the inheritance of a disease in family pedigrees assesses whether it is consistent with the dominant or recessive inheritance of a single gene, or whether several genes are acting (a **polygenic effect**). As a general principle, if a gene is recessive, the risk of the disease occurring in siblings of cases will be higher than in parents. This is because if the parents are gene carriers neither will manifest the disease. If the gene is dominant, the risk to siblings will be half that in parents because, on average, the offspring of the parents will inherit the disease-causing gene 50% of the time.

3.4 MOLECULAR GENETICS OF CANCER-PREDISPOSITION

With the exception of one cancer family syndrome so far, cancer-predisposition genes are tumour suppressor genes. At the gene locus responsible for the predisposition in an individual, one of the alleles is inactivated by mutation in every cell in the body, whereas the other normal allele acts to suppress tumour formation; hence the term tumour suppressor gene. If the normal allele is inactivated by mutation or deletion in one or more somatic cells, this constraint is lost.

3.5 DETECTING GENETIC LINKAGE IS A POWERFUL WAY OF LOCATING GENES CAUSING INHERITED CANCER-PREDISPOSITION

Genetic markers used in genetic linkage studies are sequences of DNA whose unique chromosomal locations are known. Genetic linkage is the co-inheritance of genetic markers, including alleles of known genes, with a disease phenotype. This occurs when genetic markers are close, or 'linked', to the disease-causing gene on the same chromosome (Figure 3.3).

Since genetic markers are not usually in the disease-causing gene itself, but within a variable genetic distance, co-inheritance can occur by chance. By analogy to a clinical trial where chance effects are rejected if their probability is less than 1

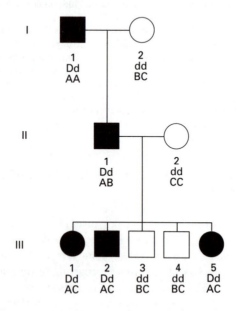

Figure 3.3 Demonstration of genetic linkage. D is the allele of the cancer-predisposition gene, and d is the normal allele at that locus. The genetic marker A is segregating with D, i.e. is linked to it (on the same arm of the same chromosome).

in 20 ($p < 0.05$), linkage to a genetic marker is considered to be statistically signifi-cant if a parameter called the Lod score is ≥ 3. Lod stands for logarithm of the odds of linkage. A Lod score of 3 ($\log_{10} 1000$) represents odds of linkage of 1000:1.

3.6 WHAT ARE THE LIMITATIONS OF LINKAGE?

Linkage data can be misleading

A disease may appear to be linked to a marker due to chance segregation of chro-mosome pairs, rather than co-segregation of genetic markers on the same chromo-some. Confidence that linkage is truly present can be increased by assessing the inheritance of other markers in the same region of the chromosome. The ideal sit-uation is to choose markers that flank the region within which the disease gene is thought to lie. The only situation in which linkage data will be misleading is if recombinations occur both above and below the disease-causing gene during meiosis, a very unlikely event.

Individuals may have the clinical features of an inherited disease without having inherited a genetic defect

Sometimes individuals in a cancer family develop the disease by chance, rather than by the inheritance of predisposition running through the family. These cases are referred to as **phenocopies**. As a general rule, one can have a high degree of suspicion as to which individuals are phenocopies because they are likely to be affected at an older age and are unlikely to have multiple primary sites of the dis-ease. However, this is only a general rule. Breast cancer families have been described where the cancer-predisposition gene has been identified, but where breast cancer has developed, for example, at age 35 in a woman who has not inher-ited the cancer predisposition in the family.

3.7 ANALYSIS OF TUMOURS IS IMPORTANT IN LOCATING THE SITES OF CANCER-PREDISPOSITION GENES

Cytogenetics

Specific changes may be seen in the chromosomal complement (karyotype) of individual tumours. In an individual inheriting a mutation in one allele of a cancer predisposition gene, the normal allele must be inactivated for tumour development to occur. Although this usually involves a submicroscopic loss of genetic material, larger losses (several to many millions of base pairs) can sometimes be seen cytog-enetically as loss of part, or all, of a chromosome arm. Observations like these help to localize the position of the gene responsible.

Allele loss (loss of heterozygosity, or LOH)

DNA polymorphisms are naturally occurring variations in DNA sequence, usually between genes (Chapter 27). Many hundreds of polymorphic loci have been identified and mapped in the human genome. There is usually at least one polymorphic site close to any gene locus. The significance of polymorphisms lies in the fact that they enable homologous (maternal and paternal) chromosomal loci to be distinguished by these minor sequence variations using simple molecular techniques. Individuals who are heterozygous for a polymorphism in a genetic marker appear homozygous if one of the alleles is lost. This can occur when the locus is near the tumour suppressor gene of interest and loss of the normal gene includes both the tumour suppresssor locus and the allele that has been tested for hetero-zygosity (Figure 3.4). LOH therefore provides a clue as to the sites of tumour suppressor genes.

3.8 CO-SEGREGATION OF A CHARACTERISTIC CLINICAL PHENOTYPE WITH CANCER HELPS TO LOCALIZE CANCER-PREDISPOSITION GENES

If an individual has an identifiable phenotype that segregates with cancer development, e.g. an abnormal skin condition, then if the position of the gene for the

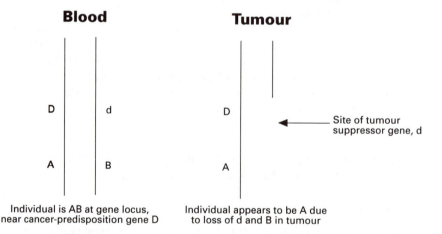

Figure 3.4 Loss of heterozygosity (allele loss). D is the mutated allele and d the normal allele of the cancer-predisposition gene. A marker near to the cancer-predisposition locus has two alleles, A and B. This individual is heterozygous for these markers (i.e. is AB). Analysis of the marker in blood DNA would therefore confirm that the individual is AB. In the tumour, loss of the normal allele, d, is accompanied by loss of B as it is nearby on the chromosome. Analysis of the tumour would therefore indicate that the individual is A, i.e. the heterozygous state has been lost in the tumour cells (so-called loss of heterozygosity or LOH).

abnormal phenotype is known, this gives a clue to the site of the cancer-predisposition gene. The *APC* gene, which predisposes to adenomatous polyposis coli (APC), was found when an individual with mental retardation and APC was found to have a deletion of the long arm of chromosome 5 on cytogenetic analysis of his lymphocytes. The *APC* gene was then localized to 5q by linkage analysis in a larger number of polyposis families.

3.9 GENES ARE CLONED AFTER THEIR CHROMOSOMAL LOCATIONS ARE DISCOVERED

The region in a chromosome where the gene lies is narrowed down to a few million base pairs using linkage analysis studies. So-called 'recombinants' are very useful in fulfilling this task; these are patients with the disease due to a predisposition gene, but who have had a recombination event between the genetic marker and the disease gene. In these individuals, recombination has occurred during meiosis in either the sperm or the egg before fertilization (Figure 3.5).

Roughly speaking, if recombination between linked marker and disease phenotype is detected in 1 in 100 meioses (patients), it represents a physical separation of about 1 million base pairs. Once the region has been narrowed down to an area of about 1–3 million bases of DNA, a 'contig' (short-hand for contiguous) map of

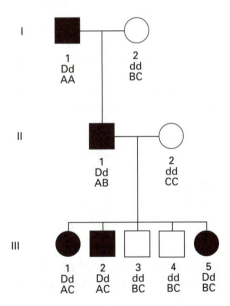

Figure 3.5 Meiotic recombination. During meiosis, crossover has occurred so that the allele of the marker linked to the disease allele, D, has changed from A to B in case III.5. This indicates that the cancer-predisposition gene is closer to the centromere than the marker locus AB. The meiotic crossover has occurred during spermatogenesis in individual II.1.

the region is constructed. The map is based on a series of DNA fragments, which usually have regions of overlap at one of their ends and which taken together encompass the region of interest. Each fragment of DNA is large and is replicated for analysis (cloned) in yeast or bacteria as a yeast artificial chromosome (YAC) or bacterial artificial chromosome (BAC) respectively. An artificial chromosome is created by splicing human DNA into a yeast or bacterial chromosome and per-suading the organism to replicate it, in order to synthesize enough human DNA for molecular analysis. An alternative is the P1 cloning system, where pieces of human DNA are cloned in bacteriophage. Systems such as BACs and P1 clones are becoming more popular than YAC cloning systems because they are more sta-ble (YACs tend to lose parts of the human DNA they contain very easily).

Genes are identified within contigs using various techniques; for example, genes often contain CpG dinucleotides, called CpG islands. Searches for CpG islands within contigs can reveal sites of genes amongst the nonsense sequences of DNA. Once genes are identified and sequenced, their wild-type (normal) sequences are compared to individuals thought to have developed cancer due to an inherited defect, in the hope of identifying cancer-causing alterations (muta-tions) within the gene being tested.

3.10 SEVERAL MOLECULAR TECHNIQUES DETECT CHANGES (MUTATIONS) IN CANCER-PREDISPOSITION GENES

Once a candidate gene in the region of a cancer-predisposition gene has been cloned, the sequence of the gene is checked in individuals who are thought to have developed cancer due to an inherited predisposition. It greatly strengthens the case for thinking a cancer-predisposition gene has been found if sequence differences are seen in affected individuals that are not present in normal controls. Several molecular methods are used to detect base-pair mismatches in wild-type and mutant DNA sequences, summarized in Table 3.1.

Every method can miss mutations, and in general these methods have a lower sensitivity than direct sequencing of DNA. The accuracy may also vary between genes; those which are GC-rich (containing a lot of GC bases) produce more arte-factual bands on a sequencing gel. This is important when considering diagnostic testing and is the reason why many laboratories confirm the presence of a genetic mutation by two methods when performing diagnostic tests or repeat the test on two occasions.

3.11 WHAT OTHER TESTS DISTINGUISH BETWEEN MUTATIONS PREDISPOSING TO CANCER AND NORMAL SEQUENCE VARIATIONS?

Expected impact of mutation on protein structure and function

Although a single gene may code for a particular phenotype (e.g. hair structure), dif-ferent coding sequences may code for normal variants of this feature (e.g. curly as

Table 3.1 Methods of comparing test mutant with normal gene sequences

Method	Principle	Advantages	Disadvantages
SSCP (single strand conformational polymorphism)	Compares conformation of single DNA strands (altered by alteration in base sequence)	Rapid Can test many samples at once	Only analyses DNA fragments up to 600 bases long Gels can be difficult to read if single strands have many conformations
HOT (hydroxylamine osmium tetroxide) or chemical mismatch cleavage	Compares mismatches between normal and mutant DNA Chemicals bind to and then cleave at mismatch	Can analyse large DNA fragments (up to 1000 bp) Indicates site of mutation	Uses hazardous chemicals
DGGE (denaturing gel electrophoresis)	DNA denatures at different temperatures/concentration of denaturant depending on the base composition	Does not use radioactivity (there are now variants of SSCP and HOT that use less dangerous or no radioactivity)	Gels complicated to construct
RNase protection	RNase cleaves at RNA/DNA mismatches at mutation sites		Less sensitive than other methods

opposed to straight hair). When mutations are found in genes that are candidates for causing cancer, a distinction has to be made as to whether the mutations are normal variants or cancer-causing. In general, the most difficult mutations to distinguish are those in which one base is altered to another (e.g. A is changed to C, a so-called **missense mutation**) where the protein product is the same size, but contains one amino acid different from the normal, or splice site mutations which confer alternative splicing, resulting in the production of a functional protein, but of a different size. Mutations which truncate or shorten the protein size because they code for stop codons or insert or delete parts of the gene are usually cancer-causing.

Segregation with the cancer phenotype

If a mutation is to be cancer-causing, it should segregate (be inherited in the same pattern) with the disease. Some members of the family may have cancer, but not have the mutation and this may arise either because the gene is not a cancer-predisposition gene or because some of the cancers are sporadic and due to chance, not to inherited predisposition. In general, if the cases are occurring at young age, it is less likely that they are sporadic.

Functional effects of the mutation

A mutation is more likely to be cancer-causing if it can be shown to alter the function of the gene. Functional assays have been developed for some cancer-predisposition genes, particularly those that are most commonly mutated by missense mutations (see above). An example is the *p53* gene; part of the action of this gene is to bind to DNA and most missense mutations occur in the DNA-binding region. The mRNA from the *p53* gene is amplified by the polymerase chain reaction and transfected into a vector in yeast. If the product is normal and binds to DNA it turns on a reporter gene downstream which has the effect of converting a colourless substance into a red substance, resulting in the growth of a red yeast colony. If the *p53* message is mutant, this cannot occur and the yeast colony is white. The problem with such assays is that mutations affecting other areas of the gene with different functions will be missed. Mutations which result in the reduced production of normal message (mRNA) will also be missed.

The incidence of a mutation in a cancer population is compared with that in an unaffected population

A polymorphism is a normal variation in a gene sequence occurring in the population at large. Conventionally, if a mutation is not present in 100 non-cancer controls representing the population at large, it is considered to be a cancer-causing mutation rather than a polymorphism. However, polymorphisms present in the normal population at a lower level than 1% may be responsible for occasional errors in interpretation.

3.12 DIFFERENT GENES ARE SOMETIMES RESPONSIBLE FOR A VERY SIMILAR PATTERN OF INHERITED CANCER PREDISPOSITION (GENETIC HETEROGENEITY)

There is often more than one cancer-predisposition gene that predisposes to a particular type of cancer running in families (genetic heterogeneity). For example, there are at least three genes (*BRCA*1, *BRCA*2 and *p53*) that predispose to breast cancer in families. The gene involved will be the same within a particular family, but different families may carry defects in different genes. One clue as to which gene is responsible in a particular family is to assess the presence of other cancers. For example, *BRCA*2 predisposes to both male and female breast cancer, whereas *BRCA*1 predisposes to breast cancer in women, but not in men (with the exception of two families described to date). If male breast cancer is also present in a female breast cancer family, it is much more likely to be due to *BRCA*2 than *BRCA*1.

3.13 SUMMARY; THE EXAMPLES OF *BRCA*1 AND *BRCA*2

The points in this chapter are illustrated by the recent advances in understanding two forms of breast cancer predisposition. Cancer pedigrees were ascertained on the basis of young age of onset, particularly with bilateral disease. Linkage analysis demonstrated the presence of at least two breast cancer genes, *BRCA*1 and *BRCA*2 on the long arms of chromosomes 17 and 13, respectively. Cloning of *BRCA*1 and *BRCA*2 was successful at the end of 1994 and in 1995, respectively. Cancer-causing mutations have been demonstrated in families due to *BRCA*1 and *BRCA*2. Many of these mutations truncate the protein because they either insert or delete bases into/out of the gene. Missense mutations have been described, but these will have to excluded as polymorphisms. The genes are very large and it is hoped that as the function of the protein products are elucidated, functional assays may be developed to avoid the necessity for complete gene analyses to provide a diagnostic test.

3.14 FURTHER READING

Eeles, R. A., Ponder, B. A. J., Easton, D. F. and Horwich, A. (eds) (1996) *Genetic Predisposition to Cancer*, Chapman & Hall, London (in press).
Lemoine, N. R. and Hurst, H. C. (1994) Genetics of cancer. *Eur. J. Cancer*, 30A.
Ponder, B. A. J., Solomon, E. and Cavenee, W. (eds) (1995) *Genetics and Cancer: A Second Look*, Cancer Surveys, vol. 25, Cold Spring Harbor Press, New York.
Ponder, B. and Waring, M. (eds) (1995) *The Genetics of Cancer*, Cancer Biology and Medicine, vol. 4, Kluwer Academic Publishers, London.

4 | Chromosomal translocations in cancer

Martin J. S. Dyer

4.1 SPECIFIC CHROMOSOME TRANSLOCATIONS ARE CHARACTERISTIC OF SOME CANCERS AND PLAY A MAJOR PATHOGENIC ROLE

The development of malignancy depends on the inappropriate, deregulated expression or inactivation of particular sets of genes. Such genetic abnormalities may arise as a consequence of specific DNA deletions or amplifications. Alternatively, genes may become deregulated by chromosomal translocations in which segments of DNA normally found on different chromosomes become juxtaposed. The outcome is inappropriate expression of a gene close to the breakpoint or the creation and expression of a new (fusion) gene from DNA sequences spanning the breakpoint.

A key feature of all these abnormalities is their association with specific tumour types: the association of some chromosome translocations with various tumours is shown in Table 4.1.

A paradigm is the translocation t(15;17)(q21;q22) seen in all cases of acute promyelocytic but no other subtype of myeloid leukaemia. (The notation t(15;17) indicates that a fragment of chromosome 15 has been translocated to chromosome 17; (q21;q22) indicates that bands 21 and 22 of the long arms (q) of chromosomes 15 and 17, respectively, are involved.) Such strong associations suggest a major pathogenic role. All the translocations shown in Table 4.1 have been molecularly cloned, i.e. the genes at both chromosomal breakpoints have been identified. All the recurrent chromosomal translocations in the acute leukaemias have been cloned.

The clinical implications of these studies have been profound. The development of the molecular cytogenetic approach to malignancy has allowed new objective approaches to diagnosis and prognosis, and strategies for the sensitive detection of residual disease, and may allow for new therapeutic approaches targeted against

Molecular Biology for Oncologists. Edited by J. R. Yarnold, M. R. Stratton and T. J. McMillan. Published in 1996 by Chapman & Hall, London. ISBN 0 412 71270 9

Table 4.1 Some chromosomal translocations in cancer

Chromosomal translocation	Disease	Genes involved	Molecular outcome
t(8;14)(q24;q32)	Burkitt's lymphoma B-cell ALL	myc (8q24) IG(14q32)	Deregulated myc expression myc mutation
t(14;18)(q32;q21)	Follicular lymphoma	bcl-2 (18q21) IG (14q32)	Deregulated bcl-2 expression bcl-2 mutation
t(9;22)(q34;q11)	Chronic myeloid leukaemia Some ALL	abl (9q34) bcr (22q11)	bcr–abl fusion protein
t(15;17)(q21;q22)	Acute promyelocytic leukaemia	PML (15q21) RARA (17q22)	PML–RARA fusion protein
t(11;22)(q24;q12)	Ewing's sarcoma	FL1I (11q24) EWS (22q12)	FL1I–EWS fusion protein
t(12;22)(q13;q12)	Melanoma	ATF1 (12q13) EWS (22q12)	ATF1–EWS fusion protein
11q23/various	Acute leukaemias	MLL (11q23)	MLL fusion proteins

This table demonstrates only a fraction of the recurrent chromosomal translocations that have been cloned in recent years. In the first example, t(8;14) indicates that a fragment of chromosome 8 has been translocated to chromosome 14; (q24;q32) indicates that bands 24 and 32 of the long arms (q) of chromosomes 8 and 14 respectively are involved. Translocation breakpoints were first cloned in the haematological malignancies because of the presence of well defined cytogenetic abnormalities and the location of cloned genes immediately adjacent to the breakpoints. The production of fusion genes is, however, common to translocations found in haematological and non-haematological malignancies. Certain genes, notably *EWS* and *MLL*, are involved with a large number of different partner chromosomes, each producing a different fusion gene. Most translocations have been found only in malignant cells; however certain translocations, such as t(14;18)(q32;q21), may be found at low frequency in apparently normal individuals.

the genetic lesions. The object of this short chapter is to describe some of the methods used in the characterization of translocation breakpoints and their clinical application.

4.2 FLUORESCENT *IN-SITU* HYBRIDIZATION (FISH) HAS TRANSFORMED CYTOGENETICS BY HELPING TO IDENTIFY THE POSITION OF TRANSLOCATION BREAKPOINTS

The starting point for all these studies has been tumour cytogenetics. The precise number of human chromosomes was only established unequivocally in 1955 and the first recurrent cytogenetic abnormality in human tumours, the Philadelphia (Ph) chromosome, was identified in 1958. This prompted the search for other recurrent abnormalities in malignancy. Progress was slow; the t(14;18)(q32;q21) seen in over 80% of follicular B-cell non-Hodgkin lymphomas was only recognized in 1978, reflecting the difficulties in obtaining and interpreting metaphases from many tumour types. Clinically advanced and chemotherapy-resistant cancers often have multiple chromosomal abnormalities which may preclude objective assessment on the basis of chromosome morphology alone.

These difficulties can now be overcome by the use of fluorescent *in-situ* hybridization (FISH) techniques, which have demonstrated a level of genetic complexity unexpected from regular cytogenetics alone (Plate 1). Traditionally, cytogenetics was based on metaphase nuclei because these offered the only opportunity for visualizing individual condensed chromosomes. Now, however, the combined use of two probes adjacent to translocation breakpoints allows the detection of translocations by the lack of comigration of signals in interphase nuclei, circumventing the need for metaphase preparations (Plate 2). Since FISH can be performed on formalin-fixed material, the development of these techniques is a major advance in the detection of translocations in routine clinical material. In this context, mention should also be made of the technique of comparative genomic hybridization, which allows the detection of DNA deletions and amplifications. The principle of this technique is that by mixing whole genomic single-stranded DNA from normal (sense strand) and cancer (antisense strand) cells, DNA that fails to hybridize identifies sequences that are either missing or over-represented in the neoplastic cells. Development of these and other methods may ultimately allow a 'global' analysis of all genetic abnormalities within a tumour, from DNA alone without the necessity of obtaining tumour metaphases.

4.3 THE DNA FLANKING TRANSLOCATION BREAKPOINTS CAN BE ISOLATED AND SEQUENCED TO IDENTIFY THE GENES INVOLVED

To understand how chromosomal translocations might contribute to neoplasia it is necessary to identify the genes located at both breakpoints. Historically, when the number of cloned human genes was low, initial studies were aided by two related observations. Firstly, the human homologues of some viral oncogenes were located at sites of recurrent chromosomal breakpoints: examples included the *myc* gene at chromosome 8q24 involved in the t(8;14)(q24;q32) seen in Burkitt's lymphoma, and the *abl* gene at chromosome 9q34, involved in the t(9;22)(q34;q11) of chronic myeloid leukaemia and acute lymphoblastic leukaemia. Both viral *myc* and *abl* genes had been identified in retroviruses and had been shown to induce leukaemias in animal models; therefore the finding that the human homologues of these genes were directly involved in translocations supported their central role in the development of human neoplasia.

Secondly, translocations in B- and T-lymphoma malignancies were found to be targeted to the immunoglobulin (*IG*) and T-cell receptor (*TCR*) loci respectively. Both *IG* and *TCR* loci have to undergo recombination during normal lymphoid development in order to produce functional antigen-receptor genes. Chromosomal translocations at these loci are thought to arise when errors occur in the recombination process, so that genes from other chromosomes become involved. Since the *IG* and *TCR* loci had been cloned it was a relatively simple matter to use these genes to clone translocation breakpoints and to isolate the genes juxtaposed from other chromosomes: the *bcl* series of genes, standing for **B**-cell **l**eukaemia/lymphoma, of which *bcl2* is the most widely known, was cloned in this fashion.

As the number of cloned human genes has increased, so it has been possible to use these genes to clone other translocation breakpoints. An example of this approach was the cloning of the t(15;17)(q21;q22) of acute promyelocytic leukaemia following the demonstration that the retinoic acid receptor α gene mapped to chromosome 17q22. This practice will increase as more of the estimated 100 000 total human genes become cloned and mapped to precise chromosomal locations.

4.4 SPECIFIC TRANSLOCATIONS CAUSE INAPPROPRIATE EXPRESSION OF A GENE CLOSE TO THE BREAKPOINT OR THE EXPRESSION OF A NEW (FUSION) GENE FROM DNA SEQUENCES SPANNING THE BREAKPOINT

The molecular consequences of chromosomal translocation include:

- **Deregulated gene expression**: a gene normally 'silenced' during normal cell differentiation may come under the influence of strong transcriptional promoters and enhancers, resulting in deregulated expression of the translocated gene. This mechanism is seen in translocations at the immunoglobulin (*IG*) and T-cell receptor (*TCR*) loci; an example is the *bcl2* gene at 18q21, involved with the *IG* gene at 14q32 in the translocation t(14;18)(q32;q21) in follicular lymphoma. The *bcl2* gene is switched off in most B-cells within the germinal centre, thereby allowing programmed cell death or apoptosis to occur. However, placement of the *bcl2* gene next to the *IG* locus, which contains a B-cell specific transcriptional enhancer, results in deregulated *bcl2* expression and prolonged

B-cell survival. Note that the coding sequences of the *bcl2* gene are not disrupted in these translocations.

- **'Ectopic' gene expression**: a variation on the above theme is the expression of genes not normally expressed in normal cell differentiation. This exemplified by some translocations in the T-cell precursor leukaemias in which various genes not expressed in normal T cells become translocated to the *TCR* loci. Again these genes are not disrupted after translocation.

- **'Fusion' gene expression**: in contrast to the two above categories, if the breakpoints occur within introns of the two genes (thereby disrupting the gene) then translocation results in the formation of novel 'hybrid' or 'fusion' genes containing components of the two. The resulting proteins consequently will have unique functions. This is a common occurrence both in leukaemias and in the non-haematological malignancies (Table 4.1). Probably the best recognized example is the *bcr–abl* fusion gene produced from the derivative chromosome 22 of the Philadelphia chromosome, t(9;22)(q34;q11). The alternate transcript derived from the derivative chromosome 9 (i.e. the *abl–bcr* transcript) is also expressed in some cases.

The majority of genes involved in chromosomal translocation encode transcription factors important in determining the early stages of development and differentiation. In the chronic leukaemias and lymphomas however some of the genes encode proteins of novel functions. The *bcl2* gene for example encodes an outer mitochondrial protein of importance in the control of apoptosis; the *bcl1* gene (cyclin D1) participates in the control of the cell cycle.

Chromosomal translocations are often associated with other genetic events, including DNA deletion and mutation. Thus, the abnormalities of chromosome 16 seen in myelo-monocytic leukaemia with eosinophilia (M4Eo) may often be associated with deletion of several megabases of DNA. Translocated *myc* genes are almost invariably heavily mutated: mutation may continue after translocation.

4.5 RAPID ASSAYS OF SPECIFIC TRANSLOCATION BREAKPOINTS IN CANCER PATIENTS WILL HELP DIAGNOSIS AND THE MONITORING OF THERAPEUTIC RESPONSE

The association of specific cytogenetic abnormalities with specific disease subtypes allows their use in diagnosis. A variety of molecular cytogenetic techniques may be used as an adjunct (or indeed to replace) conventional cytogenetics as shown with reference to some lymphoid malignancies in Table 4.2.

The choice of methods used depends on the precise molecular anatomy and consequences of the translocation. The polymerase chain reaction (PCR) may be useful for the analysis of breakpoint regions and for the detection of novel fusion mRNAs. In cases where the gene is expressed ectopically as a consequence of translocation, monoclonal antibodies may be used to detect expression. DNA electrophoresis and hybridization are useful for the detection of breakpoints dispersed

Table 4.2 Molecular cytogenetics of leukaemia and lymphoma: methods of detection (− = not used for detecting abnormality; + = useful for a subset of cases; ++ = useful for detecting the majority of cases; NA = not applicable)

Translocation	Gene	Disease (% cases with abnormality)	Clustered	FISH	PFGE	SB	PCR	MAb
t(8;14)(q24;q32)	myc	Burkitt's NHL (> 90) Diffuse NHL (20)	No	++	+	++	+	+
t(11;14)(q13;q32)	CCND1	Mz–NHL (> 80) SLVL (25) Myeloma (?)	No	+	++	+	+	++
t(14;18)(q32;q21)	bcl-2	Follicular B-NHL (80) Diffuse B-NHL (20) B-CLL (2)	Yes	–	++	++	++	+
t(3;14)(q27;q32)	bcl-6	Diffuse B-NHL (30)	?Yes	?	?	++	?	?
Trisomy 12	?	B-CCL (30)	NA	++	NA	NA	NA	?
Inv(14)(q11;q32)	TCL1 (?)	No	?	?	?	?	?	?

To detect all chromosomal aberrations associated with a particular phenotype may require a combined approach using fluorescent *in-situ* hyridization (FISH), DNA blot and antibody techniques. Breakpoints clustered within a few hundred base pairs of each other can be analysed by polymerase chain reaction (PCR). Breakpoints that are more widely dispersed are analysed by DNA blot. DNA blots are generated by pulsed-field gel electrophoresis (PFGE) which separate large DNA fragments (< 2000 kb) or conventional 'Southern' blots, which separate smaller fragments (< 50 kb) and are much easier to perform. An alternative approach is to look for deregulated gene expression by using gene-specific monoclonal antibodies (MAb).

throughout a gene or locus which are not detectable by PCR but they are labour-intensive and expensive to perform. Fluorescent *in-situ* hybridization (FISH) techniques will play an increasingly important role in routine practice.

The clinical importance of developing simple methods for the detection of chromosomal translocations may again be seen in the haematological malignancies. Firstly, some but not all translocations are of diagnostic and prognostic significance, allowing the rational introduction of 'stratified' therapy (e.g. allogeneic bone marrow transplantation in first remission for Ph$^+$ ALL with t(9;22) (q34;q11)). Cytogenetic lesions of comparable prognostic significance have been detected in CLL and lymphoma. Secondly, since these translocations are tumour-specific markers, they may be used to monitor response to therapy with unprecedented sensitivity: using PCR, malignant cells can be detected with a sensitivity of one cell in 100 000 normal cells. This technique is therefore invaluable in assessing the disease status of patients undergoing autologous and allogeneic transplantation.

4.6 FUSION GENES MAY BE TARGETS FOR TUMOUR-SPECIFIC THERAPIES

Most of the common translocation breakpoints have been cloned. Identification of the genes flanking the breakpoints has allowed new strategies in 'molecular cytogenetics' for diagnosis, and for monitoring response to therapy. The prospect arises that these genes and particularly tumour-specific 'fusion genes' may be suitable targets for the rational design of new, tumour-selective therapies. How these translocations arise, in particular how translocations become targeted to specific regions of specific genes in specific tumours, remains unknown.

4.7 ACKNOWLEDGEMENTS

I thank my cytogenetic colleagues for kindly providing illustrations for this chapter.

4.8 FURTHER READING

The reader is referred to an excellent recent review (Rabbitts, T. H. (1994) Chromosomal translocations in human cancer. *Nature*, **372**, 142–149), which provides a good 'way in' to a huge literature and to a recent book (Kirsch, I. R. (ed.) (1993) *The Causes and Consequences of Chromosomal Aberrations*, CRC Press, Boca Raton, FL) containing reviews with a historical viewpoint: see particularly chs 22–24 for a description of the early days in this field. A review of FISH techniques is found in Le Beau, M. M. (1993) Detecting genetic changes in human tumor cells: have scientists 'gone fishing'? *Blood*, **81**, 1979–1983.

Signal transduction and cancer

5

Liza Ho and Michael J. Fry

5.1 CHEMICAL SIGNALS DETERMINE CELL BEHAVIOUR

The object of this chapter is to introduce some of the key concepts of cellular signal transduction. Examples will be presented to explain how the study of signal transduction pathways is relevant to a fuller understanding of mechanisms underlying the development of cancers. So what is meant by signal transduction? All cells, from the simplest unicellular bacterium to the cells that make up the tissues of a complex multicellular organism, have to be able to respond to externally applied stimuli. In the case of a bacterium this may be a simple growth response to favourable conditions, e.g. to high concentrations of a suitable nutrient. Once favourable conditions for growth are encountered bacteria tend to proliferate until the nutrients are exhausted. For the cell from the multicellular organism the stimuli and responses are generally much more complex and subject to a greater degree of regulation. Cellular functions in multicellular organisms including cell migration, proliferation, differentiation, survival and apoptosis are controlled and coordinated in response to extracellular signals. Let us take, for example, the controlled growth and replacement of cells in a tissue following a wound. Cells must recognize that they are at the site of a wound and undergo division to replace the damaged cells. However, once the lost cells have been replaced the dividing cells must also be able to recognize this too and cease their proliferation. Failure of cells to recognize when they should cease to divide contributes to many proliferative disorders such as cancer.

The above examples require that the cells in question are able to sense their current environment and be able to mount appropriate responses to the prevailing conditions. The varied mechanisms by which external information is recognized

Molecular Biology for Oncologists. Edited by J. R. Yarnold, M. R. Stratton and T. J. McMillan. Published in 1996 by Chapman & Hall, London. ISBN 0 412 71270 9

and conveyed into the cell and processed such that an appropriate response is carried out by that cell is the essence of signal transduction. The key molecules involved in signal transduction processes are as follows. The extracellular signals, or first messengers, provide the environmental cues that essentially tell a cell how to behave. First messenger molecules include nutrients, peptide growth factors, cytokines, hormones, extracellular matrix proteins and cell surface proteins on adjacent cells (Figure 5.1).

Next are the cell surface receptor molecules which allow cells to recognize the presence of these external signals. A specific cell surface receptor is usually present for each signal that a cell is required to recognize and respond to. These receptors generally span the plasma membrane and come in a wide variety of structural types beyond the scope of this chapter. Recognition of signals received

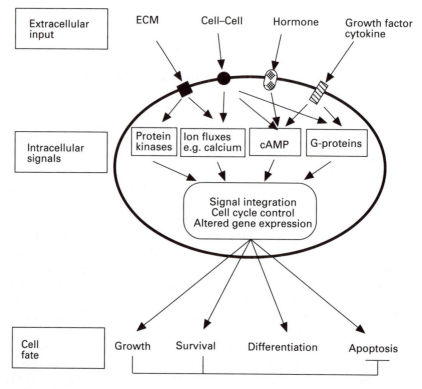

Figure 5.1 Diversity of signals and responses. Schematic cartoon of the complexity of cellular signalling. Cells continually receive multiple external inputs in the form of growth factors and hormones, and through interactions with the local extracellular matrix (ECM) proteins and adhesion molecules on adjacent cells. These ligands are recognized by specific cell surface receptor molecules on the target cell leading to the generation of intracellular signals. The cell cycle and gene transcription machinery 'interprets' the current spectrum of activated intracellular signalling pathways and on the basis of this cell fate is determined.

from outside of the cell is mediated by their binding to an appropriate cell surface receptor molecule and is then converted into the generation of second messenger molecules within the cell. This requires the function of key intracellular enzymes such as protein kinases, guanine-nucleotide binding (G)-proteins, phospholipases, etc., which couple to the activated cell surface receptors. These enzymes are responsible for generating the specific second messenger molecules and intracellular signals.

5.2 ACTIVATION OF INTRACELLULAR SECOND MESSENGER SYSTEMS LEADS TO AN INTRACELLULAR AMPLIFICATION OF THE INITIAL GROWTH FACTOR SIGNAL

Common intracellular second messenger molecules used in signalling responses in addition to protein phosphorylation include cyclic AMP, lipid mediators such as diacylglycerol (DAG) and calcium ions. The activation of these enzymes and the generation of second messengers, which are themselves able to activate additional signalling molecules, generally results in a considerable intracellular amplification of the initial signal encountered at the cell surface. This is due to the nature of the enzymes activated. In general, once an enzyme is activated it will be able to perform a number of catalytic cycles before being inactivated by regulatory mechanisms. Therefore, for example, a single molecule of epidermal growth factor (EGF) binding to its receptor will activate that receptor which will in turn recruit a number of signal generating enzymes including phospholipase C (PLC). Each activated PLC molecule will then be able to catalyse the hydrolysis of a number of membrane lipid molecules generating two second messenger molecules, DAG and inositol trisphosphate ($InsP_3$). The former will activate protein kinase C, which in turn will catalyse the phosphorylation of many substrate molecules, while the $InsP_3$ will release calcium ions from intracellular stores, which in turn will lead to the activation of another wave of calcium-sensitive enzymes. The concept of signal amplification is illustrated in Figure 5.2.

The overall effect of this is that low concentrations of an extracellular ligand can be converted into a substantial response within the cell. This allows a cell to respond to very small changes in their environment.

All of the enzymes in these signalling cascades are very tightly regulated. Their activity at any time is dictated by the balance of activating signals and inhibitory mechanisms. Activation of signalling enzymes usually results from a short lived displacement of this equilibrium state in favour of the output signal. To prevent unregulated signalling (and growth) the inhibitory mechanisms must rapidly compensate for the shift in the equilibrium brought about by the activating signal and restore the resting state (Figure 5.3). As we shall see later in the chapter alterations in the activity of any of the components in a pathway can have dramatic effects on this equilibrium state and trigger prolonged and unregulated signalling and whatever consequences arise from this.

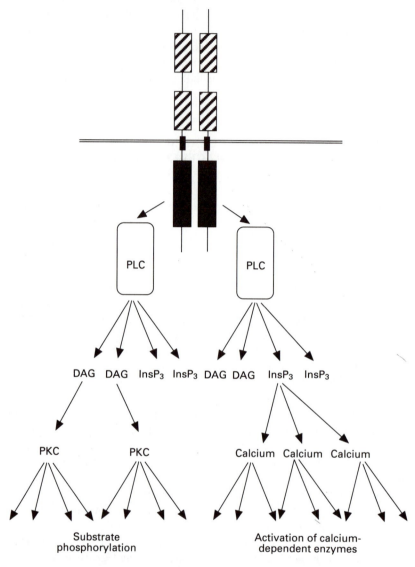

Figure 5.2 Amplification of extracellular signals by second messenger systems. The principle of signal amplification is illustrated using the example of activation of phospholipase C (PLC) by a single PTK receptor complex. This leads to the production of diacylglycerol (DAG) and inositol trisphosphate ($InsP_3$), which in turn activate protein kinase C (PKC) and release calcium stores respectively. Arrows indicate the amplification at each stage of the signalling pathway.

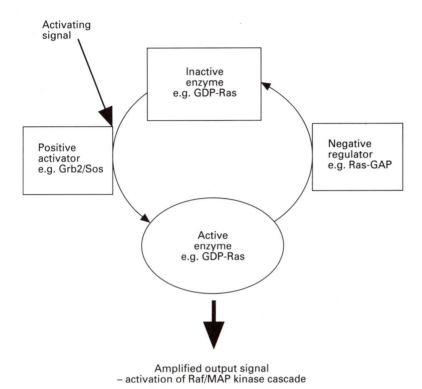

Figure 5.3 Dynamic equilibria in signalling. An isolated enzyme in a signalling cascade, which can exist in active and inactive states, is presented schematically. Using the example of Ras the principle of signalling equilibrium is illustrated. The relative activities of the activating signal and the negative regulators will determine the strength and duration of the output signal generated.

Finally, second messenger molecules trigger diverse and cell-type-specific enzymatic activities and in some cases new gene expression within the responding cell. These events ultimately lead to an appropriate reaction by the responding cell to the stimuli received. The series of events whereby a cell recognizes a particular molecule or an environmental state and initiates a particular response defines the signal transduction pathway(s) activated by that stimuli. A specific stimuli will activate both common and specific signalling components. The overall response or fate of a cell to these multiple inputs is determined by the sum of the inputs received at any given time, the duration of the stimuli and the particular subset of intracellular signals generated. Responses to specific stimuli are further modified by the individual cell context. The sum of these signals is assumed to be 'unique' to the stimuli, and allows the cell to identify the type of stimuli encountered and respond accordingly. A cell in a multicellular organism at any given moment will be receiving multiple signals simultaneously in the form of cell–cell contact, growth factors/hormones in the surrounding media and contact with extracellular matrix proteins and basement membranes. While having specific receptors, many of the different signal transduction pathways which operate in a given cell will use a number of common intracellular second messengers and enzymes. Therefore, rather than multiple linear signalling pathways leading to specific responses, signal transduction is more correctly viewed as a very complex network of intracellular events that the cell is able to decode and from which the cell's ultimate behaviour will be determined.

As many signal transduction pathways are intimately linked to processes that either positively or negatively regulate growth, differentiation and cell survival, it becomes clearer how alterations in these signalling pathways due to the mutations of its regulatory components might lead to deregulated proliferation of the type seen in many types of cancer. Cancer is a disease characterized by multiple mutations accumulated over a period of time, which either inactivate regulatory gene products (tumour suppressor genes) or activate gene products that contribute positively to cell proliferation and survival (oncogenes). Since the recognition that the products of retroviral oncogenes are homologues of normal cellular genes (e.g. v-ErbB is an oncogenic form of the EGF receptor while v-Sis is a viral form of the platelet-derived growth factor (PDGF)), it has become clear that most oncogenes are the altered products of normal cellular signalling proteins. However, cancer is more than simply a disease of accelerated cell proliferation. This disease results from alterations in the delicate balance between the rates of cell growth, cell differentiation and cell death. The signalling pathways that control cell survival and differentiation are closely linked to those implicated in regulating cell proliferation, and changes in the activity and function of specific oncogenes can alter the coordination of cell function and ultimately cell fate.

As we will see in the following sections, the intracellular components of these different signalling pathways have many common gene products, creating scope for extensive crosstalk between the different signalling cascades. From the processes that these pathways regulate, it becomes clear how aberrant signalling may result in proliferation abnormalities and the formation of populations of cancerous cells which become detached from the normal regulatory mechanisms. The remaining portion of this chapter examines how recent work defining intracellular signalling pathways has contributed to an understanding of the functioning of the oncogenes and tumour suppressor genes. This will be illustrated using the example of how receptor protein-tyrosine kinases couple to the *ras* proto-oncogene. It also describes how alterations in components of these pathways contribute to the aberrant behaviour underlying many types of human cancers.

5.3 SIGNALLING PATHWAYS ARE ACTIVATED WHEN GROWTH FACTOR MOLECULES (OR OTHER CLASSES OF LIGANDS) INTERACT WITH SPECIFIC CELL SURFACE RECEPTORS

The first step in activating signalling cascades involves the interaction of a ligand with a specific cell surface receptor. Whether a cell expresses the appropriate receptor for a ligand is therefore the first and possibly most important level of control of a particular signalling pathway along with the availability and accessibility of ligand to the target cell. Among several classes of receptors, the protein-tyrosine kinases (PTKs) will be described here since these have received the greatest attention in recent years and feature frequently in naturally occurring and experimental tumours.

PTKs fall into two major classes: membrane-spanning receptor PTKs and non-receptor or cytosolic PTKs. The latter group contains the Abl, Fps/Fer, Src and JAK (or Janus) families of PTKs. Although unable to span the plasma membrane, many of these cytosolic PTKs act as intracellular subunits for transmembrane receptors or ligand-binding proteins and therefore the term non-receptor PTK is a bit of a misnomer. Over 50 transmembrane PTK receptors falling into at least 15 subfamilies are structurally related (Figure 5.4). They all possess an extracellular glycosylated portion in which resides the ligand-binding domain capable of recognizing specific factors. This is generally linked via a single hydrophobic transmembrane segment to the catalytic PTK portion of the molecule in the cell interior. While the ligand-binding domain provides the specificity of response to extracellular signals, the intracellular domain specifies the spectrum of signalling pathways to which a receptor will couple and thus will determine the appropriate cellular response to the ligand. The response to a given ligand can, however, vary between cell types, expressing an identical receptor molecule on the basis of the intracellular signalling components present.

The first detectable response to binding of ligand is the phosphorylation of the receptor itself on intracellular tyrosine residues, a process termed autophosphorylation. Autophosphorylation sites are found both within the PTK domain and in the flanking intracellular sequences. Phosphorylation sites fall into two categories: those that up-regulate the intrinsic kinase activity of the ligand-bound receptor and those that are responsible for coupling the receptor to downstream signal generation. As we shall see in the following section autophosphorylation is critical to the propagation of the intracellular signals initiated by the receptor. Current theories of receptor activation favour a model in which the binding of ligand promotes the dimerization of receptor molecules on the cell surface, resulting in intermolecular autophosphorylation of receptor subunits and activation of the PTK domain.

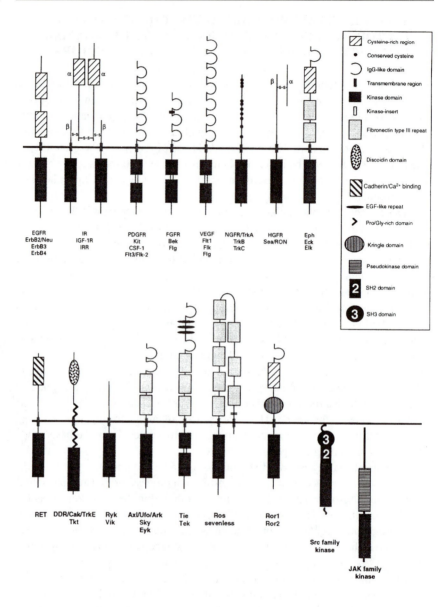

Figure 5.4 Structural topology of receptor and receptor-associated PTKs. This figure illustrates the diversity of PTKs of the receptor and receptor-associated classes. The classification is based mainly on the relatedness of their extracellular and PTK domains. The many structural motifs identified in these receptors are indicated in the key.

5.4 EXTRACELLULAR SIGNALS ARE CONVERTED INTO INTRACELLULAR SECOND MESSENGER MOLECULES BY PROTEIN-TYROSINE KINASE (PTK) RECEPTORS

Protein phosphorylation cascades, G-proteins, cAMP and calcium were long suspected to be important to intracellular signalling pathways, but it was only recently discovered how activated PTK receptors are coupled to them. One mechanism is via direct tyrosine phosphorylation of downstream proteins resulting in modulation of their activities. Such a mechanism is common for substrates of protein-serine/threonine kinases, but fewer such substrates have been identified for PTKs. A major step forward in this area was the recognition of the key role of the SH2 and SH3 domains (see below) in mediating protein–protein interactions that regulate the activation state and function of many intracellular signalling proteins. At least one of these interaction domains is found in the majority of known signalling proteins (Figure 5.5).

Other protein interaction domains are also recognized, e.g. PH domains (see below). Intracellular signalling proteins which contain these domains have been classified into two groups: adaptor proteins, which contain protein–protein interaction domains only and couple PTKs to downstream signalling proteins (e.g. Grb2); and signalling enzymes, which contain both interaction domains and catalytic activity within the same molecule (e.g. PLCγ) and which are capable of directly generating additional intracellular signals upon activation.

5.5 THE SH2 AND PTB DOMAINS ARE PROTEIN MODULES IN SIGNALLING PROTEINS WHICH RECOGNIZE AND BIND TO REGIONS CONTAINING PHOSPHOTYROSINE ON OTHER SIGNALLING MOLECULES

The best understood of these signalling domains is the SH2 domain (SH stands for src **h**omology, named after the retroviral oncogene called *src* in which this domain was first identified). The SH2 domain is approximately 100 amino acids in length and its function is the recognition of phosphotyrosine residues on target molecules when present within specific peptide sequences. Binding of SH2 domains has an absolute requirement for tyrosine phosphorylation of the target sequence. Different SH2 domains have distinct recognition sequences, which allows highly specific interactions to take place between signalling proteins in different pathways. For example, the SH2 domains of phosphoinositide 3-kinase recognize the sequence tyrosine(P)–x–x–methionine, while that of Grb2 recognizes tyrosine(P) –x–asparagine–x (where tyrosine(P) indicates phosphotyrosine and residues x are highly variable). The recognition motifs are simple enough to allow the interaction of a given SH2 domain-containing protein with several potential target proteins, thus introducing greater flexibility into the signalling pathway. The SH2 domain is of key importance in the recruitment of signalling molecules to activated PTK recep-

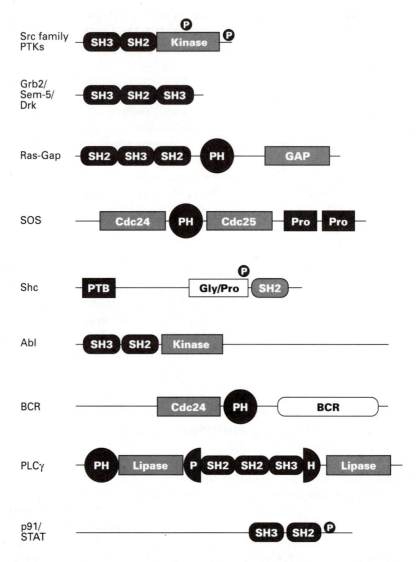

Figure 5.5 Protein–protein interaction domain containing proteins in signalling. The modular structures of the signalling proteins described in the text are shown. The homologous domains such as PH, PTB, SH2 and SH3 are boxed and labelled. Pro indicates proline rich sequences of the type shown to bind to SH3 domains. P indicates important tyrosine phosphorylation sites. Cdc24, Cdc25 are domains involved in nucleotide exchange factors. BCR domains are related to other GAP proteins and have been suggested to have activity against the Rho class of small G-proteins. 'Kinase' and 'Lipase' indicate PTK domain and phospholipase catalytic domains respectively.

tor complexes. This domain links activated receptor PTKs into phospholipid metabolism via phosphoinositide 3-kinase and PLCγ, to the regulation of the small G-protein Ras via Grb2, Shc, IRS-1 and Ras-GAP (see below and Chapter 6), and to transcription factors via the family of STAT proteins (Figure 5.5).

The PTB (**p**hospho**t**yrosine-**b**inding) domain is a second distinct phosphotyrosine-binding domain about which little is currently known (Figure 5.5). First identified in the Shc adaptor protein this domain would appear to have quite different recognition properties from SH2 domains. The Shc PTB domain recognizes the sequence asparagine–proline–*x*–tyrosine(P) with the residues providing the specificity being amino-terminal to the phosphotyrosine residue, in contrast to all known SH2 domain recognition motifs. It is currently believed that the PTB domain plays a similar role to the SH2 domain in the activatiòn of signalling molecules.

5.6 SH3 AND PH DOMAINS ARE OTHER CONSERVED AMINO ACID MOTIFS INVOLVED IN PROTEIN–PROTEIN INTERACTIONS

The function of the SH3 domain is now well understood. The SH3 domain is an independently folding protein module 50–60 amino acids in length that is found in signalling proteins with diverse functions (Figure 5.5). The targets of SH3 domains are proline-rich sequences which take up a particular type of structure, a polyproline type II helix. Two classes of ligands have been identified that bind to SH3 domains in opposite orientations. The core recognition sequences have been defined as arginine–*x*–leucine–proline–proline–*x*–proline–*x*–*x* for class I ligands and *x*–*x*–*x*–proline–proline–leucine–proline–*x*–arginine for class II ligands (where residues *x* are highly variable and thus probably determine the binding specificity for individual SH3 domains.

The final domain that we will briefly cover is the PH domain. PH stands for Pleckstrin Homology, with this motif first being identified as a repeat sequence in the protein kinase C substrate of the same name. Despite its relatively recent identification, crystal and NMR structures of this domain have already been presented. The function of this domain remains elusive, although a role in 'membrane association' has been suggested, based on largely circumstantial evidence. While phospholipids such as phosphatidylinositol-4,5-P_2 have been reported to associate with this domain, as have the β/γ subunits of heterotrimeric G-proteins, the true nature of the PH domains remains to be firmly established. However, its widespread presence in diverse, functionally unrelated proteins strongly argues for a key role in signalling processes (Figure 5.5).

These four modular protein domains are all clearly important components, which regulate the network of interactions that occur between signalling molecules. This is reflected in the frequent occurrence of these domains in known signalling proteins (Figures 5.5 and 5.6).

Their modular nature allows these domains to be inserted into proteins at vari-

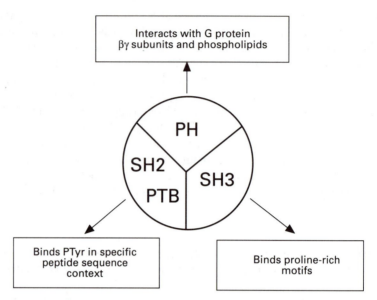

Figure 5.6 Proposed roles for protein–protein interaction modules in signalling. Proteins containing multiple SH2, SH3, PTB and PH domains can form multiple interactions and potentially networks of interacting signalling proteins.

ous points (all have their amino and carboxy termini close together on one face of the domain), and they are often found in multiple copies. It is likely that more such protein association modules await identification in signalling components that coordinate other types of signalling interactions.

5.7 INTRACELLULAR CASCADES AND THE CENTRAL ROLE OF Ras AND MAP KINASE MEDIATED SIGNALS

As indicated in Figure 5.1, receptor protein-tyrosine kinases (PTKs) couple to a wide range of downstream signalling molecules including enzymes involved in lipid metabolism (e.g. phosphoinositide 3-kinase and PLC), G-protein cascades (e.g. Ras, Rac and Rho proteins), protein kinase cascades (e.g. cytosolic PTKs, MAP kinases, protein kinase C and S6 kinases), ion fluxes (e.g. Ca^{2+}) and transcription factors (e.g. p91 and other STAT proteins). Many of the regulatory events involved require interactions between the modular domains described above and their appropriate target sequences. Figure 5.1 indicates the complex nature of the signals that can be activated by a single PTK receptor. So how do SH2 and SH3 domains coordinate to regulate signalling pathways? The best example has come from the analysis of the mechanism by which receptor PTKs activate the small G-protein Ras. The Ras protein and the pathways that it regulates seem to be central to many of the intracellular signalling pathways, including those that regulate growth, differentiation and cell survival. Experimental approaches including genetic analysis of *Drosophila* and *Caenorhabditis elegans* and tissue culture studies using murine fibroblasts lead to a fuller understanding of signalling processes in man.

Several SH2/SH3/PH domain-containing proteins have been implicated in Ras signalling and the precise mechanism of Ras regulation appears to vary with cell type. Key proteins in the activation cascade are Grb2 and Sos. Grb2 is an adaptor-type protein composed of two SH3 domains flanking an SH2 domain. The SH2 domain couples to activated receptor PTKs either directly or via additional adaptor proteins. The two SH3 domains link Grb2 to Sos via proline-rich sequences in the C-terminus of the latter protein. The Sos protein contains a guanine nucleotide exchange activity which exchanges GDP for GTP on Ras, thus activating the signalling properties of Ras. It is likely that Ras activation coincides with colocalization of all the appropriate regulatory machinery at the plasma membrane. Activated Ras binds to and activates its targets, including Raf, a protein-serine/threonine kinase that sits at the top of the MAP kinase cascade, and a phosphoinositide 3-kinase that mediates p70 S6 kinase activation. Activation of these kinases provides amplification of the signal as described earlier. Inactivation of Ras is promoted by the GTPase-activating protein (Ras-GAP), which also possesses SH2 and SH3 domains and is recruited to activated receptor PTK signalling complexes (Figure 5.7).

5.8 SOME CANCER GENES ENCODE SIGNALLING MOLECULES; MUTATIONS IN THESE GENES UPSET CELL BEHAVIOUR BY DEREGULATING SIGNAL PATHWAYS

Cell proliferation, differentiation and apoptosis are regulated by the interplay of signalling pathways. Abnormal expression of signalling molecules can lead to the deregulation of these processes, resulting in neoplasia. This is brought about by the shifting of the delicately balanced equilibrium maintained by components of the signalling pathways between their active and inactive states. The following examples of signalling molecules involved in human cancers will illustrate how overexpression, gene amplification, gene rearrangement and point mutations can disrupt signalling pathways and hence normal cellular processes. The Ras protein described in the preceding section plays an important role in many human cancers and is covered elsewhere in this book (Chapters 6 and 19). We will focus here on a couple of examples of alterations in PTKs in cancer and how this can result in aberrant signalling.

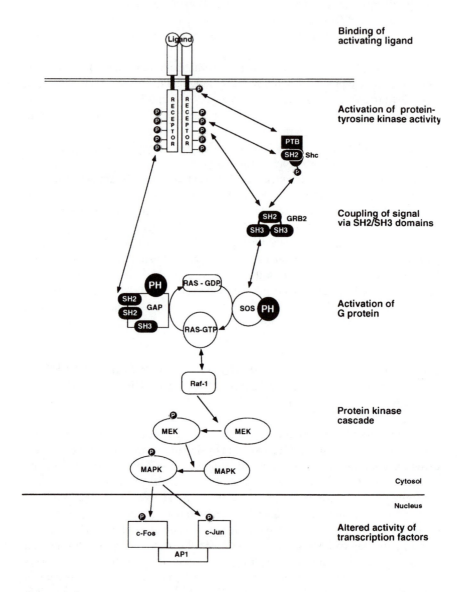

Figure 5.7 Schematic of the role of SH2/SH3 domain proteins in the regulation of Ras. Abbreviations and key to protein interaction domains are given in the caption to Figure 5.5. Double-headed arrows indicate potential protein interactions. Single-headed arrows denote enzymatic processes.

5.9 ACTIVATING MUTATIONS IN PROTEIN-TYROSINE KINASE (PTK) RECEPTORS ACTIVATE DOWNSTREAM SIGNAL PATHWAYS

In the simplest model, gene amplification or overexpression of receptor PTKs can lead to the elevation of PTK activity and the constitutive activation of downstream signalling pathways, resulting in uncontrolled mitogenesis. Examples include the EGF receptor and the related protein, c-ErbB2. Overexpression of these receptors is associated with about 30% of breast or ovarian cancers. Overexpression is thought either to result in an enhanced sensitivity to low concentrations of their ligands, or to drive ligand-independent dimerization and PTK activation.

Other means of PTK receptor activation have been described. The *ret* proto-oncogene is a receptor PTK that currently has no known ligand or substrates. However, it has recently been described as a dominant acting oncogene in three different inherited cancer syndromes, multiple endocrine neoplasia type 2A (MEN2A), MEN2B and familial medullary thyroid carcinoma (FMTC). Over 90% of MEN2A and FMTC patients have missense point mutations, resulting in the substitution of one of the five cysteine residues in the *ret* extracellular domain. These mutated *ret* proteins display enhanced receptor dimerization and elevated PTK activity compared with wild-type *ret*. It is likely that the substitution of cysteine in the extracellular domain enhances ligand-independent dimerization and autophosphorylation of receptors, producing a constitutively active Ret protein kinase (Figure 5.8).

In about 97% of MEN2B mutations Ret exhibits a methionine to threonine substitution at amino acid 918 of the catalytic core of the PTK domain altering its activity. PTK domain mutations have also recently been described in samples from some FMTC patients. A comparison of phosphotyrosyl proteins between cells expressing MEN2A-Ret and MEN2B-Ret proteins suggests that they may also have different intracellular substrates. This indicates that mitogenic signalling pathways may be deregulated in cancers by altering the response of a PTK receptor to ligand binding, by elevating its unstimulated PTK activity or possibly by changing the substrate proteins with which it interacts.

5.10 MUTATIONS OF GENES ENCODING INTRACELLULAR SIGNALLING MOLECULES ARE ALSO ASSOCIATED WITH CANCER

The final protein-tyrosine kinase (PTK) to consider is an intracellular non-receptor kinase rather than a transmembrane PTK. Hybrid proteins of *bcr* (**b**reakpoint **c**luster **r**egion) gene product and the cytoplasmic PTK Abl are found in patients with chronic myelogenous leukaemia (CML) and acute lymphocytic leukaemia (ALL). The chromosome translocation of c-*abl* from chromosome 9 and *bcr* on chromosome 22 leads to fusion of 5' *bcr* sequences to 3' *abl* sequences. This

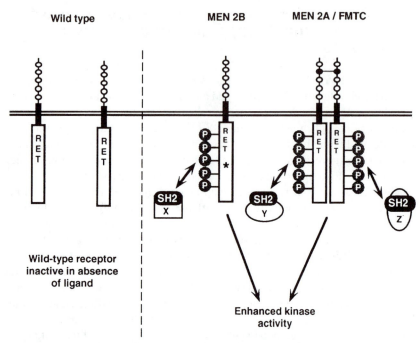

Figure 5.8 Activation of the RET PTK signalling pathway in tumours. Point mutations in the kinase domain (*) increase RET PTK activity, whereas substitution of one of the Cys residues (•) in the extracellular domain enhances dimerization, thereby increasing PTK activity. Differential coupling of the different activated forms of RET to common and specific signalling pathways is indicated by the SH2 proteins X, Y and Z.

results in the generation of p210 or p185 Bcr–Abl proteins. These two hybrid proteins are identical in their *abl* sequences, but differ in the extent of the acquired *bcr* sequences at the N-terminus of the fusion. Both p210 and p185 are transforming *in vitro* and possess deregulated PTK domains. A combination of mutagenesis and protein-binding studies have revealed that the *bcr*-derived sequence both deregulates the Abl protein kinase domain and confers novel signalling properties to the fusion protein (Figure 5.9).

Sequences in the *bcr* portion of Bcr–Abl provide novel tyrosine phosphorylation sites that are able to bind to the SH2 domain of the Abl protein to give the Bcr–Abl fusion protein elevated PTK activity required for transformation. The deletion of the Abl SH3 domain from c-Abl has also been shown to result in elevated Abl PTK activity, suggesting that other factors may be involved in controlling the transformation activity of Bcr–Abl *in vivo*, e.g. intramolecular interactions between *bcr* sequences and the Abl SH3 domain. Furthermore, the autophosphorylation of the *bcr* sequence at tyrosine residue 177 of Bcr–Abl appears to be required for transforming activity, presumably by binding to the SH2 domain of Grb2 (this tyrosine lies in a tyrosine–*x*–asparagine–*x* sequence motif) and thus

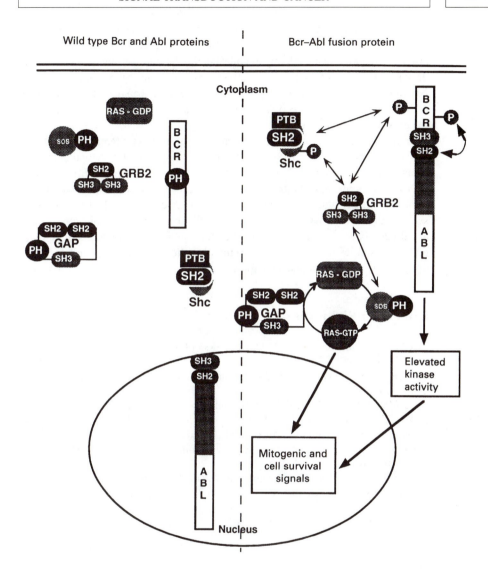

Figure 5.9 Aberrant signalling by Bcr–Abl fusion proteins in leukaemia. The wild-type Bcr and Abl proteins are localized to the cytoplasm and nucleus respectively, while the Bcr–Abl hybrid protein is cytoplasmic. The combination of increased Abl PTK activity at an inappropriate site with coupling to mitogenic signalling pathways (e.g. Ras/MAP kinase) leads to the oncogenic activity of the Bcr–Abl protein. (See Figure 5.5 for the key to the protein–protein interaction domains.)

activating the Ras/MAP kinase pathway and stimulating mitogenesis. Associations of the Bcr–Abl hybrid with phosphoinositide 3-kinase, Ras-GAP and PLCγ have also been reported that are likely to contribute to the transforming potential of this protein. In addition to the positive effects exerted on mitogenic pathways, both *bcr–abl* and the murine retroviral oncogene v-*abl* have been demonstrated to have additional inhibitory effects on apoptosis *in vitro*, which may further contribute to the growth potential of tumour cells.

In the brief space available to us in this chapter we have attempted to highlight the key concepts and, with a few relevant examples, current developments in the signal transduction field. Though incomplete, it is hoped that this will indicate to the reader how an appreciation of this difficult subject can contribute to our understanding of many types of human disease including cancer. While the networks of signalling pathways that exist in any given cell are clearly incredibly complex, we are already beginning to identify the central regulatory components. It is hoped that these studies will in time lead to the identification of new targets for novel drugs and better therapeutic strategies.

5.11 FURTHER READING

Moodie, S. A. and Wolfman, A. (1994) The 3Rs of life: Ras, Raf and growth regulation. *TIG*, **10**, 44–48.

Pawson, T. (1994) SH2 and SH3 domains in signal transduction. *Adv. Cancer Res.*, **64**, 87–110.

Pawson, T. (1995) Protein modules and signalling networks. *Nature*, **373**, 573–580.

Puil, L., Liu, J., Gish, G. *et al.* (1994) Bcr–Abl oncoproteins bind directly to activators of the Ras signalling pathway. *EMBO J.*, **13**, 764–773.

Santoro *et al.* (1995) Activation of RET as a dominant transforming gene by germline mutations of MEN2A and MEN2B. *Science*, **267**, 381–383.

van der Geer, P., Hunter, T. and Lindberg, R. A. (1994) Receptor protein-tyrosine kinases and their signal transduction pathways. *Annu. Rev. Cell. Biol.*, **10**, 251–337.

G proteins, Ras and cancer | 6

Julian Downward

6.1 PROTEINS THAT BIND GTP, INCLUDING G PROTEINS AND Ras, ARE IMPORTANT COMPONENTS OF SIGNAL TRANSDUCTION PATHWAYS

The interpretation of extracellular signals and their translation into intracellular effects is central to all biological systems and is of particular importance in complex multicellular organisms (Chapter 5). Information is gathered from outside the cell by the binding of signalling molecules such as hormones or growth factors to cell surface receptors that span the plasma membrane. The protein molecules responsible for transferring information from these receptors on to intracellular target enzymes fall into a number of functional families: one of the largest is the protein kinases, while another is the guanine nucleotide-binding proteins, the subject of this chapter.

Guanine nucleotide-binding proteins are found in many different forms. They all display very limited sequence homology, which in most cases is restricted to the sites of contact of the protein with the guanine nucleotide. They all share the ability to bind guanosine triphosphate (GTP) and catalyse its hydrolysis to guanosine diphosphate (GDP). The GTP-bound form of the protein has different properties and a different conformation to the GDP-bound form. Not all guanine nucleotide-binding proteins are involved in signalling pathways: a number are structural, such as tubulin, while others are involved in macromolecular synthesis, such as the elongation factors of protein synthesis.

The signal transducing guanine nucleotide-binding proteins are made up of two principal groups: the heterotrimeric GTP-binding proteins, generally referred to as **G proteins**, and the monomeric Ras superfamily proteins. For all these proteins the GTP-bound form represents an active state that is capable of interacting with some cellular target enzyme system, while the GDP-bound form is without

Molecular Biology for Oncologists. Edited by J. R. Yarnold, M. R. Stratton and T. J. McMillan. Published in 1996 by Chapman & Hall, London. ISBN 0 412 71270 9

biological activity. The activity of the proteins is regulated by modulating the rate at which bound GDP is replaced by GTP and the rate at which bound GTP is hydrolysed. GTP-binding proteins are often compared to a timer switch; they have an on (GTP-bound) and an off (GDP-bound) state and an inbuilt clock (the rate of the GTP hydrolysis reaction).

6.2 G PROTEINS ARE COMPOSED OF THREE DIFFERENT PROTEIN SUBUNITS

The heterotrimeric G proteins are usually located at the inner surface of the plasma membrane, held in place by covalently attached lipids (palmitylations or myristylations). They are made up of three different subunits: the α subunit, which contains the nucleotide binding site, and the ß and γ subunits. So far some 20 distinct mammalian α subunits have been characterized, along with five ß and six γ subunits. These could be combined in different permutations to give perhaps 600 distinct G proteins. The α subunits characterized so far are at least 50% identical to each other at the level of amino acid sequence, while the ß and γ subunits are even more conserved at 80% or more identity. Remarkably similar G proteins are found in all eukaryotic organisms including yeast.

6.3 G PROTEINS LINK CELL SURFACE RECEPTORS TO INTRACELLULAR ENZYMES

G proteins all interact directly with members of a class of cell surface receptor proteins that have seven membrane-spanning regions. These share a degree of amino acid sequence similarity and include the adrenergic receptors, the muscarinic acetylcholine receptors, rhodopsins, neuropeptide receptors, purinergic receptors and many more. In all, more than 100 distinct receptors have been identified that are coupled to G proteins. The G proteins only interact with activated receptors, i.e. receptors that have bound to their specific ligand. This interaction stimulates the exchange of GDP bound to the α subunit of the inactive heterotrimer for GTP (Figure 6.1).

The α subunit now acquires the active, GTP-bound conformation and dissociates from the ßγ subunits. The GTP-bound α subunit then goes on to interact with and activate its target enzyme, often referred to as an **effector**. In an interesting economy of design, the free ßγ subunits can also interact with and regulate several different effector enzymes. The signal is terminated when the α subunit hydrolyses GTP to GDP, an event in some cases catalysed by interaction with the effector, and reassociates with ßγ. Since each ligand-bound receptor can activate several molecules of G protein and each G protein can activate several molecules of effector, there is an inbuilt signal amplification. This is particularly marked in the detection of light in the retina.

The effector systems controlled by G proteins are almost as diverse as the

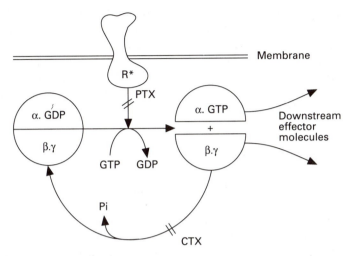

Figure 6.1 The G protein cycle. Activated receptor (R*) causes αβγ trimer to release GDP and take up GTP. This complex is unstable and breaks up to give α.GTP and ß.γ, both of which can control various effector enzymes. GTP on α is hydrolysed to GDP, resulting in rebinding of α.GDP to ß.γ and signal termination. Cholera toxin (CTX) blocks GTP hydrolysis on α; while pertussis toxin (PTX) inhibits the interaction of activated receptor with G protein trimer.

receptors involved. They include several types of adenylyl cyclases, phospholipases C and A$_2$, cyclic GMP phosphodiesterase and many different ion channels. These enzymes all are capable of causing changes in the intracellular concentration of small molecules, or ions, involved in cellular signalling.

6.4 G PROTEINS ARE TARGETS FOR BACTERIAL TOXINS

Recently it has been established that G proteins play an important role in a number of human tumours. Their original identification, however, was due to their involvement in two well known infectious diseases. *Vibrio cholerae* produces severe disease through the production of cholera toxin; this multisubunit toxin delivers an ADP ribosylase to target cells. This ADP ribosylase is specific for the α subunit of G$_s$, a G protein that normally functions to positively regulate the activity of adenylyl cyclase and hence cyclic AMP levels. Cholera toxin ADP ribosylates G$_{s\alpha}$ at arginine 201 and causes a very large reduction in the rate at which bound GTP is hydrolysed. The G protein therefore remains locked in the active, GTP-bound state, leading to the continuous activation of its effector, adenylyl cyclase, and the generation of abnormally high levels of cyclic AMP. It is the uncontrolled production of cAMP in the cells lining the gut that leads to the characteristic symptoms of cholera.

Another organism, *Bordetella pertussis*, produces a G-protein-specific ADP ribosylase. In this case the target is $G_{i\alpha}$, a closely related subgroup of G proteins that also regulate adenylyl cyclase, but in an inhibitory manner. Pertussis toxin ADP ribosylates $G_{i\alpha}$ on cysteine 350; this results in an inhibition of the interaction of the G protein with its activating receptor. The G protein cannot therefore be stimulated to a GTP-bound state which would inhibit adenylyl cyclase, but instead remains inactive. As with cholera, a rise in the intracellular levels of cAMP results.

6.5 G_s CAN ACT AS AN ONCOGENE

In 1989 the discovery was made that heterotrimeric G proteins can act as oncogenes and be involved in the development of tumours (Table 6.1).

Analysis of $G_{s\alpha}$ in eight pituitary tumours revealed that half had undergone point mutations, either at codon 201 (the site of ADP ribosylation by cholera toxin) or at codon 227. Both mutations caused strongly reduced GTPase activity and the tumour cells contained elevated cAMP levels; intracellular cAMP concentration is known to correlate with growth in this cell type. More recently, similar mutations have been found in $G_{s\alpha}$ in a number of thyroid tumours. As with all known oncogenes, the mutation of $G_{s\alpha}$ is likely to represent only one of several steps required to achieve a full malignant phenotype. The limited number of tumour types in which mutations in $G_{s\alpha}$ have been found may reflect the fact that cyclic AMP is only growth-stimulatory in a limited number of cell types. In addition, activating mutations in $G_{s\alpha}$ have been found to be responsible for McCune–Albright syndrome and rare cases of gonadotrophin-independent precocious puberty, or testotoxicosis. Inactivating mutations in $G_{s\alpha}$ are responsible for a number of pseudohypoparathyroidisms and Albright's hereditary osteodystrophy.

Table 6.1 Activated G protein oncogenes in human tumours

Tumour type	Incidence (%)	Predominant oncogene
Pituitary adenoma (growth-hormone-secreting)	43	$G_{s\alpha}$
Thyroid carcinoma	4	$G_{s\alpha}$
Ovarian stromal tumours	30	$G_{i2\alpha}$
Adrenal cortical tumours	27	$G_{i2\alpha}$

6.6 G$_i$ CAN ALSO ACT AS AN ONCOGENE

A second type of heterotrimeric G protein, G$_{i2\alpha}$, has also been shown to be capable of acting as an oncogene. A subset of adrenal cortical and ovarian tumours carry mutations at sites cognate to those mutated in G$_{s\alpha}$. These mutations again give rise to a protein that is constitutively activated due to decreased GTPase activity. While it has been proven that mutant G$_{i2\alpha}$ can transform Rat-1 fibroblast cells in culture, it is not clear by what mechanism this occurs; the documented targets for G$_{i2\alpha}$, adenylyl cyclase inhibition, potassium channels and phospholipase A$_2$, have not been correlated with growth stimulation in adrenal cortical or ovarian cells. Given the rapid progress in this field, it is quite likely that mutation of other G proteins will be found to be involved in the genesis of a number of other tumour types.

6.7 Ras PROTEINS MAKE UP A SUPERFAMILY OF RELATED GTP-BINDING PROTEINS

Like the heterotrimeric G proteins, members of the Ras superfamily of proteins bind GTP and hydrolyse it to GDP, being active in the GTP-bound state and inactive in the GDP-bound state. However, they are monomeric and only about half the size of a G protein α subunit, hence the term 'low molecular weight GTP binding protein' which is sometimes used to describe them. They possess only limited homology to G proteins; how analogous their functions are is still a matter of debate.

Some 60 distinct mammalian Ras superfamily proteins have been identified. They are all at least 30% related to each other and can be grouped into three or more distinct families (Table 6.2) in which each member is at least 50% homologous to every other member.

Table 6.2 The mammalian *ras* superfamily

ras *family*	rho *family*	rab *family*
H-*ras*	*rhoA*	*rab1A*
K-*ras* (*A* and *B*)[a]	*rhoB*	*rab1B*
N-*ras*	*rhoC*	*rab2*
	rhoG	*rab3A*
rap1A[b]	*rac1*	*rab3B*
rap1B	*rac2*	*rab4*
rap12	*TC10*	*rab5*
	CDC42 HS	*rab6*
R-*ras*		*rab7*
ralA		*BRL-ras*
ralB		*rab8*
		rab9
		rabs10–17

[a] Products of alternative splicing
[b] Also known as *Krev-1* and *smg p21*

The main families within the Ras superfamily are those named after, and typified by, the Ras, Rho and Rab proteins. Ras proteins were the first to be identified, due to their role in human cancer (see next section); the others were subsequently identified due to their sequence homology to Ras. Three very closely related proteins in the Ras family, H-Ras, K-Ras and N-Ras, are primary regulators of cell growth and frequently mutationally activated in tumours. The function of other members of the Ras family (R-Ras, TC21, Ral, Rap) is not known, though there is some evidence that Rap may act antagonistically to Ras. TC21, and to a lesser extent R-Ras, may be able to act in a similar way to Ras and are capable of weakly transforming some cells in culture.

6.8 Ras PROTEINS REGULATE CELL PROLIFERATION

The mechanisms by which the three oncogenic Ras proteins control cell growth has been the subject of intense study. In the last couple of years considerable progress has been made towards a full understanding of their regulation and function. At least one of the three oncogenic Ras proteins is expressed in all eukaryotic cells, and in most cases disruption of the function of these proteins results in an arrest of cell growth. In a few cell types, particularly neuronal lineages, Ras proteins appear to regulate a differentiation response rather than cell growth.

6.9 GTPase ACTIVATING PROTEINS (GAP) SPEED GTP HYDROLYSIS AND Ras INACTIVATION

Most of the progress made in understanding the regulation and function of Ras proteins has been through the study of proteins that interact with them. Unlike G proteins, Ras proteins have very slow endogenous rates of GTP hydrolysis: the first protein, $p120^{GAP}$, that was found to interact with Ras was a GTPase activating protein (GAP) that speeded up the hydrolysis reaction and hence the inactivation of the Ras proteins (Figure 6.2).

At least one other GAP exists for the oncogenic Ras proteins: neurofibromin, the product of the Type 1 neurofibromatosis gene (see below). These GAPs act as negative regulators of Ras; in some cell types the activation state of Ras is regulated by control of the activity of these proteins.

6.10 EXCHANGE FACTORS ARE PROTEINS THAT PROMOTE Ras ACTIVATION

A more ubiquitous control mechanism for Ras proteins is through regulation of the rate of guanine nucleotide exchange. Various exchange proteins have been identified which are likely to be key activators of Ras (Figure 6.2). These have

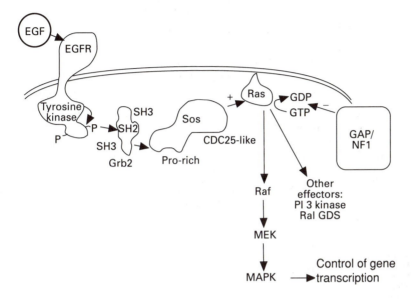

Figure 6.2 The Ras cycle. Activated cell surface receptor binds exchange factor (Sos) via adaptor protein (Grb 2). GDP-bound Ras is stimulated to exchange GDP for GTP by interaction with exchange factors (Sos). Ras.GTP interacts with effectors such as Raf to transmit its growth signal to the cell. The signal is terminated when GTP on Ras is hydrolysed to GDP, a reaction catalysed by GAP and neurofibromin (NF1).

been named CDC25 and Sos after their homologues in yeast and fruit fly, which were discovered first. Ras is activated when cells are treated with any of a number of growth factors, especially those that regulate tyrosine kinases.

6.11 ADAPTOR PROTEINS LINK ACTIVATED CELL-SURFACE RECEPTORS TO EXCHANGE FACTORS PRIOR TO Ras ACTIVATION

The simplest version of the regulation of the activation state of Ras in response to growth factor stimulation of cells is this: growth factor binding to receptor tyrosine kinases at the cell surface causes activation of their kinase activity through dimerization. A direct result of this is autophosphorylation on tyrosine of the intracellular part of the receptor. The tyrosine phosphorylated sequences on the receptor form specific binding sites for proteins containing the Src-homology-2 (SH2) class of domains. SH domains are amino acid sequences that recognize and bind specific sequences in other proteins. A subset of these proteins is known as the adaptor proteins; they contain only protein binding amino acid motifs, such as SH2, which binds tyrosine phosphorylation sites, and SH3, which binds proline rich sequences, and do not possess any catalytic activity. The regulation of Ras involves tyrosine phosphorylated receptors binding to the adaptor protein Grb2, which is made up solely of one SH2 and two SH3 domains. This in turn is bound to the Ras-specific guanine nucleotide exchange factor Sos. Thus, upon growth factor binding to its receptor, the exchange factor Sos is brought into a multicomponent complex at the plasma membrane via Grb2 (Figure 6.2). It is likely that the activation of Ras results from the localization of Sos, which is normally cytosolic, to the plasma membrane where Ras is located. Further complexity is found in this signalling system in many instances; one of the most common is the involvement of another adaptor protein, Shc, as an intermediate between the receptor tyrosine kinase and Grb2. Shc is phosphorylated by activated tyrosine kinases and then binds to the SH2 domain of Grb2.

6.12 Raf IS AN IMPORTANT DOWNSTREAM TARGET OF ACTIVATED Ras

Great advances have also been made in understanding the growth signalling pathway downstream of Ras. Recent years have seen the identification of not just one but several direct target enzymes for Ras. The best characterized of the Ras 'effector' enzymes is the cytosolic serine/threonine protein kinase Raf. The amino terminal regulatory region of this protein interacts directly with Ras in its GTP-bound active conformation, but not the inactive Ras-GDP. Interaction of Ras with Raf brings the kinase to the plasma membrane, which appears to be essential for its function. It may be that there are additional activation events that occur once Raf has bound to Ras at the membrane.

6.13 ACTIVATED RAF PHOSPHORYLATES A MITOGEN-ACTIVATED PROTEIN CALLED MAP KINASE KINASE

The activated Raf kinase phosphorylates another enzyme, MAP kinase kinase, which, as its name indicates, phosphorylates an enzyme called MAP kinase (from mitogen-activated protein). This kinase cascade serves to amplify the signal from Ras and facilitates communication with other signal pathways. The activated MAP kinase has several target proteins, including various proteins in the nucleus involved in the regulation of gene expression. In this manner the growth signal is transmitted from outside the cell to the cell nucleus. Unfortunately, matters may not be a simple as this single linear pathway suggests. A number of other proteins have been identified that bind directly to activated Ras, including the lipid kinase phosphatidylinositol 3-OH kinase and Ral-GDS, an exchange factor for the Ras-related protein Ral. At present it is unclear exactly how much these pathways contribute to the normal or transforming function of Ras, but there is good evidence that the Raf/MAP kinase pathway alone is insufficient to account for all the effects of Ras on the cell.

6.14 Rho PROTEINS CONTROL THE STRUCTURE OF THE ACTIN CYTOSKELETON

The other main families of Ras-related proteins have very different, but equally critical, functions. Members of the Rho family of proteins are involved in controlling the structure of the actin cytoskeleton. Rho controls actin stress fibre formation while Rac controls actin-mediated plasma membrane ruffling and Cdc42 regulates the formation of filopodia, cell surface actin processes. Like Ras, they are activated by various mitogenic growth factors. GAPs and exchange factors have also been identified for members of this family; intriguingly, some of these have been previously identified as oncogenes in their own rights (Vav, Dbl). Indeed, some Rho family proteins, particularly Rac, can act as weak oncogenes, and synergize well with certain other oncogenes such as Raf in causing neoplastic transformation. The Rab family of Ras-related proteins is involved in the regulation of secretion and intracellular vesicle traffic, with a different Rab protein being associated with each of the intracellular vesicular compartments, possibly as a marker or sorting protein. GAPs and exchange factors have also been identified for these proteins.

6.15 Ras PROTEINS ARE ACTIVATED BY POINT MUTATIONS IN HUMAN CANCERS

Ras proteins are best known for their involvement in human cancer. The Ras family members that are capable of causing malignant transformation are H-Ras, K-Ras and N-Ras. Of these, H- and K-Ras were first discovered as the products of the retroviral oncogenes of the Harvey and Kirsten murine sarcoma viruses. In the late 1970s it was recognized that retroviruses had picked up their transforming oncogenes from the host genome. In a turning point for the understanding of the molecular basis of cancer a transforming gene was discovered in a human tumour which was almost identical to the retroviral H-*ras* oncogene. The ability to transfer the *ras* gene from one cell to another by calcium-phosphate-mediated transfection led to the discovery that a large number of human tumours carried activated forms of the H- or K-*ras* genes, or the closely related N-*ras* oncogene. The difference between the oncogenic *ras* genes found in tumours and the normal *ras* genes was a point mutation that resulted in a constitutively active form of the protein. Almost all the mutations found in tumours occur at codons 12, 13 or 61: knowledge of the three-dimensional structure of the Ras protein shows that these residues all lie at points of contact of the Ras protein with the phosphate groups of

GTP. The mutations result in a reduced rate of GTP hydrolysis and a resistance to the inhibitory effects of GAPs. The mutant Ras proteins are therefore locked into an active GTP-bound state and transmit a constitutive growth signal to the cell.

The advent of the polymerase chain reaction has lead to the examination of enormous numbers of human tumours for mutations in *ras* and other oncogenes (Table 6.3).

Ras mutations appear in nearly 30% of all tumours examined, with extremely high rates in pancreatic carcinoma (95%), thyroid tumours (60%) and adenocarcinoma of the colon (50%). Analysis of the time course of activation of *ras* has shown it to be a relatively early event in the genesis of many colon carcinomas. *Ras* is known to require other cooperating activated oncogenes, such as *p53* or *myc*, to cause full transformation in various model systems.

Table 6.3 Activated *ras* oncogenes in human tumours

Tumour type	Incidence (%)	Predominant oncogenes
Pancreatic carcinoma	95	K-*ras*
Thyroid carcinoma	60	N-*ras*
Colon adenocarcinoma	50	K-*ras*
Colon carcinoma	50	K-*ras*
Seminoma	40	K-*ras*, N-*ras*
Lung adenocarcinoma	40	K-*ras*
Myelodysplastic syndrome	30	N-*ras*
Acute myelogenous leukaemia	30	N-*ras*
Keratocanthoma	30	H-*ras*
Melanoma	30	N-*ras*
Squamous cell carcinoma	25	H-*ras*
Bladder carcinoma	15	H-*ras*
Breast carcinoma	< 5	
Cervical carcinoma	< 5	
Ovarian carcinoma	< 5	
Stomach carcinoma	< 5	

6.16 NEUROFIBROMATOSIS TYPE 1 INVOLVES DEFECTIVE *Ras* INACTIVATION BY GAP

An unexpected involvement has recently emerged of *ras* in a hereditary disease that is characterized by a high incidence of tumours: type 1 neurofibromatosis

(NF1). This is caused by inheritance of a defective allele of the *NF1* gene, which has been shown to encode a GTPase activating protein (GAP) for *ras*. Since GAPs are negative regulators of *ras*, NF1 patients may be especially susceptible to development of certain tumours due to reduced ability to switch off *ras*, especially in cells where the single healthy allele of *NF1* has been lost. Increased tumour incidence is confined to cells of neural crest origin, with benign neurofibromas being found in almost all patients. It has been suggested that the *NF1* gene product might also be involved in delivering a differentiation signal from Ras to neural crest cells; failure to differentiate, or dedifferentiation, is often associated with neoplasia.

As yet there is little evidence for involvement of other members of the *ras* superfamily in the formation of human tumours. As for other diseases, it is possible that the ADP ribosylation of Rho protein by *Clostridium botulinum* exoenzyme C3 could contribute to the illness caused by that organism.

6.17 THERE ARE INTERESTING PROSPECTS FOR THERAPIES

The rapid advances in understanding the role of GTP binding proteins in the formation and maintenance of the transformed phenotype have not yet led to improvements in therapy or, with rare exceptions, diagnosis. One case in which *ras* has been used for diagnosis is the use of polymerase chain reaction to identify mutations in *ras* from stool samples of patients with colorectal tumours. Considerable efforts are being made by a number of pharmaceutical companies to develop drugs that will interfere with the function of Ras proteins, with the hope that the removal of the Ras-dependent growth signal from tumour cells might cause their growth arrest or death (Chapter 19). Two aspects of Ras function have been targeted for drug development; one is the essential post-translational modification of Ras by farnesylation that is required for its localization to the plasma membrane and subsequent biological activity and the other is its interaction with its cellular effectors.

Drugs that inhibit the interaction of Ras with its effectors have proved to be hard to develop, although the known three-dimensional structure of Ras is available for modelling. However, the enzymes involved in the post-translational modification are now well characterized and a number of good pharmacological inhibitors of these are now available. As with all drugs, a major worry with the use of such agents is the lack of specificity: they would be very likely to effect other farnesylated proteins, such as nuclear lamins, and would inhibit normal Ras in healthy tissues. However, use of a number of farnesyl transferase inhibitors in animal studies indicate that they have very low toxicity. Indeed, testing of these compounds on transformed cells in culture and Ras-induced tumours in animals have produced some very encouraging results. It is no exaggeration to say that farnesyl transferase inhibitors hold out a very good prospect of being an effective and non-toxic treatment for several types of cancer, including some very common ones, in the not too distant future. If this turns out to be true, it will be the first case in which basic-science–driven rational drug design has led to an effective cancer therapy.

Recently, clinical trials have been carried out of experimental gene therapies to treat disseminated lung cancer with vectors that would express *ras* antisense messenger RNA. These have not been particularly successful as yet. As with all gene therapy approaches to cancer treatment, the major difficulty is likely to be the delivery of the therapeutic DNA to all the affected cells. However, as gene therapy protocols and, probably more importantly, drug design improve, the prospect of targeting mutant *ras* as an effective way of treating human cancer becomes increasingly promising.

6.18 FURTHER READING

Gibbs, J. B., Oliff, A. and Kohl, N. E. (1994) Farnesyltransferase inhibitors – *ras* research yields a potential cancer therapeutic. *Cell*, **77**,175–178.

Neer, E. J. (1995) Heterotrimeric G proteins – organizers of transmembrane signalling. *Cell*, **80**, 249–257.

Pawson, T. (1995) Protein modules and signalling networks. *Nature*, **373**, 573–580.

Cell cycle control and cancer | 7

Antony M. Carr

7.1 ENTRY INTO AND EXIT FROM THE CELL CYCLE ARE CONTROLLED BY GENES WHICH ARE DEREGULATED IN CANCER

The cell cycle is the name given to the string of events that results in the duplication of the genetic material and the segregation of these two copies into two daughter cells. The molecular nature of the cell cycle is conserved in all eukaryotic cells. Many of the mechanisms and pathways that control the cell cycle in mammalian cells evolved before the appearance of multicellular organisms and therefore studies of cell cycle control in unicellular eukaryotes such as the yeasts *Saccharomyces cerevisiae* and *Schizosaccharomyces pombe* have greatly facilitated our understanding of how all cells control their division.

In human beings, cancer can be described as the inappropriate division of a clone of cells at an inappropriate time. Differentiation and programmed cell death usually ensure that our bodies do not continue to increase in mass and determine that only those cells that should divide (such as stem cells) proliferate. Following the introduction of errors into the genes that control the cell cycle (either by mutation, misreplication or aberrant chromosome segregation) the complex controls that govern entry into and exit from the cell cycle can fail, and then the resulting cell can acquire the potential to proliferate in an uncontrolled manner and ultimately may form a tumour which can kill the organism.

An understanding as to how the cell cycle is controlled, and how these controls are compromised in cancer cells, is therefore essential to our understanding of oncogenesis. In this chapter, the cell cycle and its control is introduced and linked to our current understanding of cancer biology.

Molecular Biology for Oncologists. Edited by J. R. Yarnold, M. R. Stratton and T. J. McMillan. Published in 1996 by Chapman & Hall, London. ISBN 0 412 71270 9

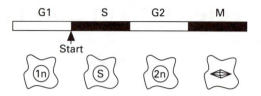

Figure 7.1 The cell cycle. 'Start' is the point at which cells make a decision whether to divide or to enter an alternative developmental state such as differentiation. The G2/M transition is a point at which cells are able to arrest should circumstances (DNA damage, for example) change during the cell cycle. Passage through both 'Start' and the G2/M transition is controlled by the activity of cyclin-dependent kinases.

7.2 DNA SYNTHESIS AND MITOSIS ARE THE MAJOR EVENTS OF THE CELL CYCLE

The two major events of the cell cycle are **DNA synthesis**, when the genetic material is copied by the replication enzymes, and **mitosis**, when the two copies of the genetic material are segregated to the daughter cells. These two key events are separated by 'gap' periods known as G1 and G2 (Figure 7.1).

Therefore, the cell cycle is usually represented as being composed of four stages, G1–S–G2–M, which occur in a cyclical manner in rapidly dividing cells. Many of the molecular events that comprise the cell cycle (DNA replication, mitosis, etc.) have been elucidated and describe a complex machine, which undertakes the many specific yet distinct events which together result in the creation of two identical daughter cells. This machine contains an intrinsic 'programme' that determines the correct order of these events and ensures that each cycle produces two daughter cells. The control of when this machine is allowed to run, and how the 'programme' can be interrupted to allow flexibility when mistakes are encountered is clearly important to the control of cell proliferation and to its accuracy.

7.3 THERE ARE TWO MAJOR CONTROL POINTS IN THE CELL CYCLE: 'START' AND THE G2/M TRANSITION

Even in simple eukaryotes, decisions on whether to proceed with a new round of cell division are made in response to environmental signals. In yeast these are mainly criteria such as nutrient availability and the presence of mating pheromone (a signalling peptide that diffuses from cells of the opposite mating type). In more complex organisms, where a steady supply of nutrients is maintained by the circulatory system, the equivalent environmental signals are the presence of a multitude of growth factors and inhibitors and cell-to-cell contact.

There are two major decision points within the cell cycle. The first is called 'Start', and defines the point at which cells are committed to exit from G1 and enter the S phase of the cycle. In mammalian cells this is the major transition point in the cell cycle (also known as the **restriction point**) and it is often described as the point at which cells become committed to a complete cell cycle. This is because cells that are not cycling usually exist in a specialized state (referred to as G0, or stationary phase) which they enter from the G1 period of the cycle. Growth-stimulating signals therefore induce passage through 'Start' and growth-inhibitory signals prevent passage through 'Start'.

A second decision point exists prior to the onset of mitosis, at the G2/M border. In some instances, such as in liver tissue, cells can enter stationary phase from G2, before transition through the G2/M decision point, but this is unusual. The usual function of the G2/M control point seems mainly to ensure that the DNA is free of errors and is ready to be separated at mitosis. Since mitosis is an irreversible act, it is vital that the DNA is completely replicated and undamaged when mitosis occurs. The pathways that ensure that this is the case, and that act to delay mitosis when problems are detected (i.e. prevent passage through the G2/M transition point and thus temporarily interrupt the cell cycle 'programme') are known as **checkpoints**.

7.4 PROGRESSION THROUGH THE CELL CYCLE IS DETERMINED BY THE ACTIVITY OF CYCLIN-DEPENDENT KINASES

Work on the control of the cell cycle has identified many of the proteins that coordinate and control the cycle in all eukaryotic cells. Key regulators of cell cycle progression are the cyclin-dependent kinases (CDKs), which consist of two major subunits, a catalytic subunit and a targeting (cyclin) subunit. In mammalian cells there are at least six members of the catalytic subunit family, while the simpler yeast cells contain a single such protein ($p34^{cdc2}$). The kinase activity of all CDK catalytic subunits is dependent on their association with cyclins. All cells contain a large number of cyclin proteins which act during the different stages of the cell cycle. Different cyclins are required at different points in the cell cycle and are thought to target the CDK activity to particular substrates. A distinctive feature of cyclins is that they fluctuate in level during cell cycle progression. Different cyclins peak in level at different stages of the cell cycle (i.e. when they are required) and are then destroyed once that point is passed. Hence some cyclins are known as G1 cyclins, others as G2 cyclins and yet others as S phase cyclins (Figure 7.2).

Different combinations of $p34^{CDK}$ and cyclin subunits form kinase complexes which act to control progress through different points in the cell cycle. The association of some human CDK catalytic subunits with G1 cyclins form the kinase complexes responsible for passage through 'start' and entry into S phase. Different CDK–cyclin complexes are then required to coordinate DNA synthesis. Finally,

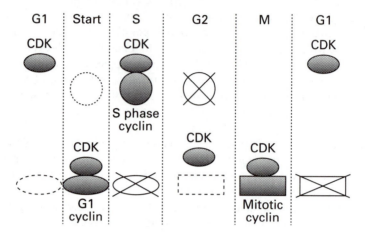

Figure 7.2 Different cyclins complex with CDK catalytic subunits at different cell cycle stages. Different cyclins are present at different times in the cell cycle, and are destroyed by proteolysis once they are no longer required. The different cyclins define the different stages in the cell cycle and their periodic appearance and disappearance helps to maintain the intrinsic 'programme' that determines the order of the different events that comprise the cell cycle.

further CDK catalytic subunits form complexes with the mitotic cyclins and initiate and coordinate mitosis.

Therefore, the association of CDK catalytic subunits with the different cyclin subunits at each of the different stages of the cell cycle helps coordinate the 'programme' that is intrinsic to the cell cycle. However, in order to stop or pause the cell cycle at either of the two major control points, further methods of regulating the activity of these CDK complexes are required. Two such methods, phosphorylation of CDK proteins and specific CDK inhibitors, are discussed briefly below.

7.5 THE ACTIVITY OF CDKS CAN BE REGULATED BY TYROSINE PHOSPHORYLATION

The catalytic activity of CDK complexes are regulated at the G2/M transition by tyrosine phosphorylation at amino acid position 15. CDK proteins which are phosphorylated on this residue are inactive, and the cells therefore remain in G2. Passage through the G2/M transition and subsequent entry into mitosis is achieved by dephosphorylation of the tyrosine 15 residue. A balance of kinase and phosphatase activities determines the tyrosine 15 phosphorylation state of the CDK

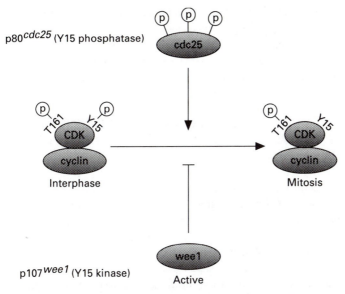

Figure 7.3 Regulation of CDK activity by tyrosine phosphorylation. The activity of CDK–cyclin complexes at mitosis is partly determined by the phosphorylation status of the tyrosine 15 residue (Y15) on the catalytic subunit. Phosphorylated CDK is inactive, and dephosphorylated CDK is active. Thus the kinases that phosphorylate Y15 (the Wee1 family of kinases) inhibits mitosis and the phosphatase which dephosphorylates Y15 (the Cdc25 family) activates mitosis.

complex and in this way determines the timing of mitosis. A family of CDK-specific protein kinases (p107^{wee1}) tyrosine phosphorylates CDK complexes (and inhibits the onset of mitosis), while a family of CDK-specific tyrosine phosphatases (p80^{cdc25}) acts antagonistically to p107^{wee1} kinases to activate CDK complexes by dephosphorylation (Figure 7.3).

A further phosphorylation event, at the threonine residue at position 161 of the catalytic subunit, can also regulate CDK complex function, both at 'Start' (the G1/S transition) and at the G2/M transition. This phosphorylation event is required for CDK activity and is thus another potential mechanism for regulating the activity of CDK complexes.

7.6 CDK COMPLEXES CAN ALSO BE REGULATED BY SPECIFIC INHIBITORS

Recent research has identified a class of molecules that directly bind to and inhibit cyclin-dependent kinases. Thus far at least five such inhibitors have been found in mammalian cells. Some appear to inhibit specific CDK–cyclin complexes, while others appear to be more general in their activity. The best understood of these inhibitors are the p16 and p21 proteins. One of the functions of p16 protein is to arrest cells in G1 when the cell receives specific hormonal signals to cease proliferation. Similarly, p21 is found expressed at high levels in some differentiated tissues such as muscle, suggesting that it has a role in preventing entry into the cell cycle in these circumstances.

7.7 CELL CYCLE CHECKPOINTS ACT TO INHIBIT CDK ACTIVITY AND ALLOW INTERRUPTION OF THE CELL DUPLICATION PROGRAMME

Clearly, the control of the cell cycle machine is complex, and there are many ways to regulate the CDK–cyclin complexes that coordinate progress through the cell cycle. Once a cell has embarked on the cell cycle, it will usually complete the cycle unless an important event (e.g. DNA synthesis) is interrupted or an error is detected. In the event of such an interruption, signalling mechanisms, known as checkpoint pathways, delay passage through the major cell cycle control points ('Start' and/or the G2/M transition) until the defective event is completed or the error is corrected.

Two examples serve to illustrate the action of checkpoints.

- If DNA synthesis is interrupted (for example by chemical inhibitors) then cells do not proceed through the G2/M transition into mitosis until the block is removed and DNA synthesis is completed. This inhibition appears to act through tyrosine 15 phosphorylation of CDK catalytic subunits. In yeast cells, mutations in a number of genes have been identified which abolish this checkpoint. In such cells, a block to DNA synthesis results in the catastrophic attempt to segregate the unreplicated chromosomes and subsequent cell death.
- If the DNA is damaged during the cell cycle, this is recognized and signals are sent that inhibit both the G2/M transition (mitosis) and 'Start' (DNA synthesis). The prevention of passage through 'Start' appears to act through activation of inhibitors of CDK function.

Attempting to either replicate or segregate damaged DNA is likely to lead to the fixation and accumulation of mutations, or genomic instability, both of which are precursors to cancer. Several tumour suppressor genes (see below) are involved in the correct operation of the checkpoints, attesting to the important role the checkpoints play in preventing oncogenesis by maintaining the fidelity of DNA replication and chromosome segregation.

7.8 TUMOUR SUPPRESSOR GENES PROVIDE A LINK BETWEEN CANCER AND THE CELL CYCLE CONTROL PROTEINS

Most tumours are a result of a number of sequential mutations. Typically these include two distinct types:

- the inappropriate activation of proto-oncogenes involved in signalling the presence of growth stimulatory factors;
- the inactivation of tumour suppressor genes (Chapter 1).

Tumour suppressors are often involved in preventing cell cycle progression. Of the various loci that have been identified as tumour suppressors, the major ones are the retinoblastoma (*Rb*) gene, the *p53* gene and the *p16* gene. The exact roles of these tumour suppressor genes in controlling inappropriate proliferation is not fully understood, but it is clear that they ultimately interact with cyclin-dependent kinases. Each of these examples provides useful insight into the links between the cell cycle control proteins and cancer.

The Rb protein provides a link between the CDK/cyclin cell cycle machinery and transcription. Rb acts as a negative regulator of a transcription factor, E2F, which is active during the transition from G1 into S phase of the cell cycle and is required (and probably regulatory) for entry into the cell cycle. Rb protein binds to, and thereby inactivates, E2F. During normal proliferation Rb is phosphorylated (probably by CDK/cyclin kinase) which causes it to disassociate from E2F, which then becomes active. When Rb protein is lost, this leads to loss of regulation of E2F transcription activity and thus loss of one method by which cells can prevent the accumulation of errors and inhibit inappropriate cell proliferation.

The *p53* tumour suppressor gene is involved in a number of pathways that influence cell cycle progression. p53 protein is a transcription factor and is often mutated in human tumours. In response to DNA damage or other perturbations in DNA metabolism, p53 appears to direct the transcription of the *p21* gene, which is an inhibitor of cyclin–dependent kinases. This suggests that p53-dependent p21 induction is responsible for arresting the cell cycle in G1 following DNA damage. p53 protein is also required to direct cells into apoptosis when the DNA is damaged, a method of removing from the population those cells which are most likely to carry mutations. Loss of p53 causes genomic instability since p53-null cells are unable to prevent cell proliferation or activate apoptosis when the integrity of the DNA is compromised. It is possible that the role of p53 as a tumour suppressor protein reflects its crucial role in maintaining the integrity of the genome and that p53 null cells are simply more likely to accumulate the mutations required for complete tumorigenesis.

The *p16* gene is also found to be mutated in many human tumours, and presumably plays a role in inhibiting cell cycle progression (and thus growth) in normal cells. It is known that *p16* is activated by some signalling molecules that prevent cell proliferation, and that p16 protein can bind to cyclin-dependent kinases during G1 to prevent passage through 'Start' into the cell cycle. It is not yet clear how *p16* mutations contribute to the cancerous state, but they presumably do so through loss of the CDK inhibitory activity that prevents inappropriate proliferation.

7.9 SUMMARY

In all eukaryotic cells the cell cycle and its control is essentially the same. A variety of cyclin-dependent protein kinase complexes act at the different stages of the cell cycle to coordinate the independent events that comprise the cell cycle. In mammalian cells, where the number of developmental fates is large, a series of signalling pathways interact with the cyclin-dependent kinases and restrain or promote entry into the cell cycle when it is appropriate. When errors occur, either in the genetic material itself or through a conflict in signalling processes, eukaryotic cells are prevented from progressing through the cell cycle and, in some circumstances, are lost completely through programmed cell death. Many of the mutations that contribute to carcinogenesis involve activating mutations in oncogenes and inactivating mutations in tumour suppressors. Activated oncogenes ultimately promote activation of CDK complexes and subsequent cell proliferation. Inactivating mutations in some tumour suppressor genes apparently disable the checkpoint pathways by which cells inhibit the activity of cyclin-dependent kinases and prevent the accumulation of genetic damage. Others may participate directly preventing inappropriate progression through the cell cycle in otherwise potentially neoplastic cells.

7.10 FURTHER READING

Hunter, T. and Pines, J. (1994) Cyclins and cancer II: cyclin D and *cdk* inhibitors come of age. *Cell*, **79**, 573–582.

Marx, J. (1995) Cell cycle inhibitors may help brake growth as cells develop. *Science*, **267**, 963–964.

Nurse, P. (1990)Universal control mechanism regulating onset of M-phase. *Nature*, **344**, 503–507.

Sheldrick, K. S. and Carr, A. M. (1993) Feedback controls and G2 checkpoints: fission yeast as a model system. *BioEssays*, **15**, 775–782.

Human papillomaviruses and cancer of the cervix

Emma S. Hickman, Rachel C. Davies and Karen H. Vousden

8.1 EPIDEMIOLOGICAL EVIDENCE LINKING CANCER OF THE CERVIX AND HUMAN PAPILLOMAVIRUS INFECTION

For over a century, cancer of the cervix has been recognized as a sexually transmitted disease. In 1976, human papillomavirus (HPV) was first implicated in the development of cervical tumours and since then a large body of evidence from both epidemiological and molecular biological studies has accumulated to support this observation. Studies of biopsy samples collected from a diverse range of patients has shown that HPV DNA is found in around 90% of all cervical tumours, although a small percentage arise without apparent HPV infection. The genital HPVs can be broadly described as 'high-risk' and 'low-risk' depending on their ability to contribute to malignancy; development of tumours most frequently involves infection with HPV types 16 or 18 (high-risk types) whereas other HPV types that commonly infect the genital tract, such as HPV6 or 11, are almost invariably associated with benign lesions and are thus considered low-risk. Despite the clear association of HPV with cervical malignancies, the long interval between infection and the development of a tumour indicates that high-risk HPV infection does not directly produce malignancies, although it appears entirely sufficient for the development of cervical intraepithelial neoplasia (CIN), a premalignant lesion. Indeed, many women infected with HPV never develop cervical cancer and additional oncogenic events must occur during progression of the disease. Hence HPV and cervical tumour formation fits in with the well established view that malignant progression is a multistage process. Evidence has emerged that the continuous expression of HPV-encoded oncoproteins is needed for both development and maintenance of the malignant phenotype, providing important targets for the

Molecular Biology for Oncologists. Edited by J. R. Yarnold, M. R. Stratton and T. J. McMillan. Published in 1996 by Chapman & Hall, London. ISBN 0 412 71270 9

design of both therapeutic and preventative treatments. In addition, the study of how viral proteins interfere with the negative growth signals is providing valuable insights into the normal control of cell proliferation.

8.2 PAPILLOMAVIRUS GENOME STRUCTURE AND BIOLOGY

The human papillomaviruses are small DNA viruses that consist of a 7.9 kb double-stranded circular DNA genome enclosed within a 55 nm viral capsid. The viral genome, with its nine major open reading frames, is depicted in Figure 8.1, although the complexity of expression is increased by multiple splicing patterns.

Though studies of both human and bovine papillomaviruses have assigned at least one role to the proteins encoded by each of these open reading frames (Table 8.1), our inability to grow these viruses efficiently in culture has hampered research significantly.

HPVs are epitheliotropic viruses and viral gene expression is dependent on the growth and differentiation of the host cell, late gene expression and viral particle production only occurring in terminally differentiated keratinocytes. This stage of differentiation is very difficult to mimic in tissue culture and no simple experimental system for viral growth is currently available, resulting in a very poor understanding of the natural life cycle of the virus.

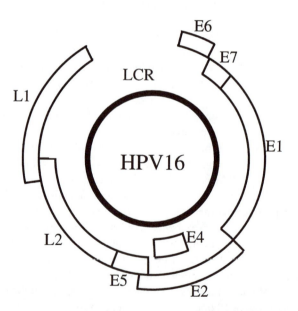

Figure 8.1 Genome of HPV16 showing the position of the major open reading frames with the potential to encode proteins and the long control region containing transcription and replication control elements. Functions of the proteins encoded by the various open reading frames are described in Table 8.1.

Table 8.1 Some functions of high-risk genital HPV-encoded proteins

Early proteins	
E1	Control of replication
E2	Control of transcription
E5	Transformation through cell surface receptors (not expressed in cancers)
E6	Immortalization/transcription control
E7	Immortalization/transcription control
Late proteins	
E4	Disruption of cell keratin cytoskeleton
L1	Structural protein
L2	Structural protein
Non-coding region	
LCR	*cis*-acting transcription and replication control

8.3 IDENTIFICATION OF HPV ONCOPROTEINS

Introduction of the entire HPV16 genome into primary human genital ker-atinocytes results in immortalization of cells that normally show a finite life span in culture. Keratinocytes infected with HPV16 are unable to differentiate normally and resemble cells of premalignant cervical lesions, supporting the association between infection with high-risk HPV types and the development of CIN. Although these cells are not tumorigenic, they can be induced to progress to malignancy either spontaneously or following introduction of activated oncogenes. Cells expressing high-risk HPVs show some degree of genomic instability, which could itself contribute to further malignant progression. Low-risk HPVs do not encode such activity, leading to the proposal that high-risk HPVs may act as soli-tary carcinogens, requiring only time for the accumulation of additional oncogenic genetic events, while low-risk viruses need additional carcinogenic events to give rise to a malignant lesion.

Genetic analysis of the HPV genome revealed that only the *E6* and *E7* genes are required for the immortalization of primary human genital epithelial cells and that both proteins show independent transforming and immortalizing activities in rodent cells. The products of *E6* and *E7* are small proteins (100–150 amino acids) showing limited similarity to each other, each containing conserved C-terminal zinc-binding cysteine motifs that mediate oligomerization and contribute to the stability of the proteins (Figure 8.2).

The relevance of E6 and E7 transforming activity to the development of human cancer is supported by three observations. Firstly, the E6 and E7 proteins of the high-risk viral types function much more efficiently in these assays than the low-

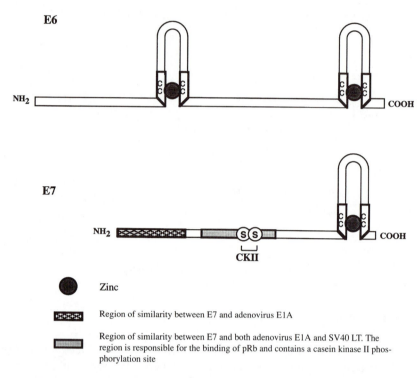

Figure 8.2 The structures of the HPV E6 and E7 proteins showing the zinc-binding cyteine motifs and the regions of homology to other viral oncoproteins. No significant homology between HPV E6 and any other protein has yet been found, although the protein shares the ability to bind p53 with both adenovirus E1B and SV40 large T antigen.

risk proteins, mirroring the apparent malignant potential of the respective viruses. Secondly, keratinocytes expressing E6 and E7 remain phenotypically untransformed and are not tumorigenic, in keeping with both the contribution of HPV to tumour development and the multistep model of carcinogenesis. Lastly, almost all cervical tumours retain expression of E6 and E7, down-regulation of which impairs the growth of the tumour cell.

8.4 INTEGRATION OF VIRAL DNA INTO THE HOST GENOME MAY BE A CRUCIAL STEP IN CARCINOGENESIS

In benign productive lesions such as warts, the viral DNA exists as an episome in the nucleus of the host cell, often at high copy number. Although several events additional to HPV infection are likely to contribute to full malignant transformation, in many tumours integration of the viral DNA within the host chromosomes appears to contribute at least one of these steps. The site of integration on the chromosome does not seem to be consistent and yet the viral genome is almost invariably disrupted in the *E2* and *E1* open reading frames, with consistent retention of *E6* and *E7* gene expression. In trying to understand how integration could lead to malignancy, it is of interest to note that the E2 and E1 proteins are involved in the control of transcription and replication and is possible to imagine how loss of the *E2* and *E1* genes could lead to deregulated expression of the E6 and E7 oncoproteins. Indeed, in the context of the full length genome it has been shown that mutation of either *E1* or *E2* genes can increase the immortalizing activity of HPV16 DNA.

8.5 THE E6 PROTEIN OF HIGH-RISK HPVS TARGETS THE p53 TUMOUR SUPPRESSOR PROTEIN FOR RAPID PROTEOLYTIC DEGRADATION

Although the E6 protein does not share sequence homology with any other viral oncoprotein, a clear functional similarity, the ability to bind the cellular protein p53, is shared between HPV E6, adenovirus E1B and SV40 large T antigen. Studies have revealed that many tumours contain deletions, rearrangements or mutations in both alleles of the *p53* gene and the germline *p53* mutation carried by Li–Fraumeni patients confers a predisposition to a wide range of tumours. In addition, *p53*-null mice develop tumours at a very early age, although these mice develop normally, supporting a role for p53 as a tumour suppressor protein. Consistent with this role for p53, it has been shown that the wild-type protein can inhibit transformation by various oncogenes, including *E7*, and is capable of suppressing the growth of p53-negative tumour cells. This simple picture is complicated, however, by the existence of dominant oncogenic *p53* mutants capable of binding and inactivating the wild-type protein in cells containing only one mutant *p53* allele.

Recently, the clarification of the roles of wild-type p53 protein has been the focus of a considerable research effort (Chapter 9). After DNA replication, the protein is involved in both a G2 phase and a spindle checkpoint, arresting the cell cycle and preventing the separation of damaged chromosomes during mitosis. In addition, p53 protein appears to be essential prior to DNA replication in the cellular response to DNA damage and can participate in both the G1 phase cell cycle arrest and the induction of apoptosis. Studies in murine haematopoietic cells show that cells can undergo either pathway upon DNA damage, depending on such factors as the presence or absence of cytokines, providing that functional p53 protein is expressed. In the normal cell the p53 protein is subject to rapid turnover, becoming stabilized upon DNA damage. The ability of p53 to act as a transcriptional activator is probably critical to its function in the G1 arrest pathway, since its target genes include *p21*, encoding a potent cyclin-dependent kinase (CDK) inhibitor and GADD 45, involved in DNA repair. The p21 protein also plays a direct role in preventing DNA replication through interaction with the proliferating cell nuclear antigen (PCNA). In the event of DNA damage, the interaction of p53 protein with both cell cycle and DNA replication machinery has been proposed to allow time for repair to be carried out before DNA synthesis is resumed. At present it is not clear how p53 functions in apoptosis but in both pathways there seems to be a common theme, that p53 acts to prevent DNA damaged cells from progressing through S phase and thus accumulating potentially tumorigenic mutations.

The HPV E6 protein contributes to abnormal cell growth by binding and rapidly targeting the p53 protein for degradation via the ubiquitin pathway, thus releasing the cell from p53-mediated negative growth control. This interaction involves another cellular protein, E6AP, and the E6–E6AP complex functions as a ubiquitin ligase, an enzyme that transfers ubiquitin to targeted substrates. HPV16 E6 has been shown to disrupt the p53-induced G1 arrest, the transcriptional transactivation associated with p53 and p53-mediated apoptosis. The contribution of E6 to malignant progression is likely to be related to the removal of this checkpoint that ensures genomic integrity. Studies of genomic *p53* sequences from HPV-positive anogenital tumours reveal that, unlike most epithelial tumours, HPV-positive malignancies rarely contain mutant p53, highlighting the importance of E6/p53 protein interactions.

8.6 THE E7 PROTEIN INTERFERES WITH THE FUNCTION OF THE RETINOBLASTOMA GENE PRODUCT

The amino terminal portion of E7 contains regions of similarity with oncoproteins of other DNA tumour viruses, adenovirus E1A and SV40 large T antigen, and through this region, all three proteins are shown to interact with the cellular retinoblastoma gene product (pRb) and other members of the pRb family such as p107 and p130. Mutations within this region of E7 have shown that both pRb binding and phosphorylation of the casein kinase II site (Figure 8.2) are required for efficient transformation of rodent cells. Although mutational analysis shows that E7 protein plays additional roles in transformation, the ability to interfere with pRb function is probably central to its contribution to malignant development. Like p53, the pRb protein acts as a tumour suppressor and the presence of a mutant allele of *Rb* in the germline of an individual is associated with a predisposition to retinal tumours (Chapter 1). The ability of pRb to regulate cell cycle progression is demonstrated by the reversion of malignant phenotype of several pRb-negative tumour cell lines upon the re-introduction of a wild-type *Rb* gene.

A clue as to how pRb might function in growth control lies in its ability to bind to and inhibit various transcription factors such as E2F-1 and c-Myc, proteins that activate many genes, including some required for DNA synthesis (Chapter 7). As the normal cell cycle progresses, pRb is subject to sequential phosphorylation, resulting in its inactivation and the release of active transcription factors (Figure 8.3).

Several lines of evidence point to the involvement of cyclin–CDK complexes in the phosphorylation of pRb and it seems likely that pRb is the principal substrate of the cyclin-D-containing kinases. Interestingly, cyclin D is known to contain the LXCXE motif, which is shared with E7, E1A and large T antigen and is essential

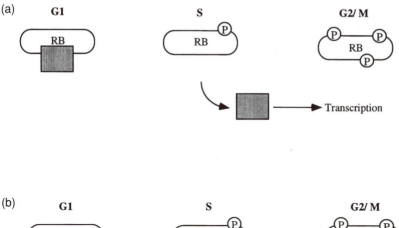

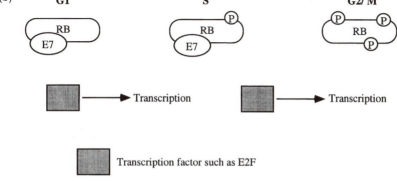

Figure 8.3 pRb phosphorylation during the cell cycle. **(a)** The hypophosphorylated form of pRb binds to and sequesters transcription factors involved in cell cycle progression during G1 phase. Upon phosphorylation of pRb at the G1/S boundary, these factors are released and the transcription of the genes involved in DNA replication is initiated. **(b)** In the HPV-infected cell, pRb is constitutively inactive as a result of binding the E7 protein. The inappropriate release of transcription factors allows deregulated expression of the genes that normally control cell division.

for pRb binding. This motif could bring cyclin D–CDK4 complex into close proximity with pRb during G1, allowing phosphorylation to occur and cell cycle progression upon the release of transcription factors. In the HPV-infected cell, pRb-dependent control of cell proliferation appears to be overcome by E7. The E7 protein of high-risk but not low-risk viruses binds to the under-phosphorylated form of pRb, interfering with its ability to complex transcription factors (Figure 8.3). The consequent availability of transcriptionally active forms of these factors at inappropriate stages during the cell cycle is thought to contribute to a shortening of G1 phase and deregulation of cell growth found in HPV-positive tumours.

8.7 THE E6 AND E7 PROTEINS OF HIGH-RISK HPV TYPES MAY TARGET THE SAME GROWTH CONTROL PATHWAY IN CELLS – IMPLICATIONS FOR SURVIVAL OF THE VIRUS

The discovery of the p21 CDK inhibitor and the ability of p53 protein to induce its expression has provided a potential link between the activation of p53 and the inhibition of the kinases responsible for pRb phosphorylation. Such a pathway would explain how p53 activation upon DNA damage leads to cell cycle arrest and places the target of E7 function downstream of the target of the E6 protein (Figure 8.4).

This notion is supported by several studies demonstrating that E7 is capable of overcoming a DNA-damage-induced, p53-dependent G1 arrest in various tissue culture systems. During normal lytic infection, the abrogation of this pathway would drive the infected cell through the cycle, enabling efficient production of viral progeny. In the premalignant lesion, the loss of this control point would allow DNA-damaged cells to enter into division, possibly accumulating potentially tumorigenic mutations. This hypothesis leaves the question, why should HPV evolution have selected for two proteins to target a single cellular pathway? Recent reports show that deregulated expression of E2F in quiescent rat fibroblasts leads

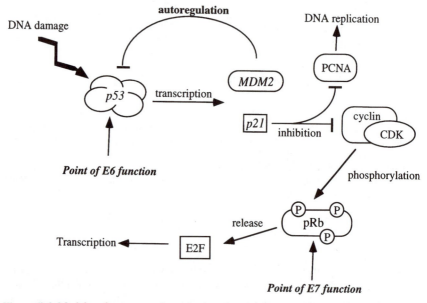

Figure 8.4 Model to demonstrate how the functions of the two HPV oncoproteins may be linked. In the normal cell, DNA damage leads to the stabilization of p53 and the subsequent transcriptional activation of many genes including the cyclin-dependent kinase inhibitor, *p21*. Inactivation of these kinases would lead to an accumulation of the hypophosphorylated form of pRb and hence cell-cycle arrest in the G1 phase. In a cell expressing the high-risk HPV oncogenes this DNA damage response would be disrupted at two points; firstly the E6 protein, by targeting the p53 protein for degradation, would prevent transcriptional activation of the kinase inhibitor and secondly, expression of E7 would lead to constitutive inactivation of pRb and hence deregulation of E2F.

to S phase entry and apoptosis in a p53-dependent manner. In addition, transgenic mice expressing the E7·protein show apoptosis in the retina whereas those containing E7 in a *p53*-null background develop retinal tumours. The ability of E6 to rescue this phenotype suggests that targeted degradation of p53 is required to avoid apoptotic cell death upon E7-mediated release of E2F.

8.8 THERAPEUTIC APPROACHES TO TARGET HPV INFECTION AND MALIGNANT PROGRESSION

The significant correlation between HPV infection and the development of cervical cancer has prompted considerable interest in the use of antiviral agents in both treatment and prevention of the disease. Prophylactic and therapeutic vaccination against the high-risk HPVs, using virus-like particles as an antigen, is an exciting prospect and much research is currently under way to repeat the success of animal model studies in humans.

Since it is known that continual expression of E6 and E7 is required for maintenance of the transformed phenotype, an alternative strategy would be to target these oncoproteins directly by gene therapy. It has been demonstrated that bovine papillomavirus E2 can inhibit proliferation when introduced into cervical carcinoma cell lines, presumably due to its ability to down-regulate the papillomavirus promoter (LCR) and hence switch off expression of the oncoproteins. Other ideas include targeting E6 function or expression, with the hope that deregulated expression of E7 in the absence of E6 activity would drive infected cells into apoptosis. Since the targets of each of these approaches are not found in uninfected human cells it is hoped that such therapy would be specific for tumour cells and would not interfere with normal cellular function.

8.9 ACKNOWLEDGEMENTS

Research sponsored by the National Cancer Institute, DHSS, under contract with ABL.

8.10 FURTHER READING

Bosch, F. X., Manos, M. M., Munoz, N. *et al.* (1995) Prevalence of human papillomavirus in cervical cancer: a worldwide perspective. *J. Nat. Cancer Inst.*, **87**, 796–802.

Hawley-Nelson, P., Vousden, K. H., Hubbert, N. L. *et al.* (1989) HPV16 E6 and E7 proteins cooperate to immortalize human foreskin keratinocytes. *EMBO J.*, **8**, 3905–3910.

Pan, H. and Griep, A. E. (1994) Altered cell cycle regulation in the lens of HPV-16 E6 or E7 transgenic mice: implications for tumor suppressor gene function in development. *Genes Dev.*, **8**, 1285–1299.

Picksley, S. M. and Lane, D. P. (1994) p53 and Rb: their cellular roles. *Curr. Opin. Cell Biol.*, **6**, 853–858.

Zur Hausen, H. (1976) Condyloma acuminata and human genital cancer. *Cancer Res.*, **36**, 794.

9 | The *p53* gene in human cancer

Michael R. Stratton

9.1 INTRODUCTION; HOW p53 WAS DISCOVERED

Certain DNA viruses can induce tumours in experimental animals. The transforming activity of one of these, the small DNA virus SV40, is attributable to a
portion of the viral genome that encodes a protein known as the large T antigen.
In order to understand the mechanism by which large T induces oncogenesis,
attempts were made in the late 1970s to detect the host cell protein to which large
T binds. Immunoprecipitation of large T from cells infected by SV40 revealed a
host cell protein of 53 kDa (known as p53) in addition to large T itself. p53 did not
appear to be bound directly to the antibody, but was precipitated because it was
associated with large T.

The first indication that mutation of the *p53* gene might be a common step in
the development of human cancer emerged from investigations into the role and
location of tumour suppressor genes. Comparison of germline and tumour DNAs
using polymorphic probes revealed that loss of one or other parental allele (loss of
heterozygosity) on the short arm of chromosome 17 occurs at high frequency in
many types of neoplasm (Chapter 3). This type of result is usually interpreted as
indicating the presence of a tumour suppressor gene in the vicinity. The *p53* gene
had previously been localized to this region and fine mapping of heterozygous
deletions in colon carcinoma indicated that the common deleted area includes the
p53 locus. Prompted by this clue, the *p53* gene was sequenced in two colon carcinomas and subsequently in other tumours showing loss of heterozygosity on chromosome 17p. In most cases single base substitutions were detected in the
remaining *p53* allele, indicating that the *p53* gene was likely to be the tumour suppressor gene in this region at which the loss of heterozygosity was directed.

Molecular Biology for Oncologists. Edited by J. R. Yarnold, M. R. Stratton and T. J.
McMillan. Published in 1996 by Chapman & Hall, London. ISBN 0 412 71270 9

9.2 STRUCTURAL ALTERATIONS IN THE *p53* GENE ARE COMMON SOMATIC EVENTS IN HUMAN CANCER

A substantial body of data contributed by several groups now indicates that the *p53* gene is the most commonly mutated gene known in human cancer. Mutations are present in all major histogenetic groups and indeed in most subtypes of cancer. Thus they are found in epithelial, mesenchymal, haemopoietic and lymphoid neoplasms and in tumours of the central nervous system. The proportion carrying *p53* mutations is high in some types of cancer and low in others, but in no tumour type does it reach 100%. Mutation of *p53* therefore appears unlikely to be an obligate step in oncogenesis.

9.3 SOMATIC MUTATION OF THE *p53* GENE IN HUMAN CANCER IS OFTEN NOT THE INITIATING EVENT IN ONCOGENESIS

Mutation of the *p53* gene constitutes an intermediate or late step in the sequence of genetic alterations required for the development of many tumours. For example, mutations usually occur around or during the transition between adenoma and carcinoma of the colon. Similarly, they are found in blast crisis of CML but not in the earlier chronic phase. However, *p53* gene mutations often occur early in the development of gliomas and are sometimes present in the *in-situ* phase of breast cancer. Exceptions to all these patterns exist and therefore it may be that mutation of *p53* occurs at different stages in different tumour types and that adherence to a predetermined sequence of genetic events is not as important as the final accumulation of the necessary set of mutations.

9.4 THE TYPE AND PATTERN OF MUTATION IN THE *p53* GENE SUGGESTS INACTIVATION OF A TUMOUR SUPPRESSOR GENE

The majority of somatic mutations of the *p53* gene found in human tumours are single base substitutions (point mutations) resulting in replacement of one amino acid by another in the p53 protein (missense mutations) (Chapter 2). A minority result in abnormalities of mRNA processing, frame shifts or premature termination of translation. Less commonly, complete or partial deletions and other gross rearrangements of the gene are detected. Point mutations of *p53* are scattered through numerous codons but are mainly confined to four regions of the gene that are conserved through evolution, located in exons 5–8 (Figure 9.1).

Within these conserved regions there are four codons that are 'hot spots', which together account for approximately 30% of mutations in the gene. This diversity of mutation, coupled with the fact that some must result in absence or truncation of the protein, suggests that inactivation is the functional outcome. Comparison of the types of mutation found in different cancers suggests that patterns of mutation

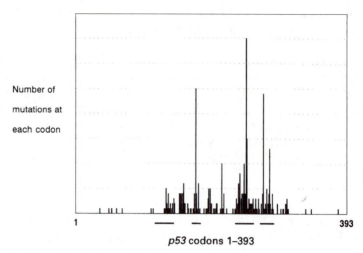

Number of mutations at each codon

p53 codons 1–393

Figure 9.1 Somatic mutations in the *p53* gene. Mutations are widely scattered over a large number of codons. However, they are preferentially found in regions that show sequence conservation during evolution (black horizontal bars). Moreover, there are hot spots that account for over 30% of mutations. The data are collated from several reports and reflect mutations in many different types of cancer.

differ. It seems plausible that at least part of this variation may reflect exposure to particular carcinogens and hence may generate an insight into aetiological mechanisms (considered in more detail in Chapter 14).

In most normal cells, the level of p53 protein is low and usually undetectable by immunocytochemistry. However, many p53 proteins with single amino acid substitutions have a much longer half life than the wild type and are consequently present at higher levels in the cell. Thus detection of the p53 protein in tumours by immunohistochemical methods is often used to provide indirect evidence of a mutated gene.

9.5 GERMLINE MUTATIONS OF THE *p53* GENE PREDISPOSE TO THE DEVELOPMENT OF SEVERAL TYPES OF CANCER

The Li–Fraumeni syndrome (LFS) is a rare familial predisposition to cancer that is transmitted in an autosomal dominant manner. The syndrome is characterized clinically by sarcomas in children (of bone and soft tissue) associated with a high incidence of early onset breast cancer in female relatives. Leukaemia, brain, lung and adrenocortical tumours constitute less common features of the disease.

It transpires that some LFS patients carry one mutated and one wild-type *p53* allele in their germline and that the predisposition to cancer is transmitted with the mutant allele. Many of these mutations are at the same sites as those documented as somatic events in sporadic tumours, including some of the 'hot spots' described above. It is therefore thought that LFS patients are predisposed to cancer because one *p53* allele is inactivated in the germline and therefore only the remaining allele needs to be altered by somatic mutation before a cell can escape from the tumour suppressor activity of the p53 protein. In normal individuals developing a sporadic tumour, both *p53* alleles must be inactivated in the same cell by somatic mutation. On this model the genetics of *p53* are rather similar to those of the prototype tumour suppressor, the retinoblastoma (*Rb*) gene (however, see below for complications).

Further studies now indicate that germline *p53* mutations account for only a proportion of LFS families, and that the range of tumours developing in individuals carrying such mutations is wider than that classically associated with LFS. Germline mutations in *p53* account for approximately 5% of cases of osteosarcoma and of children developing two independent primary cancers. A much smaller proportion of early onset breast cancer (< 1%) is attributable to germline *p53* mutations. In contrast to familial retinoblastoma the new mutation rate appears to be low. Cancer screening to provide early diagnosis and prophylactic treatment may be appropriate for some. However, the limited beneficial effects of these may have to be weighed against the detrimental effects of testing upon lifestyle and economic status. Some of the problems associated with counselling are exemplified in the kindred shown in Figure 9.2.

9.6 IS *p53* A DOMINANT TRANSFORMING GENE OR A TUMOUR SUPPRESSOR GENE?

Early in the 1980s several groups demonstrated that overexpression of *p53* can immortalize or transform cells either on its own or in collaboration with other

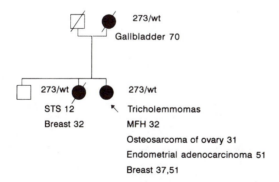

Figure 9.2 A cancer family due to a germline *p53* mutation at codon 273 (Arg to Cys). The pedigree illustrates the wide spectrum of tumours that can be induced and how variable the penetrance can be in individuals with germline *p53* mutations. The mother of the index case lived to age 70 before developing a carcinoma of the gallbladder, while the index case (arrowed) suffered multiple cancers before the age of 50. Both women and the sister of the index case who developed breast cancer and a sarcoma carry the mutation.

oncogenes such as *ras*. This pattern of activity was regarded as similar to that of genes such as c-*myc*, which are conventionally regarded as dominantly acting oncogenes (i.e. they can transform cells in the presence of a normal allele and the proto-oncogene usually requires activation to contribute to oncogenesis). Doubt was cast upon this interpretation when it emerged that the *p53* cDNA clones used in all these experiments were mutated rather than wild-type. Moreover, the pattern of structural alterations found in tumours (widely scattered point mutations and various types of homozygous deletion) is suggestive of a tumour suppressor gene (or recessive oncogene in which both alleles require inactivation) rather than of a dominantly acting gene. Further experiments have recently revealed that wild-type *p53* does indeed suppress the transformed phenotype in a number of cell systems.

However, if *p53* were to behave as an orthodox tumour suppressor gene, then mutated versions should theoretically be inactivated and thus not affect the phenotype when introduced into normal cells. Nevertheless *p53* clones with several different mutations stubbornly act to transform cells. Reconciling the respective transforming and suppressor activities of mutated and wild-type *p53* has been problematic. One possibility is that p53 protein functions in the form of protein oligomers. Oligomers composed exclusively of wild-type protein function normally. Oligomers composed of mutant protein are inactivated. However, hybrid oligomers of mutant and wild-type protein take the mutated (i.e. inactivated) phenotype (Figure 9.3).

On this model a mutation in a single *p53* allele may provide the cell with a proliferative advantage without alterations of the remaining allele. Nevertheless, the

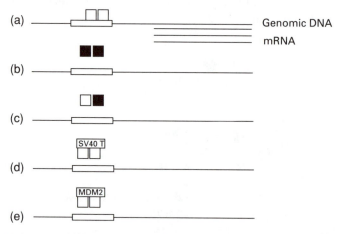

Figure 9.3 Schematic representation of p53 function and its regulation by other proteins. (a) Wild-type (wt) p53 protein binds to specific DNA sequences and regulates the expression (transactivates) of other genes. (b) Mutated p53 cannot bind and transactivate. (c) Hybrid oligomers of mutated and wt p53 cannot bind and transactivate. (d) SV40 large T antigen associates with p53 and prevents it binding or transactivating. (e) MDM2 protein associates with p53 and prevents it binding or transactivating.

overall effect is inactivation of the protein, a dominant negative or dominant loss of function effect.

There is now substantial evidence for a dominant loss of function effect of some mutated forms of p53 protein. This dominant effect should only be observed if there is normal p53 protein in the cell with which the mutated form can oligomerize. However, some studies have shown that introduction of mutated p53 protein into cells without any functioning p53 protein at all still alters the cellular phenotype, making the cells behave in an even more transformed way. These results suggest that, in addition to the inactivation by mutation of certain p53 functions, and in addition to the dominant negative effect, some *p53* mutations have a genuine 'gain of function' effect. Indeed, a single mutation may have all three effects! This complexity illustrates the problems of attempting to construct simple classifications of genes and how they act.

9.7 INTERACTIONS OF p53 AND Rb PROTEINS WITH DNA VIRUS ANTIGENS ARE CENTRAL TO THE TRANSFORMING ACTIVITIES OF MANY DNA VIRUSES

p53 protein was discovered through its association with SV40 large T. It also binds to other DNA virus transforming proteins including the E1B protein of adenovirus and the E6 protein of papillomavirus (Chapter 8). Interaction with these DNA tumour virus proteins inactivates many functions of p53 protein, hence mimicking the effects of mutations in the gene. Interestingly enough it is now known that the protein encoded by the retinoblastoma tumour suppressor gene also binds to SV40 large T, to the E1A protein of adenovirus and the E7 of papilloma virus. These interactions with DNA tumour virus antigens are also believed to result in inactivation of the Rb protein. There is therefore a remarkable symmetry between *p53* and *Rb*. *Rb* was detected and isolated as a gene involved in a familial cancer syndrome, was subsequently found to be a target for somatic mutation in many tumours and finally was shown to be a bound and inactivated by DNA virus transforming proteins. Conversely, *p53* was originally discovered as a protein bound to DNA virus proteins, was subsequently shown to be a target for somatic mutations and finally was implicated in a familial cancer syndrome.

9.8 THE PRODUCT OF THE *MDM2* GENE, WHICH IS AMPLIFIED AND OVEREXPRESSED IN SOME HUMAN CANCERS, BINDS TO AND INHIBITS p53

The observation that DNA tumour virus proteins bind to p53 and to the product of the *Rb* gene inevitably rekindled interest in putative eukaryotic homologues of these oncoproteins which might also interact and inhibit tumour suppressor proteins. Direct isolation and characterization of proteins associated with p53 revealed that one of these is the product of the *MDM2* gene. *MDM2* was originally identified as a gene amplified in spontaneously transformed BALB/c mouse cells and has been shown to have transforming activity when overexpressed in NIH3T3 cells. Interestingly, MDM2 protein appears to inhibit the transcriptional regulatory activity of p53 in a similar way to SV40 large T and mutated p53. The *MDM2* gene is located on the long arm of chromosome 12 and is frequently amplified in sarcomas. Therefore, amplification and increased expression of MDM2 in sarcomas inhibits the normal function of p53 and represents an alternative to *p53* gene mutation as a means by which p53-dependent pathways can be disrupted in human cancers.

9.9 NORMAL p53 PROTEIN REGULATES THE EXPRESSION OF OTHER GENES BY BINDING TO A SPECIFIC DNA SEQUENCE

Normal p53 protein binds to a specific DNA sequence and can regulate the transcription of (transactivate) genes in the vicinity of such sequences. Mutated forms of p53 found in human cancers, however, cannot bind or regulate transcription through this DNA element. Moreover, mutated p53 (presumably through the formation of hybrid p53 protein oligomers described above) and SV40 large T antigen can inhibit the transactivation function of normal p53 protein. Together, these studies support the contention that p53 exerts its biological effects (at least partially) through regulation of transcription and that abrogation of this activity by SV40 large T and mutated p53 may (at least partially) account for their transforming activities.

9.10 ONE OF THE MAJOR TARGET GENES TRANSCRIPTIONALLY REGULATED BY p53 IS *P21/WAF1/CIP1*

Some of the genes and proteins which are transcriptionally regulated by p53 and hence mediate its effects have been identified. The *p21* gene (also known as *WAF1* or *CIP1*) was identified by several independent approaches including as a mRNA that is expressed in cells that contain functional p53 but not in cells without functional p53. It turns out that p21 inhibits cyclin-dependent kinases, proteins that control entry into the cell cycle. Therefore, when p53 is functioning normally, p21 is expressed, it inhibits cyclin-dependent kinases and prevents progress through the cell cycle. When p53 is non-functional (due to mutations or other mechanisms) p21 protein is not expressed, cyclin-dependent kinases are active and cells can proceed through the cell cycle without hindrance. This unfettered progress through the cell cycle is a characteristic feature of many neoplastic cells and is probably a key outcome of inactivation of p53 protein. p21 protein also inhibits the function of a protein known as proliferating cell nuclear antigen (PCNA), which stimulates DNA replication and repair.

In addition to *p21*, p53 is known to regulate the expression of several other genes involved in cell proliferation and apoptosis. Indeed, rather than being a component of a 'pathway', p53 is probably at the centre of a 'network' or 'web' of interactions, the complexity of which is likely to defy description in simple words or pictures.

9.11 WILD-TYPE p53 PROTEIN CAN INDUCE APOPTOSIS AND HAS A ROLE IN THE REGULATION OF THE CELL CYCLE

What are the ultimate biological effects of mutation in p53 protein and disruption of expression of the target genes mentioned above? At least two biological mechanisms play a role. Firstly, wild-type but not mutated forms of p53 can induce the form of programmed cell death known as apoptosis. Secondly, wild-type but not mutated p53 arrests the transition from G1 to S phase of the cell cycle. Although these effects could just be interpreted as influencing the rate at which cells accumulate, a more subtle consequence of disruption of these biological functions is now strongly supported.

G1 arrest is important in optimizing the extent of DNA repair after exposure to DNA-damaging agents by delaying the onset of DNA synthesis, hence allowing more time for repair. This reduces the amount of DNA damage that is converted into fixed mutations in DNA. G1 arrest is induced in cells that have sustained DNA damage. Cellular levels of p53 are also dramatically increased after induction of DNA damage. It is now thought that DNA damage results in the increased p53 level that in turn induces G1 arrest. Therefore inactivation of p53, and hence absence of G1 arrest and reduction of the opportunity to repair DNA, renders tumour cells more prone to the acquisition of mutations in the genome. Indeed, there is direct evidence that introduction of mutated p53 or SV40 large T can allow gene amplification in cell types that were previously resistant to this type of change. Thus, wild-type p53 prevents acquisition of mutations and has therefore been evocatively dubbed 'the guardian of the genome'.

An alternative way of preventing mutations accumulating is to let cells with DNA damage die, and it has been recognized for some time that exposure to DNA-damaging agents often does induce apoptotic cell death. Since wild-type p53 is known to induce apoptosis this is a plausible alternative or additional way in which p53 could act as 'guardian' or perhaps 'garbage disposer' of the genome. The elegant experiments described in the following paragraphs suggest strongly that this is the case.

9.12 *p53* IS NOT REQUIRED FOR NORMAL DEVELOPMENT OF THE MOUSE

Using the technology of homologous recombination, mice without any functional *p53* gene have been produced. These survive embryogenesis, develop normally postnatally, and are fertile. It is remarkable that a gene believed to be so intimately involved in cell division, differentiation and death should be redundant in normal development. Mouse carrying a single functioning *p53* gene (called *p53*-heterozygous or *p53* –/+ mice) are at a high risk of developing tumours, predominantly sarcomas of bone and soft tissues. Mice without any *p53* gene (called *p53*-null or *p53* –/– mice) develop a different spectrum of cancers, mainly thymic lymphomas, and develop them at an earlier age than heterozygous mice.

9.13 APOPTOSIS IN RESPONSE TO DNA DAMAGE IS INHIBITED IN *p53*-NULL MICE, BUT THE APOPTOTIC RESPONSE TO OTHER STIMULI REMAINS INTACT

Several types of insult can induce cells to implement the apoptotic programme of cell death. In mouse thymocytes apoptosis can be induced *in vitro* by induction of DNA damage (for example by ionizing irradiation or drugs such as etoposide) and by steroids. Thymocytes from *p53*-null mice can implement fully the apoptotic response to steroids but cannot apoptose in response to DNA damage. This result is consistent with the idea that one of the major functions of p53 is to stop cells accumulating too much DNA damage, and that one of the ways it does this is to encourage cells accumulating DNA damage to commit suicide. Therefore p53 is directly implicated not in the apoptotic programme itself, but in the pathway stimulated by DNA damage that can turn on apoptosis. Presumably, as a result of the loss of this pathway, *p53*-null mice are much more susceptible to development of tumours in response to irradiation.

9.14 CANCERS WITHOUT FUNCTIONAL p53 PROTEIN ARE RESISTANT TO DNA-DAMAGING CHEMOTHERAPEUTIC AGENTS

It has become clear that one of the ways in which chemotherapeutic agents used in treating human cancer exert their effects is by inducing DNA damage, hence triggering apoptosis and causing cells to commit suicide. If p53 is critical in the apoptotic response to DNA damage then cancer cells without functional p53 could be less responsive to this type of chemotherapeutic agent. In fact this is precisely what is observed in experimental systems. Indeed, many of the human cancers that are most sensitive to DNA-damaging agents, for example testicular cancers, have a very low proportion carrying *p53* mutations. Moreover in certain cancers that are generally sensitive to chemotherapy and do not usually carry *p53* mutations, the minority of tumours that do carry a *p53* mutation are considerably less sensitive to chemotherapy (for example B-CLL and Wilms tumour). Indeed, studies of several types of cancer have shown that those cases that carry *p53* mutations have an overall worse prognosis than those that do not. Identification of mutated *p53* as a gene/protein that is involved in resistance to chemotherapy by preventing apoptosis is of major importance, as it may suggest rational approaches to overcoming chemoresistance in cancer.

9.15 p53 MAY PLAY A ROLE IN THE REGULATION OF ANGIOGENESIS IN TUMOUR CELLS

There has been considerable emphasis upon the role of p53 protein in preventing and disposing of DNA damage in recent years. There is some evidence, however, that p53 may serve other biological functions. In order for a fast growing neoplasm to survive it must attract a blood supply (Chapter 22). The nature of the signals that perform this function have been under scrutiny recently. p53 regulates the expression of thrombospondin-1, a protein that may serve as a signal to suppress angiogenesis. When p53 is inactivated thrombospondin-1 expression is apparently reduced, which should encourage the development of new vessels. Indeed, the ability of gliomas to induce florid angiogenesis (a key feature of the diagnostic classification of these tumours) appears to be related to the presence or absence of functional p53. Therefore in addition to its role in prevention and repair of DNA damage, p53 may be implicated in many other cellular functions, of which we know little at present.

9.16 CONCLUSION

There has been a veritable explosion of information concerning the roles and functions of *p53*. Study of this gene and its protein have led to fundamental advances in understanding of viral and human carcinogenesis, the cell cycle, programmed cell death, the response to DNA damage, resistance to chemotherapy, development of blood vessels and environmental exposures to mutagens. All these insights can be traced back to the original experiments in which p53 was detected, carried out in a relatively obscure experimental animal system. This story should serve to remind us that it is not always straightforward to predict what types of scientific endeavour are likely to bear fruit. We should remember this when evaluating what is and what is not relevant to current research on human cancer.

9.17 FURTHER READING

Baker, S. J., Markowitz, S., Fearon, E. R. *et al.* (1990) Suppression of human colorectal carcinoma cell growth by wild-type p53. *Science*, **249**, 912–915.

Clarke, A. R., Purdie, C. A., Harrison, D. J. *et al.* (1993) Thymocyte apoptosis induced by p53-dependent and independent pathways. *Nature*, **362**, 849–852.

Donehower, L. A., Harvey, M., Slagle, B. L. *et al.* (1992) Mice deficient for p53 are developmentally normal but susceptible to spontaneous tumours. *Nature*, **356**, 215–221.

El-Deiry, W. S. *et al.* (1993) WAF1, a potential mediator of p53 tumor suppression. *Cell*, **75**, 817–825.

Hollstein, M., Sidransky, D., Vogelstein, B. and Harris, C. C. (1991) p53 mutations in human cancers. *Science*, **253**, 49–53.

Kern, S. E., Kinzler, K., Bruskin, A. *et al.* (1991) Identification of p53 as a sequence specific DNA binding protein. *Science*, **252**,1708–1711.

Lane, D. P. and Crawford, L. V. (1979) T antigen is bound to a host protein in SV40 transformed cells. *Nature*, **278**, 261–263.

Lowe, S., Bodis, S., McClatchey, A. *et al.* (1994) p53 status and the efficacy of cancer therapy *in vivo*. *Science*, **266**, 807–810.

Malkin, D., Li, F. P., Strong, L. C. *et al.* (1990) Germ line p53 mutations in a familial syndrome of breast cancer, sarcomas and other neoplasms. *Science*, **250**, 1233–1238.

Cell adhesion, motility and cancer

10

Ian R. Hart

10.1 CELL–CELL AND CELL–MATRIX CONTACTS HAVE IMPORTANT STRUCTURAL AND FUNCTIONAL ROLES RELATING TO CELL DIFFERENTIATION, PROLIFERATION AND MIGRATION

Tissues in multicellular organisms are formed by organized collections of cells which interact and cooperate with one another via cell–cell junctions. The cells of a tissue are also in contact with the extracellular matrix via cell–matrix junctions. Extracellular matrix, which consists of secreted extracellular protein and polysaccharide macromolecules, is produced locally by various cell types and provides the scaffold upon which tissues form. The ways that cells dissociate, migrate, arrest and subsequently reassociate with like or unlike cells are central events in developmental biology and normal physiology. Specialized cell–cell and cell–matrix junctions regulate these interactions and thereby control processes such as haemostasis, wound healing, immune response and morphogenesis.

10.2 CELL INVASION AND METASTASIS IN CANCER INVOLVES DEREGULATION OF CELL–CELL AND CELL–MATRIX CONTACTS

Deregulation of cell adhesion and motility probably plays a minor role in the initial stages of tumorigenesis but those events possibly are crucial in the later stages of tumour invasion and dissemination. The process of tumour metastasis involves the detachment and migration of neoplastic cells from their sites of origin to dis-

Molecular Biology for Oncologists. Edited by J. R. Yarnold, M. R. Stratton and T. J. McMillan. Published in 1996 by Chapman & Hall, London. ISBN 0 412 71270 9

tant secondary sites. It involves changes in adhesion between cells and changes in cell adhesion to the proteins and polysaccharides that make up the extracellular matrix. These include collagens, elastins, laminins, fibronectins and proteoglycans. The interactions must be dynamic in order for the invading tumour cell to pass through and over other tissues. This means that the molecules involved in regulating normal cell–cell or cell–substrate adhesion are likely to be of major significance in modulating the metastatic behaviour of tumours.

Tumour invasion and metastasis can be broken down into a series of sequential interactions, which involve release of normal cellular contacts within the primary tumour, invasion through the underlying basement membrane, destruction of and passage through extracellular matrix and eventual penetration into minor blood or lymphatic vessels. The survival of tumour cells that have gained access to small vessels is partly dependent on the outcome of interactions with host immune effector cells. Circulation to a distant site is followed by transendothelial extravasation and movement through the extracellular matrix with colonization and growth at a secondary site.

The changes in cell–cell and cell–matrix interactions that underlie the metastatic sequence are regulated not just at the level of the individual cell but also at different subcellular anatomic sites. Thus, while adhesive cell–cell or cell–matrix contacts may be created at the invading or frontal edge of the cell, the trailing or the posterior edge of the cell may be engaged in de-adhesion from either cell neighbours or surrounding extracellular matrix proteins. Cancer cell mechanisms of adhesion and motility utilize or usurp the function of specific cell adhesion receptors found on the surface of many normal cells. There are four major families of cell adhesion receptors containing numerous members. Representatives of each of these receptors, or their ligands, have been found on all tumour cells examined and the four major receptor families will be considered individually (Figure 10.1). Most tumour cells express many different members of these families and there may well be redundancy, i.e. overlap of function.

10.3 CADHERINS ARE CELL SURFACE MOLECULES WHICH RECOGNIZE AND BIND TO MOLECULES OF THE SAME KIND ON ADJACENT CELLS

Cadherins are cell surface membrane glycoproteins that mediate cell–cell adhesion in a calcium-dependent fashion The ligands for these receptors generally are identical cadherin molecules expressed on adjacent cell types. Since like molecules bind to like, these associations are termed **homotypic** (cf. **heterotypic**, in which different molecular species bind, e.g. binding of cell surface integrin receptors to immunoglobulin superfamily members). Although many different cadherins have now been identified there are three classical cadherins, best characterized as E- for epithelial, N- for neural and P- for placental, which possess common structural and functional motifs (Figure 10.1). E-cadherin, expressed by epithelial cells,

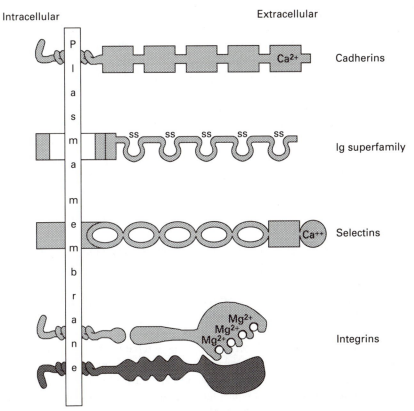

Figure 10.1 Schematic representation (not to scale) of four major families of cellular adhesion receptors.

is the most important member of the family with regard to human cancers.

Loss of E-cadherin expression has been shown to lead to the dissociation of cells, an important prelude to their invasion into adjacent tissue. When cDNAs encoding for E-cadherin are introduced into non-cadherin-expressing tumour cell lines in tissue culture, the expression of this adhesion molecule can have a striking effect on cell behaviour. The presence of E-cadherin at the cell surface restricts the capacity to invade in a matrigel invasion assay and inhibits the motility of these modified cells. In contrast, the treatment of E-cadherin-expressing cell lines with anti-E-cadherin monoclonal antibodies or with peptides that bind to and block the homotypic binding site results in cellular dispersal and abolition of the epithelial morphology. There appears to be an inverse quantitative correlation between the invasive ability of a range of cell lines derived from human carcinomas and the relative levels of E-cadherin expression. Low expressers are more invasive than high expressers. Analysis of fresh human cancer confirms this relationship and has shown that in carcinomas generally there is an inverse relationship between

E-cadherin expression and tumour grade. Poorly differentiated tumours monitored for expression of the protein at the cell surface frequently exhibit a reduced or absent reactivity. Using *in situ* hybridization to measure levels of mRNA in archival material obtained solely from Duke stage B tumours we were able to show recently that E-cadherin expression may serve as a useful prognostic marker in colorectal cancer patients. It is important to remember, though, that down-regulation at the protein level, as a consequence of down-regulation of messenger RNA, is not the only means whereby E-cadherin function may be diminished, since the number of molecules is not necessarily predictive of functional activity. Mutations in the homotypic binding site of the molecule or in catenins, proteins that link the cytoplasmic tail of cadherin molecules to the actin cytoskeleton of the cell, can result in functional down-regulation of adhesive activity. Such alterations in the integrity of this molecule have been detailed in recent analyses of human material.

10.4 INTEGRINS MEDIATE CELLULAR INTERACTIONS WITH THE EXTRACELLULAR MATRIX

Released from cohesion within the primary tumour mass, disseminating cancer cells interact with the extracellular matrix by releasing proteolytic enzymes that degrade the major protein components (including collagen, fibronectin, laminin and elastin) to allow easy movement through the intercellular spaces. Most cell interactions with the extracellular matrix are regulated by members of the second family of adhesion receptors, the integrins. Integrins are a family of transmembrane glycoprotein heterodimers which have a large extracellular component and a short cytoplasmic tail. The heterodimers are composed of an α and a β chain non-covalently associated and are responsible for many cell–matrix interactions. Specificity of ligand recognition on extracellular matrix molecules arises from the association of a particular α chain with a particular β chain. This is the case even though binding to the extracellular matrix proteins frequently occurs through recognition of a short peptide sequence common to all integrins, arginine–glycine–aspartic acid (referred to as an RGD sequence, the code letters for the three amino acids involved). Seminal work in the late 1980s suggested that the integrins have an important part to play in tumour metastasis. In a series of elegant experiments, Martin Humphries admixed linear peptides containing the RGD sequence with murine melanoma cells and coinjected this mixture into the tail veins of mice. He was able to show that RGD, and only the RGD sequence, was capable of reducing the level of experimental metastasis by greater than 80%. These results, though having little implication in terms of treatment protocols, demonstrated that the integrins and their interaction with extracellular matrix proteins must be involved in the mechanisms of tumour spread.

A great deal of effort has been expended on documenting the presence or absence of specific integrin subunits in clinical material. These immunocytochemical analyses have produced varied and conflicting results. Epithelial tumours frequently lose expression of many integrin subunits as they become more aggressive, whereas other tumour types such as melanoma often show an increase in integrin expression with progression. Contrasting results of morphological analyses may result from the fact that measuring protein levels *per se* actually gives little indication as to the functional status of the expressed integrins. It is known from *in vitro* studies of normal tissue cells that integrin activation status may vary independently of protein expression. Thus the functional status of the integrin examined, not just the amount expressed, determines the potential of these molecules to modulate cell behaviour.

Experiments in model systems, where integrin cDNAs were introduced into cells that lack expression of specific subunits, resulted in the modified cells exhibiting profound behavioural changes. Introduction of integrin α5β1, which functions as a fibronectin receptor, into cells deficient in this receptor resulted in down-regulation of tumorigenic capacity. Conversely, when α2 was introduced and expressed in a rhabdomyosarcoma line there was an impressive elevation in metastatic activity. As stated before, it is not just the presence of a specific integrin but its functional state that is important. In a series of human melanoma cell lines we showed that the α4β1 integrin exists in a constitutively active state, in contrast to normal leucocytes, where the α4β1 heterodimer requires an activation stimulus before it can bind to its appropriate ligand. Thus, α4β1, which is responsible for binding both to fibronectin and to the immunoglobulin superfamily member V-CAM, is always in the active form on the surface of melanoma cells. This could have implications for tumour cell behaviour. Since one of the ligands for α4β1 is V-CAM, an immunoglobulin superfamily member which is expressed on activated endothelial cells, it seems probable that the mere presence of the integrin on melanoma cells could play an important part in mediating the arrest of disseminating tumour cells in distant capillary beds.

Integrins have been shown to cluster at the contact sites of cellular interaction with extracellular matrix proteins. Also localized to these sites, which are termed **focal adhesion plaques**, are a number of molecules involved in regulating second-messenger production whose stimulation results, eventually, in altered gene expression. Included in this collection of molecules is the so-called focal adhesion kinase, capable of phosphorylating a variety of substrates in response to integrin occupancy. Thus the binding of integrins may result in changes in gene expression. It is of interest that some cell types have been shown to up-regulate levels of metalloproteinase activity as a consequence of integrin binding. Metalloproteinases are members of the endopeptidase class of enzymes responsible for digesting extracellular matrix molecules. They are thought to be the major effectors of proteolytic activity in malignant tumours and this activity could be of major importance. The activation of metalloproteinase activity could be especially significant when integrins are involved in the adherence of disseminating cancer cells to the endothelium at distant metastatic sites. Thus the α4β1 V-CAM interaction described above may result in stimulation of proteolytic activity which would be responsible for the degradation of the underlying basement membrane and this

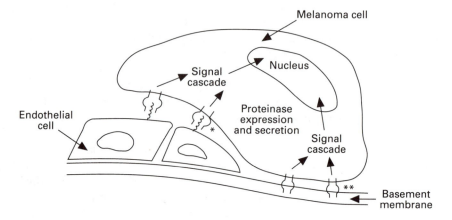

Figure 10.2 Schema of melanoma cell in a capillary binding to vascular endothelium via integrin/V-CAM interactions (*) or to basement membrane via integrin/fibronectin interactions (**). These lead to the induction of signalling events that increase metalloproteinase expression at sites of integrin attachment and facilitate extracellular matrix breakdown. The melanoma cell is then able to migrate out of the capillary into the neighbouring tissues.

could lead to facilitated extravasation of the arrested cancer cells (Figure 10.2).

The observation that motility over specific substrates has been linked to integrin-mediated binding suggests that the activity of these heterodimers could affect a number of changes in addition to cell–matrix adhesion that underlie the metastatic process.

10.5 MEMBERS OF THE IMMUNOGLOBULIN SUPERFAMILY ARE IMPORTANT CELL ADHESION MOLECULES (CAM)

Members of the immunoglobulin superfamily of adhesion molecules are recognized by the presence of several immunoglobulin-like domains in the extracellular portion of the molecule. The family contains numerous members, loss of which by gene deletions or down-regulation has been postulated to play an important part in tumour progression. In colorectal carcinomas inactivation of the *DCC* (**d**eleted in **c**olorectal **c**arcinoma) gene has been shown to occur in approximately 70% of cases. The *DCC* gene amino acid sequence predicts for a protein with strong homology to the cell adhesion molecule I-CAM 1 and other members of this family of adhesion molecules. Though the function of the gene is not fully understood it has been proposed that elimination of the activity of this molecule results in the cancer cells escaping the constraints of cell–cell cohesion and permitting the initial events of tumour cell migration to occur.

Carcinoembryonic antigen (CEA), which is expressed in colorectal cancer by colonic mucosal cells, has been shown to function as an immunoglobulin super-family adhesion receptor. Pretreating mice with CEA, to saturate the hepatocytes, increased the number of liver metastases resulting from the introduction of weakly metastatic human colon cancer cell lines. This increase in metastatic activity was presumed to result from the increased tumour cell adhesion in the liver mediated by a CEA–CEA interaction. If a comparable mechanism is operative in human cancer then the importance of CEA may relate not just to its function as a tumour marker but also to its potential as a target for therapy.

10.6 SELECTINS ARE ADHESION RECEPTORS FOUND ON LEUCOCYTES, VASCULAR ENDOTHELIUM AND CANCER CELLS

The selectins are a group of adhesion receptors with three members, E-, P- and L-selectin, which are expressed on a variety of blood cells and/or endothelial cells. E-and P-selectin are expressed by endothelial cells, where many cytokines cause dramatic up-regulation of expression, while L-selectin is expressed by leucocytes. The calcium-dependent lectin domain at the amino terminus of this long molecule binds to sugar groups expressed on different cell types. When leucocytes extravasate during inflammation, relatively weak selectin–sugar interactions serve to draw leucocytes to the endothelium from the laminar flow blood, initiating a process known as 'rolling'. The name derives from the appearance of leucocytes, which roll along the luminal surface of the endothelial monolayers before stronger integrin-mediated leucocyte–endothelium attachments occur.

These binding events are generally between white blood cells and endothelial cells as part of the processes of lymphocyte homing and lymphocyte trafficking but the expression of selectin ligands by carcinoma cells suggests the involvement of comparable mechanisms in tumour spread. Some colorectal cancers have been shown to express the ligand for E-selectin, also called sialylated Lewis X antigen, which correlates with the capacity to form distant metastases. It is tempting to extrapolate from the normal physiological behaviour of leucocyte trafficking to the situation in tumour cell dissemination but this may be misleading. The relative dimensions of the disseminating cancer cell and the arresting capillary bed may well mean that there is no need to attract cells out of a laminar flow situation. The large-diameter neoplastic cell is enclosed within a smaller-diameter capillary, which may permit the immediate involvement of integrin-mediated attachment.

10.7 ADHESION RECEPTORS MAY BE INVOLVED IN THE INDUCTION OF NEOVASCULARIZATION

Until very recently the events of cell adhesion and angiogenesis (the process of new blood supply development) were considered to be vital but unrelated processes in tumour development. However, the recent demonstration that soluble (secreted) members of the selectin and immunoglobulin families of adhesion receptors can stimulate angiogenesis serves to link these two events (Koch *et al.*,1995). Perhaps the induction of these soluble receptors by tumour cells plays a comparable role in stimulating angiogenesis to that proposed to occur via the release of angiogenic cytokines such as IL-8, TNF-α, VEGF or FGFs.

10.8 CD44 IS IN A CELL ADHESION CLASS OF ITS OWN

One cell adhesion molecule which has been ascribed a major role in determining metastatic spread is CD44, which does not fit into any of the four major families of adhesion receptors detailed above. Originally identified as a molecule involved in determining lymphocyte adhesion to high endothelial venules, it has been shown that expression of a novel form of this glycoprotein by carcinoma cells correlated with their ability to metastasize. The structure of the CD44 molecule is extremely variable according to which exons are retained or spliced out of the primary transcript. Production of the different isoforms that result from this variation in exon usage is complicated further by different glycosylation patterns, which result in the ability to generate an almost limitless number of forms of this molecule. However, recent clinical results from human tumours show that utilization of certain exons is correlated with increased tumour progression, suggesting that the molecule may play an important part in determining tumour spread. CD44 exemplifies an adhesion molecule that was first identified as being involved in tumour cell behaviour in experimental systems and has now been demonstrated to play a similar part in the behaviour of naturally occurring cancers through the analysis of clinical material.

10.9 TUMOUR CELL MOTILITY FACTORS PROMOTE METASTASIS

It has long been known from histopathological analysis of tumour tissue that the occurrence of isolated tumour cells or small emboli of tumour cells at sites distant from the primary tumour mass is probably indicative of independent motility by the neoplastic cells. This motility may be as a consequence of the binding of specific mitogenic cytokines, also known as **motility factors**, to requisite receptors on the surface of these cancer cells. Motility factors, such as autocrine motility factor, migration stimulating factor and scatter factor, have all been shown to be produced by both normal fibroblastic cells and various tumour cells, and they may function in an autocrine or paracrine fashion. As a consequence of the binding of these motogenic cytokines to the tumour cells, there is initiation of postreceptor signalling, which results in increased locomotion. As discussed above, such increased locomotion may be further stimulated by the binding of cells to various substrates.

What is not understood is precisely how, in biochemical terms, these various stimulators of motility actually increase locomotion and whether in cancer cells such mechanisms are defective or perhaps even constitutively activated. Many normal cells, including leucocytes, fibroblasts, endothelial cells and trophoblasts, also are capable of invasive activity and motile behaviour. Analysis of the driving mechanisms behind motile behaviour in these normal cells should shed light on those mechanisms operative in the cancer cells.

In conclusion, cell adhesion and motility are important factors in regulating cancer spread. Many of the molecules involved in this process in neoplastic cells appear to be very similar or identical to those involved in regulating the normal patterns of tissue migration or cell spread. Information derived from analysis of normal cells may assist in determining behaviour in cancers.

10.10 FURTHER READING

Dorudi, S., Hanby, A. M., Poulsom, R. *et al.* (1995) Level of expression of E-cadherin in RNA in colorectal cancer correlates with clinical outcome. *Br. J. Cancer*, **71**, 614–616.

Gunthert, U., Hofmann, M., Rudy, W. *et al.* (1991) A new variant of glycoprotein CD44 confers metastatic potential to rat carcinoma cells. *Cell*, **65**, 13–24.

Humphries, M. J., Olden, K. and Yamnada, K. M. (1986) A synthetic peptide from fibronectin inhibits experimental metastasis of murine melanoma cells. *Science*, **233**, 467–470.

Hynes, R. O. and Lander, A. D. (1992) Contact and adhesive specificities in the associations, migrations and targeting of cells and axons. *Cell*, **68**, 303–322.

Koch, A. E., Halloran, M. M., Haskell, C. J. *et al.* (1995) Angiogenesis mediated by soluble forms of E-selectin and vascular cell adhesion molecule 1. *Nature*, **376**, 517–519.

11	# Differentiation and cancer

Malcolm D. Mason

11.1 NORMAL TISSUE RENEWAL REQUIRES THAT MATURE CELLS ARE REPLACED BY CELLULAR DIFFERENTIATION FROM A POOL OF UNDIFFERENTIATED 'STEM CELLS'

Normal cell renewal is usually not accomplished by the division of mature cells. Instead, normal tissue growth and renewal is accomplished by division and differentiation from a pool of undifferentiated cells called **stem cells**. Stem cells are the origin of the mature cells that characterize an individual tissue. Such a system certainly exists in the haemopoietic system and in most common epithelia, and may even exist in some other non-epithelial tissues, although this is not firmly established.

Stem cells are unique in that, when they divide, their two daughter cells are different (i.e. cell division is asymmetric). One cell, identical to the parent stem cell, replaces it in the tissue pool to maintain the overall number of stem cells. The other is said to be **committed**, as it embarks on a chosen pathway of differentiation and will eventually give rise to a mature cell whose phenotype is characteristic of the tissue in which it arises. What governs the decision by a daughter cell to self-renew or to differentiate is one of the most fascinating questions in cell biology. In an organized tissue, the presence of stem cells may only be inferred from their effects on tissue growth and renewal. For example, the existence of a bone marrow multipotent stem cell was suggested many years ago, but its formal identification has been difficult.

11.2 DIFFERENTIATION IS LINKED TO PROLIFERATION

It is axiomatic that, in order to renew cells lost by death or injury, cell division must occur and feed a pool of new differentiating cells into the system. The

Molecular Biology for Oncologists. Edited by J. R. Yarnold, M. R. Stratton and T. J. McMillan. Published in 1996 by Chapman & Hall, London. ISBN 0 412 71270 9

population undergoing this cell division is initially in the stem cell compartment of a tissue. At the other extreme is the fully mature cell, for which cell division is not only unnecessary for tissue renewal, but is thought to be impossible in most tissues that have a stem cell system of organization. Such mature, fully differentiated cells are said to be **postmitotic**, but the mechanisms by which mitosis is blocked are not fully understood. However, in order to become postmitotic, a cell must exit permanently from the cell cycle, and attention is now being focused on the molecular mechanisms of cell cycle arrest. Also implicit in the stem cell model is that in normal tissues between the undifferentiated stem cell stage and the postmitotic end cell stage there are a number of intermediate stages, during which a committed cell may have limited proliferative capacity but yet not be completely differentiated. It seems therefore that, as a generalization, a cell must either be a stem cell or be engaged in the process of differentiation in order to retain proliferative capacity. The phenotypic diversity of the intermediate cells could go some way towards explaining the corresponding diversity of tumours that arise from a single tissue if such tumours were caricatures of normal differentiation. For the postmitotic cell, the only destiny is cell death. This is as true of terminally differentiated cells arising from malignant stem cells as it is of postmitotic cells from normal stem cells (Figure 11.1).

11.3 CELL CYCLE ARREST OCCURS AS CELLS MATURE AND AGE, DUE TO A SHORTENING IN TELOMERE LENGTH

How are differentiation and proliferation so elegantly linked? One mechanism may be to link the genes which control differentiation to those which control cell cycle arrest. A closely studied example of this is the retinoblastoma gene (*Rb*), described

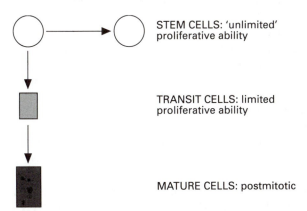

STEM CELLS: 'unlimited' proliferative ability

TRANSIT CELLS: limited proliferative ability

MATURE CELLS: postmitotic

Figure 11.1 Schema of cell hierarchy. Although the characteristic morphology of a tissue depends on the mature cells that make it up, it is usually the most primitive cells that have the ability to divide, while the mature cells themselves have lost this capacity and are said to be postmitotic.

in Chapters 1 and 7, and this will be discussed later. Another possible mechanism involves the cell's ageing process. If a certain number of cell divisions are required to reach full maturity, can a cell tell how many divisions it has undergone? One way by which this is achieved is simply by measuring the shortening of the cell's telomeres with age. Telomeres are the non-coding ends of chromosomes and, because of their structure (which includes the repeating nucleotide sequence TTAGGG), they shorten with each round of cell division. This can be circumvented by the enzyme **telomerase**, which resynthesizes telomeres to their full length. Normally, telomerase activity is strictly repressed in committed cells, so that, as the cell ages with each round of mitosis, telomere length shortens, until it reaches a critical length which triggers cell cycle arrest. If telomerase is reactivated, the results can be catastrophic, as cells can no longer tell when their lifespan is over, and cell-cycle arrest (and hence the postmitotic state) cannot be achieved. There is now evidence that this is precisely what happens in many tumours. If this is true, inhibitors of telomerase activity might one day have some use in cancer therapy.

11.4 UNDER NORMAL CIRCUMSTANCES CELLS REMAIN FAITHFUL TO ONE LINEAGE PATHWAY OF DIFFERENTIATION, AND THEIR DESTINY IS DEFINED BY DETERMINATION DURING DEVELOPMENT

The term **determination** refers to a heritable undertaking by a cell during embryonic development to follow a particular pathway of specialized development at some stage in the future. Once differentiated, a cell acquires certain structural and functional characteristics that endow it with the ability to undertake a specialized task, e.g. to carry oxygen or to absorb nutrients. Differentiation may involve, for example, the secretion of certain specialized molecules that are not produced by the undifferentiated cell. Thus, determination precedes differentiation. According to classical dogma, once a cell is committed it will remain faithful to one lineage pathway of differentiation only: a committed intestinal epithelial precursor cell will not turn into a red blood cell, for example. Some cells, particularly very undifferentiated cells, must have the option of following several alternate pathways, the best example of which is the bone marrow pluripotent stem cell which may differentiate along erythrocytic, leucocytic or megakarocytic pathways. However, it is also clear that such options in normal tissues are severely restricted.

11.5 SOME MALIGNANT TUMOUR CELLS RETAIN THE ABILITY TO DIFFERENTIATE ALONG ALTERNATIVE LINEAGE PATHWAYS

Fundamentally, the dogma of fidelity to one pathway is almost certainly correct. However, certain features of malignant cells imply that early on in differentiation

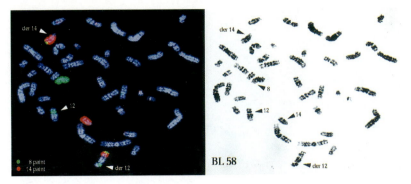

Plate 1 Chromosome 'painting' of a Burkitt's lymphoma cell line. Chromosome-specific 'paints' for fluorescent *in-situ* hybridization (FISH) can be produced from individual human chromosomes purified by flow sorting on the basis of chromosome size. The DNA may be labelled with reporter molecules of different colours and hybridized on to tumour metaphases to aid identification of 'marker' chromosomes of ambiguous morphology. Here, chromosome 14 (red) and chromosome 8 (green) paints have been used together with a chromosome 12 centromere-specific probe (green). Normal copies of chromosome 8, 12 and 14 are arrowed (right hand panel). A derivative chromosome 14 is identified which has material from chromosome 8 at the telomere, typical of the t(8;14)(q24;q32) of Burkitt's lymphoma. Unexpectedly however, a derivative chromosome 12 is identified which also has material from chromosomes 14 and 8. This chromosome was interpreted as der(12)(12;14;8)(q24;q22;q24). (Source: courtesy of Dr Lynne Hiorns, Academic Department of Haematology and Cytogenetics, ICR-RMH.)

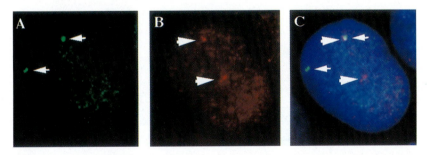

Plate 2 Interphase FISH of Ewing's sarcoma with t(11;22)(q24;q12). Ewing's sarcoma is associated with translocation t(11;22)(q24;q12) resulting in the fusion of the *FL1I* gene at 11q24 with the *EWS* gene at 22q12. This translocation is of value in diagnosing Ewing's sarcoma from other round cell tumours. Here, interphase nuclei have been hybridized with probes which lie either side of the breakpoint in the *EWS* gene on 22q12. (a) shows the signal from a centromeric probe (green) and (b) the signal from a telomeric probe (red). Both probes should lie adjacent to one another unless a translocation disrupting the *EWS* gene is present. (c) shows the position of these two markers relative to one another: the dissociation of one of the pair of red-green doublets is indicative of a *EWS* translocation. (Source: provided by Drs Aidan McMannus and Janet Shipley, Molecular Cytogenetics Laboratory, Sections of Paediatrics and Cell Biology and Experimental Pathology, ICR-RMH.)

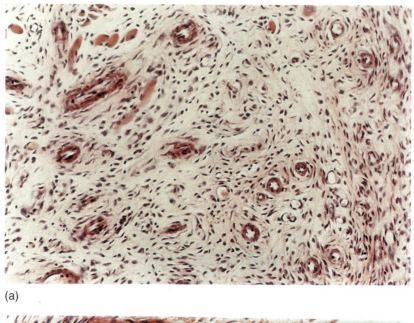

(a)

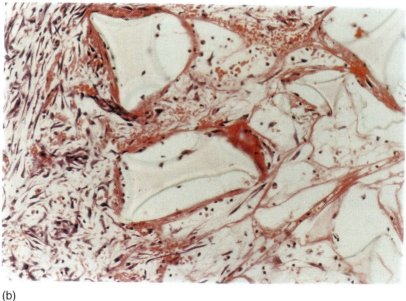

(b)

Plate 3 Vessel formation in rat sponge model following treatment with thymidine phosphorylase. (a) Thymidine-phosphorylase-treated sponge. (b) Control sponge. Cannulated polyester sponges were engrafted subcutaneously into the backs of laboratory rats. Daily incorporation of thymidine phosphorylase via the cannula results in the development of a regular vascular supply within the sponge (a). In contrast, the control sponge shows no evidence of having acquired an analogous vascular supply. This correlates with increased clearance of radioactive xenon from the sponge treated with thymidine phosphorylase. (Source: reproduced by permission of Dr R. Bicknell, Oxford).

a cell may keep its options open with regard to which of several possible lineage pathways it intends to follow, and primitive cells may carry markers characteristic of more than one lineage (sometimes referred to as 'lineage promiscuity'). Indeed, there is evidence that the system can sometimes be flexible to the point that 'impossible' transitions, such as from epithelial cell to fibroblastoid cell, do in fact, occur, but phenomena such as this should be seen as exceptions and not the rule. Such 'epithelial–mesenchymal transformations' are common events in embryogenesis, and may yet change our perception of a cell's fidelity to one lineage pathway.

Germ cell tumours provide the most dramatic model of cancer cell differentiation. The diversity of cell types present in such tumours can be understood in the light of evidence that the embryonic carcinoma (EC) stem cell in non-seminomas is very similar to a pluripotent embryonic stem cell, i.e. it is capable of differentiating along multiple lineage pathways. In the mouse, EC probably corresponds to cells of the late blastocyst. When human EC cells of the cell line NTERA-2 are treated *in vitro* with retinoic acid (RA) they undergo somatic differentiation into neurone-like cells. By contrast, cells of the mouse EC cell line F9 treated with RA undergo extraembryonic differentiation into parietal endoderm. Treatment of NTERA-2 cells with another differentiation-inducing agent, hexamethylene bisacetamide (HMBA) induces them to a differentiate into a large, flat cell type with entirely different phenotypic markers to the RA-treated cells. The outcome of an interaction between a cell and a differentiation-promoting agent clearly varies according to the cell and the stimulus.

11.6 NORMAL DIFFERENTIATION IS OFTEN BLOCKED IN MALIGNANCY, BUT 'DEDIFFERENTIATION' IS USUALLY A MISNOMER

The central theme of this chapter is the concept that cancers are merely caricatures of normal differentiation, and that the primary characteristic of a malignant cell is that the process of normal differentiation has somewhat been blocked, leading to a cell whose maturation has been 'frozen'. Human lymphoid malignancies provide the most comprehensively understood model of the relationship between normal differentiation and the phenotype of cancer cells. The phenotypic characteristics of normal lymphoid cells along their pathways of differentiation have been elaborated to a high degree of sophistication. What is striking about lymphoid malignant cells is that their phenotype is often remarkably similar to that of normal cells somewhere in the pathway between primordial stem cell and mature cell. The real problem in haemopoietic malignancies and perhaps in some other cancers may not be so much that abnormal differentiation is taking place, but rather that the freezing of normal differentiation at an intermediate stage leads to the clonal expansion of a cell type that should be transient and therefore extremely rare. The motive of altered gene expression in this setting would be the stabilization of this transient cell type. Some specific genomic abnormalities may merely reflect the method whereby this is achieved. For example, the t(14;18) translocation in lymphomas involving the *bcl-2* gene may be important because *bcl-2* codes for a protein that is an inhibitor of programmed cell death. The t(15;17) translocation seen in acute promyelocytic leukaemia results in the fusion of a retinoic acid receptor gene to a different site, which is interesting given that RA promotes the differentiation of acute promyelocytic leukaemia cells.

In haemopoietic malignancies, the block in normal differentiation can some-times be overcome: injection of leukaemia cells into the placentae of mouse embryos sometimes results in normal mice whose haemopoietic tissues are chimeric. Therefore, all the leukaemia cells injected into these mice were able to participate in normal differentiation. Differentiation of erythroleukaemia cells can also be induced by HMBA.

Germ cell tumours are caricatures of normal development in which **somatic differentiation** gives rise to the embryo and subsequently to the animal itself, and **extraembryonic differentiation** gives rise to supporting tissues such as yolk sac and trophoblast, which do not form the animal proper. By understanding what sort of cell the malignant teratoma stem cell thinks it is supposed to be, we can begin to understand certain aspects of its behaviour. Experiments by Kleinsmith and Pierce demonstrated that malignant teratoma cells could differentiate into 'benign' (i.e. postmitotic) tissues. They found that it was possible to inject a single terato-carcinoma cell into a mouse and generate a tumour comprising both malignant ter-atoma and benign differentiated tumour. Even more dramatic were the experiments by Brinster, in which a malignant teratoma cell was introduced into normal mouse blastocysts, which were then transferred to a foster mother. The healthy offspring resulting from this blastocyst were chimeric mice, in which the tissues were derived partly from normal mouse embryonic cells and partly from teratocarci-noma cells. The molecular processes favouring the survival of malignant terato-carcinoma stem cells are, indeed, 'subservient to the normal differentiation programme'. In one chimera, the cells derived from the teratoma included those of the germinal epithelium, and it was possible to propagate a second generation of apparently normal mice whose 'father' was, in a sense, a teratocarcinoma cell! 'Once a cancer cell, always a cancer cell'? Clearly not, as these experiments dra-matically illustrate. Interestingly, the transplantation of a normal mouse blastocyst into an extrauterine site led to the formation of a malignant teratocarcinoma. Although it is not possible to perform such an experiment in human teratomas, it is clear that, as in the mouse, human EC cells are pluripotent. Whether the differ-entiated cells produced in this context are truly benign and postmitotic is open to question; the clinical phenomenon of sarcomatous degeneration of differentiated testicular teratoma deposits argues that such tissues are not always stable.

Malignant tumours, far from being in a state of total anarchy as is sometimes portrayed, may therefore be subject to a 'tissue' system of organization based on a stem cell compartment. In some tumours, the stem cell population will predominate and no differentiated elements will be apparent to the light-microscopist. In others, mixtures of undifferentiated cells and more mature cells give a malignant tumour its characteristic morphology. It was Pierce, working on teratocarcinomas, who postulated that malignant tumours arose from stem cells, that they were the target for carcinogenic stimuli and not the mature end-cells that give a tissue its morphology.

In this view of cancer as a caricature of normal differentiation, malignant stem cells give rise to a larger pool of similar malignant stem cells by self-renewal and also to committed cells that have a variable capacity for differentiation, including terminal differentiation. It is implicit from what has been said that differentiation is a one-way process. Usually, normal mature postmitotic cells do not revert to undifferentiated stem cells and give rise to cancer by 'de-differentiation'. This term is used loosely by clinicians without any regard for the enormous biological implications that such a label carries. Cell biologists now believe that limited de-differentiation may occur in some specific cell types and in special circumstances but the use of the term in the clinicians' ordinary vocabulary should be banned. The evolution of a low-grade tumour into a high-grade tumour can be viewed as resulting from a shift in emphasis from differentiation to self-renewal, and not from the supposed 'de-differentiation' of well-differentiated tumour cells.

11.7 THE GENETIC CONTROL OF NORMAL DIFFERENTIATION IS BETTER UNDERSTOOD IN SIMPLER ORGANISMS LIKE *DROSOPHILA* THAN IN MAMMALS

Much of our understanding of normal differentiation comes from studies of embryonic development in lower animals, and comparatively little is known about normal adult mammalian cell differentiation. It may seem surprising to the oncologist that non-mammalian systems are relevant to human cancer, but such is indeed the case. Mutations of *Drosophila* have added to the studies on mammalian cells in identifying genes that are of crucial importance in normal development and differentiation. These include homeotic genes encoding DNA-binding proteins that are responsible for implementation of the body 'plan' in *Drosophila* development, instruct cells on their spatial position in the body and hence implement their proper pathway of differentiation. Homeotic genes include the highly conserved **homeobox** genes, a class of gene that has been found in other organisms, including mammals and even plants. It is noteworthy in this context that the action of retinoic acid (RA), a potent inducer of differentiation in several tumour systems *in vitro*, appears to involve the action of several homeobox genes.

11.8 SOME HIGHLY CONSERVED GENES THAT CONTROL NORMAL DIFFERENTIATION ARE DISRUPTED IN MALIGNANCY

Several homologues of non-mammalian genes from *Drosophila* and from the nematode *Caenorhabditis elegans* are involved in differentiation of both normal

Table 11.1 Conserved genes involved in differentiation/development in non-mammalian systems that are also found in mammalian cells and in malignant tumours

Mammalian/cancer cell	Non-mammalian cell
gli	kruppel
int-1	wingless
TGF-β	decapentaplegic
EGF	notch, lin-12
EGF receptor	torpedo
RA receptor	knirps
PBX1	extradenticle

and malignant mammalian cells (Table 11.1).

At first sight these examples of homology are confusing because they do not present a clear picture of the relationship between normal differentiation and malignancy. Our understanding will be enhanced in the future as the functions of these genes in normal development are elaborated, and already their patterns of expression are being described. For example, the *int-1 (Wnt-1)* gene, expressed in mouse mammary tumours, is not expressed in the normal mouse mammary gland, although it is in the developing central nervous system. However, several members of the *Wnt* family of genes, which have substantial homology to *Wnt-1*, are expressed in mouse mammary glands during pregnancy and lactation. As another example, the candidate Wilms tumour gene, *WT1*, is normally expressed in the early stages of nephron development in the fetal gonad and fetal mesothelium. Other cancer-associated genes will undoubtedly be shown to have a role in normal development; for example, mice that lack the *Rb* gene do not develop normal nervous systems. Tempting as it is to suggest that every important cancer-associated gene will have a role in normal development and differentiation, there will be important exceptions. The tumour suppressor gene coding for p53 has been implicated in a wide variety of tumours. Rather surprisingly, mouse embryos that are genetically engineered so that the individuals carry a deficiency of *p53* are developmentally normal, yet they are highly susceptible to the development of malignant tumours.

One reason why so many cancer-associated genes appear to be important in differentiation and embryonic development is that many molecules that regulate growth and differentiation have multiple and possibly diverse functions. However, in order to understand how a biologically active molecule influences carcinogenesis or the behaviour of malignant cells, we may first have to understand its functions in normal embryological development because the two may share many common mechanisms. One of the best models for developing this understanding is normal muscle differentiation.

11.9 MASTER REGULATORY GENES CONTROL DIFFERENTIATION IN MAMMALIAN CELLS AND MAY HELP TO CONTROL EVENTS SUCH AS CELL CYCLE ARREST

Homeobox genes may be an example of a class known as **master regulatory genes**. Such genes are extremely potent, in that they initiate a clear signal that, in effect, switches on a given pathway of differentiation. A master regulatory gene has now been identified in cells of muscle lineage, where it has been named *MyoD*. The protein produced by *MyoD* is a DNA-binding protein of the helix–loop–helix class, with sequence homology to the Myc family of proteins. The MyoD protein will directly or indirectly regulate all of the other genes whose activation or repression make up that phenotype of skeletal muscle cells (Figure 11.2).

So powerful is the *MyoD* gene that, if it is transfected into non-muscle cells such as amniocytes, and even if it is transfected into some tumour cells, it will over-ride the cells' prevailing phenotype and induce terminal differentiation into muscle cells! However, this does not mean that *MyoD* cannot itself be over-ridden, as we shall see.

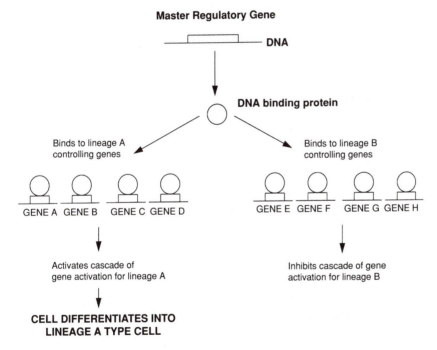

Figure 11.2 Schematic representation of how a master regulatory gene might control cellular differentiation. In this example a cell has two alternative lineage pathways, A and B. The product of the master regulatory gene is a protein that binds to many genes, in this example activating those responsible for the features of lineage A while inactivating those responsible for lineage B. Different master regulatory genes might use different combinations of activation and inactivation to achieve their objective.

Although abnormalities of *MyoD* itself may not be sufficient to induce tumours, the mechanisms by which it acts shed some light on the processes of normal differentiation and the ways in which differentiation could be disrupted in cancer. As *MyoD* becomes active, it seems to induce the expression of retinoblastoma protein in its active, unphosphorylated form in which it induces cell cycle arrest. Conversely, and of interest in relation to tumour formation in other systems, *MyoD* gene expression can be down-regulated by the expression of both c-*myc* and certain cyclins. It will be of great interest to see whether similar processes can be demonstrated in other cell types, and whether they are of importance in tumour induction.

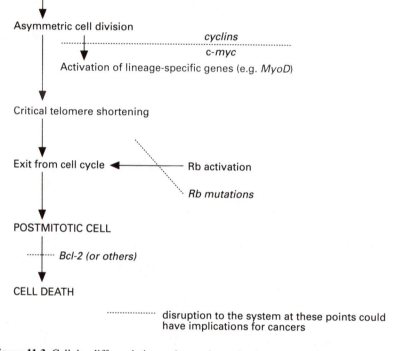

Figure 11.3 Cellular differentiation and tumorigenesis. A summary of some of the events that must take place as a stem cell matures to become postmitotic. There are several steps in this process where abnormalities could play a part in tumorigenesis.

11.10 FURTHER READING

Greaves, M. F. (1986) Differentiation-linked leukemogenesis in lymphocytes. *Science*, **234**, 697–704.

Hall, P. A. and Watt, F. M. (1989) Stem cells: the generation and maintenance of cellular diversity. *Development*, **106**, 619–633.

Kim, N. W., Piatyszek, M. A., Prowse, K. R. *et al*. (1994) Specific association of human telomerase activity with immortal cells and cancer. *Science*, **266**, 2011–2015.

Pierce, G. B. and Speers, W. C. (1988) Tumors as caricatures of the process of tissue renewal: prospects for therapy by directing differentiation. *Cancer Res.*, **48**, 1996–2004.

Riley, D. J., Lee, E. Y. P. and Lee, W. H. (1994) The retinoblastoma protein: more than a tumor suppressor. *Annu. Rev. Cell Biol.*, **10**, 1–29.

12	# Multistage carcinogenesis

Andrew H. Wyllie

12.1 INTRODUCTION

Carcinogenesis means the acquisition of those cellular properties that render a cell competent to grow into a tumour. Once tumours have formed, further new properties continue to appear in the cells within them, some carrying selective growth advantage and hence of sinister import for the patient. Properly speaking, this evolution of phenotype within tumours is referred to as **tumour progression**, while the term carcinogenesis refers to the events leading to the appearance of a tumour in the first place. Both sets of events, however, represent changes that lead from normality to aggressive tumour growth, and they are considered together in this chapter.

Evidence from many different sources indicates that the events involved in carcinogenesis and tumour progression are multiple. This chapter describes this evidence, and gives an account of some of the different events known to participate in the development of common human tumours. Finally, certain significant implications of multistage carcinogenesis are discussed.

12.2 EVIDENCE FOR MULTIPLICITY OF CARCINOGENIC EVENTS

Examination of tumour pathology has often suggested that tumours evolve by a series of discrete transitions. Thus there are clearly recognizable morphological distinctions between benign and malignant tumours. Similarly, the terms atypical hyperplasia, intraepithelial neoplasia and invasive carcinoma reflect different degrees of cytological aberration in structure and behaviour. Often these lesions occur in isolation, but there are also many observations of, for example, invasive carcinoma coexisting and contiguous with intraepithelial carcinoma, of atypical

Molecular Biology for Oncologists. Edited by J. R. Yarnold, M. R. Stratton and T. J. McMillan. Published in 1996 by Chapman & Hall, London. ISBN 0 412 71270 9

hyperplasia apparently merging with areas of intraepithelial neoplasia, or of carcinoma arising within a pre-existing benign tumour. The most obvious interpretation of these histological patterns is that new behaviour is acquired in a sequential, stepwise fashion. There is now molecular evidence in support of this view, some of which will be discussed later, but it is much more difficult to prove that a similar stepwise process is a feature of **all** tumour development.

Multistage carcinogenesis was also recognized in experimental reconstructions of the carcinogenic process. In experiments involving chemical carcinogens, Berenblum distinguished between tumour **initiators** – substances capable of causing tumour growth on their own – and tumour **promoters**, which do not cause tumour growth on their own but enhance the effect of inducers. In a series of classical experiments, he showed that initiators altered cells in a permanent manner that might nonetheless remain undetected by morphological observation until proliferation of the altered cells was stimulated by a promoter. Promoters had no observable effect if applied before initiators, but could dramatically increase the yield of tumours if applied afterwards, even after an interval of many months. Essentially similar observations have since been made with a great many carcinogens and a wide variety of tissues. Most initiators are now known to be mutagens, whereas the most widely used promoters, extracted from croton oil, contain tetradecanoyl phorbol 13-acetate (TPA), an analogue of diacylglycerol and a potent stimulator of protein kinase C.

Experiments of the same design as Berenblum's have been repeated with a specific focus on mutations in critical cancer-related genes. In experimental mammary and skin cancer of rodents (Figure 12.1), the tumours were shown to be

1. INITIATION
   Single systemic treatment with carcinogen, eg DMBA, MNNG, NMU.

2. PROMOTION
   Local treatment with promoter (TPA to skin) or physiological hormone stimulus (ovarian hormones to mammary gland).

3. TUMOUR DEVELOPMENT
   At site of promotion, but with activating mutations in *ras* genes that are characteristic of initiator.

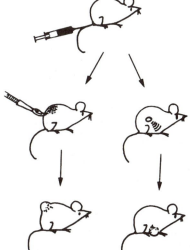

Figure 12.1 Diagram of experiments that demonstrate initiation and promotion at the molecular level.

clonal expansions of cells bearing *ras* oncogenes activated by mutations.

The mutations were of the specific types expected from cellular exposure to the initiating agents that had been applied long before. Moreover, it could be shown that the initiator had induced mutations characteristic of its interaction with DNA in cells that contributed only to hyperplasia. Further events were apparently required before a subclone from these altered cells would emerge as carcinoma.

From an entirely different angle, epidemiological data have shown that the age-related increase in human tumour incidence, with few exceptions, follows a steeply rising curve. The shape of this curve is amenable to mathematical modelling, and one model interprets the slope – following log–log transformation – to indicate the number of separate events involved in carcinogenesis. For the great majority of adult-onset carcinomas, this slope suggests the existence of five or six such independent events. For some childhood tumours the slope is less, but always indicates that carcinogenesis involves more than a single cellular event.

Most of the rest of this chapter is devoted to unravelling what these multiple carcinogenic events might be, and how they interact with each other. It has a strong bias towards irreversible genetic events because there is now a solid factual basis for the importance of these in carcinogenesis. Other types of event also occur and are unquestionably important – such as modulation in the level of transcription of important genes – a potentially reversible change that does not imply structural alteration in the genes themselves. An example might be the expression of the cell attachment molecule E-cadherin, which appears to be preferentially depressed in many metastatic tumours. As much less is known about such modifications in the cellular levels of gene products, or about how they might occur in cancer, they are not further discussed here.

12.3 GENETIC CHANGES IN HUMAN CANCERS

Many genetic abnormalities are known to occur in tumours. Some of these are not site-specific. Thus a generalized hypomethylation has been described in many tumours, both benign and malignant. Very little is known of the regulation of this change in the methylation status of the genome, but it does not appear to be specific to particular sets of genes. It has been postulated that this hypomethylated DNA in premalignant cells is more susceptible to mutagenesis, perhaps because of less tight chromatin packing and correspondingly enhanced carcinogen access.

Similarly, a high proportion of all malignant tumours are aneuploid, or show other evidence of karyotypic instability, such as the presence of bizarre marker chromosomes, double minutes or truncated telomeric sequences. Unlike hypomethylation, aneuploidy is seldom a feature of benign tumours. A superficial assessment of the karyotypes of aneuploid tumour cells reveals a wide variety of abnormalities and a broad range of DNA content. However, two large groups can often be discerned: 'near-diploid' tumours that differ from normal to a relatively small degree, perhaps by virtue of a single extra chromosome or part of a chromosome arm; and 'near-tetraploid' tumours that contain double sets of most chromosomes. These patterns suggest that different mechanisms may be at work in the genesis of tumour aneuploidy, some based on damage to or non-disjunction of individual chromosomes and others involving an endoreduplicative event in which two rounds of chromosome replication occur without intervening cell division. Despite the fact that karyotypic instability in tumours has a large number of possible outcomes in terms of gene number, integrity and expression, it now appears probable that it may result from disorder in only a few critical genes, which are discussed later in this chapter.

A third general feature of tumour genomes is telomere erosion. The nucleotide sequence of telomeres consists of tandem arrays of many hundreds of short motifs (in man TTAGGG). These subserve two functions: to bind a special set of chromosome proteins that presumably have to do with identification of chromosome ends within the nucleus, and to solve a problem intrinsic to the mechanism of DNA replication. DNA replication always requires a priming event at the 5' end of the replicated sequence (or 3' – i.e. downstream – to the template DNA). It is thought that the telomeric array provides a resource of DNA to permit this, lying well downstream of the most 3' portion of essential genomic DNA near the end of the chromosome. In each consecutive round of DNA replication one element of the tandem array is lost. In normal cells and tissues this is of no consequence, because the total number of replications within the lifetime of an individual is less than the number of telomeric tandem repeats. Germ cells possess an enzyme – telomerase – which reconstitutes the tandem array to its original length, and so the truncation of telomeric repeat arrays in the lifetime any one individual is not cumulative for the next generation. It is possible, however, that the number of replication events in the history of a tumour stem cell may exceed the number of telomeric repeats. This may lead to loss of the specific telomere binding proteins, progressive loss of non-redundant subtelomeric sequence at each subsequent round of replication, and the appearance within the nucleus of a 'free-end' of double helix. Such free ends are potent signals for a variety of nuclear events to be described later, but one important consequence is inappropriate recombination. Here the free end of one chromatid can fuse with that of its complementary chromatid, or with damaged telomeres on the ends of other chromosomes (Figure 12.2).

At the next mitosis, these interchromatid fusions prevent normal chromosome disjunction; for completion of cytokinesis, the fused chromatids must break (not necessarily at the original fusion site) so producing yet another set of free ends. This fusion–bridge–breakage cycle (originally described by the pioneer geneticist and Nobel laureate Barbara McClintock) may have much to do with the aneuploidy of tumours cells described earlier. Telomere erosion does not appear to progress indefinately, however. Recent observations show that the enzyme telomerase – normally silenced in somatic cells – is reactivated in many tumour types, even those in which some telomere erosion is known to be common.

Specific genetic alterations are acquired in carcinogenesis in oncogenes and oncosuppressor genes. Human colorectal tumours, and to a lesser extent tumours

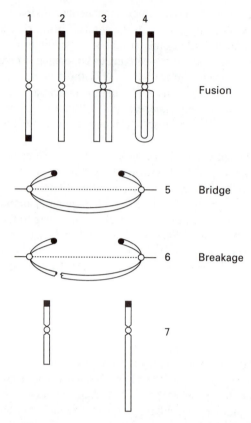

Figure 12.2 The fusion–bridge–breakage cycle. In this stylized diagram, a chromosome is shown (2) in which the telomeric sequence originally present at the end of the long arm (1 – represented by the black-shaded area) has been eroded. Following replication (3) the free ends of the chromatids fuse (4). A mitotic bridge forms (5) and breaks (6), although not necessarily at the original fusion site. This partitions to each of the daughter cells a chromosome lacking a telomere, so re-instating the cycle at position (2) in these cells.

at other sites such as the breast, bladder, cervix, ovary and lung, have been studied for such changes in some detail. About 40% of all colorectal tumours include clonal expansion of cells that carry a K-*ras* gene activated by point mutation. Around 50–60% bear a mutated *p53* gene, and between 60% and 80% have nonsense or stop-codon mutations in *APC*, the gene responsible for transmission of familial adenomatous polyposis. Allele deletions that involve the *p53* locus on chromosome 17p are found in around 70%, the *DCC* locus on chromosome 18q in 70–80%, and the *APC* locus on chromosome 5q in about 50%. Further allele deletions, presumably indicating the location of further, but as yet undiscovered oncosuppressor genes, are observed in chromosome 8p, 1p and 22q. Very few colorectal carcinomas fail to show clonally expanded alterations at any of these sites, and most have multiple defects. However, adenomas and carcinomas with proven better prognosis tend to carry fewer genetic defects than carcinomas with poor prognosis.

Rather similar principles appear to apply to carcinomas at other sites. Thus breast carcinoma frequently shows – in addition to mutation and allele deletion of *p53* – amplification of a region in chromosome 17q that includes the proto-onco-gene c-*erbB2*, and allele deletions at a wide variety of other sites, including loci on chromosome arms 1q, 3p, 6q, 16q, 17q and 18q. The critical deleted loci in chromosome 17q and 13q.12 correspond to the *BRCA1* and *BRCA2* genes, which are inherited in defective form in hereditary types of breast cancer. In lung carci-nomas, the L-*myc* oncogene is often amplified whilst a proportion show activat-ing K-*ras* mutations. Allele deletions are also common at many loci – including *p53*, the *Rb* (retinoblastoma susceptibility) oncosuppressor gene at 13q14, and still anonymous loci in 3p and 11p. Pancreatic carcinomas, like colorectal carcinomas, show mutations in K-*ras* and *p53*, but at much higher frequency, close to 95%. Ovarian cancers show a different spectrum of allele loss (4p, 6p, 7p, 12p, 19q as well as 17p and 17q), while in bladder tumours the commonest sites of allele loss are 8p, 9p, 9q, 11p, 13q, 17p and 4p. Interestingly, cervical carcinoma, in the majority of which oncogenic human papillomavirus (HPV) genomes are present, seldom show alterations at the loci commonly involved in other tumours, includ-ing mutation and deletion of *p53*. Presumably the viral oncogenic proteins are ini-tiating disruption of cell function that would otherwise arise only through loss or distortion of expression of endogenous genes.

Different tumours tend to show differing but overlapping combinations of fre-quently affected genes. This pattern of multiple gene abnormality in tumours raises several questions. Does each abnormality represent a distinct stage in tumour evo-lution, or is the critical factor in tumour progression merely the total number of defects? Do the genetic events become added sequentially in time, or do they occur simultaneously in those cells that are the targets of carcinogen attack? Do the events interact with each other? Are there alternative routes to the same end-point? The answers to these questions bear on the whole concept of multistage car-cinogenesis and tumour progression, and are discussed below.

12.4 MULTIPLE GENETIC EVENTS: STAGE-SPECIFIC OR CUMULATIVE?

Early anecdotal observations in colorectal tumours suggested that some genetic lesions could be acquired at almost any stage of tumour evolution, and the state-ment is often reiterated that it is the total number of defects rather than their nature that determines tumour behaviour. Much subsequently gathered data, however, indicate that certain genetic defects preferentially appear at specific stages of tumour progression, presumably because they initiate or permit the clonal expan-sion of cells with characteristic neoplastic behaviour (Figure 12.3).

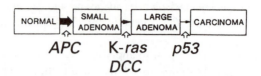

Figure 12.3 Sequence of pathological changes in colorectal carcinogenesis with stages at which several acquired gene abnormalities probably exert their major effects.

Thus K-*ras* mutations in colorectal cancer appear in the larger and more villous adenomas with the same incidence as in carcinomas, but are relatively rare in small, tubular adenomas. Where a K-*ras* mutation appears in a carcinoma that bears a contiguous fragment of adenoma (from which, presumably, it has derived) the same mutation is often found in the adenoma. Thus the evidence supports the view that presence of a mutationally activated K-*ras* gene confers growth advantage on cells within the larger adenomas, but offers little to the cells of established carcinomas.

Point mutations in *APC* also occur with the same frequency in adenomas and carcinomas, but here even small adenomas are involved. Indeed, point mutations in *APC* appear also in the tiny 'aberrant crypt foci' which may represent one early (and possibly reversible) step towards neoplasia. In the Mendelian dominant condition of familial adenomatous polyposis (FAP), heterozygous mutation in the *APC* gene is inherited, and the colorectal mucosa, over the first 20 years of life, becomes studded with hundreds to thousands of adenomas. Although some controversy remains, it is probable that the residual normal *APC* allele is lost prior to adenoma appearance.

Mutations in *p53* show an entirely different pattern in colorectal carcinogenesis. They are rare in adenomas of all kinds, even those that are undergoing transition to carcinoma, but are relatively common in carcinomas. Thus, *p53* mutation appears to confer no growth advantage on cells evolving from normal to adenoma, but provides a potent advantage to carcinoma cells. Moreover, the advantage seems to be limited to cells early in the process of growing as carcinoma: divergent stemlines of carcinomas that bear no *p53* mutation tend not to acquire one. The reason for this distinctive relationship is becoming clearer. A major function of *p53* is to respond to the appearance of DNA double strand breaks in the nucleus. Under conditions in which such 'free ends' appear (e.g. following ionizing radiation, genotoxic damage by mutagens, or chromosome breakage in aberrant mitosis) *p53* is stabilized and sets in train either G1 arrest or the initiation of apoptosis. G1-arrested cells have the opportunity to repair their DNA damage before replication. Hence, whichever pathway is chosen (and this choice appears to be conditioned by the cell type involved) the result of *p53* activation is removal from the tissue of cells with the potential to replicate damaged or altered DNA. In contrast, cells that lack functional p53 protein tend both to accumulate DNA damage and to engage in premature DNA replication. At least some

of them may survive following chromosome breakage and are therefore susceptible to abnormal recombination events and entry to the McClintock
fusion–bridge–breakage cycles described above. They may also undergo, more
readily than normal cells, inappropriate replication of short DNA segments leading to gene amplification. Gene amplification is itself a known stimulus for
abnormal recombinational and deletional chromosomal events. Cells with defective p53 protein function might also be expected to become tetraploid through
endoreduplicative mitosis. In these ways, mutation in *p53* may facilitate the
development of aneuploidy.

12.5 MULTIPLE GENETIC EVENTS: SEQUENTIAL OR SIMULTANEOUS?

There is much to support the view that genetic abnormalities are acquired in pre-
sumptive tumour cells in a sequential fashion with time. Foremost is evidence
from persons who inherit a tumour susceptibility gene, such as a mutated copy of
APC, *Rb*, *p53* or the genes responsible for ataxia telangiectasia. Such individuals
are usually born tumour-free, but tumours develop during childhood or early adulthood, as further genetic lesions are acquired in the cells of the tissues at particular
risk of carcinogenesis.

Nonetheless, there are difficulties in accepting that each genetic event in a carcinoma occurs sequentially in time and independent of the others. For example, it
is not unusual to find carcinomas of the colorectum that bear five or six of the
defects mentioned earlier (mutation of K-*ras*, mutation and deletion at the *p53*,
DCC and *APC* loci, deletion of 8p alleles, aneuploidy). Tumours with only two or
three of these lesions are also common. The question therefore arises how
sufficient concentrations of carcinogen can be made available to colorectal stem
cells so as to deliver five consecutive hits to one, without ensuring that many more
cells receive two or three. Yet five-defect carcinomas can occur as single tumours;
colorectal specimens bearing multiple carcinomas are relatively rare.

There are several possible solutions to this paradox. First, recent work has
demonstrated the importance of mismatched repair genes in carcinogenesis.
These genes code for proteins that recognize, excise and repair mismatch
nucleotides that arise through errors in DNA replication. Heterozygous inactivating mutations in these genes are responsible for most cases of HNPCC, where
it is clear that they permit a 'mutator phenotype' in which critical mutations in
the familiar cancer genes discussed above persist unrepaired at much greater frequency than normal. Patients carrying such defective genes may be mosaics for
APC mutation, e.g. a mutation in *APC* having arisen and persisted uncorrected
during early embryonic life. Such patients do not show the classical clinical picture of familial polyposis, presumably because the crypts bearing *APC* mutation
are scattered among normal crypts and embedded in normal stroma. Nonetheless
these patients are at greater risk of multiple additional carcinogenic mutations in

the dispersed *APC*-mutant cells, and hence at greater risk of colorectal cancer and even the extracolonic manifestations commonly associated with Gardner's syndrome (e.g. mandibular osteomas).

A second type of explanation for concordance of multiple genetic changes within single cells that then become the founder cells of cancers relates to the status of genes involved in xenobiotic metabolism. Such genes (e.g. class M glutathione S-transferase – *GSTM1*) are not essential for normal growth or development; hence a substantial proportion of the population are homozygous for null (functionless) mutations. There is some evidence that such null individuals are at greater risk of certain types of cancer (e.g. carcinoma involving the proximal colon). Presumably this is because the absence of this enzyme permits their cells to accumulate higher active concentrations of potential carcinogens than would appear in normal individuals with the same environmental exposure. It is possible, though unproven, that persons who are heterozygote for the null mutation undergo somatic mutations (perhaps in early life, so generating mosaicism) of their residual normal *GSTM1* gene. The cells that are thus rendered homozygous null for the enzyme would therefore be at greater risk than their neighbours of developing multiple carcinogenic mutations, because of the abnormally high concentration of active carcinogen they would experience.

12.6 POTENTIAL INTERACTIONS BETWEEN GENETIC EVENTS

Experiments with cultured rodent fibroblasts reveal that carcinogenesis requires cooperation between certain groups of oncogenes. Transformation *in vitro* and tumorigenicity *in vivo* occur in diploid embryo fibroblasts when they are forced (by gene transfer) to express a mutated *ras* oncogene together with mutated *p53*, c-*myc* or the adenoviral oncogene *E1B*. None of these genes on its own produces a transformed phenotype in such cells. In contrast, the *ras* oncogenes are highly effective in transforming rodent fibroblasts that have been immortalized through prolonged culture *in vitro*. Presumably the process of prolonged culture permits or selects for changes that have equivalent effects to expression of the cooperating genes described above. It has been pointed out that this cooperation between oncogenes often involves interaction between a protein involved in transduction of signals received at the cell surface (e.g. Ras) and a transcription factor (e.g. Myc, p53), but this association may be fortuitous.

Further evidence for gene cooperation in carcinogenesis comes from studies with transgenic animals – in which foreign genes are inserted into the embryonic stem cells and so appear in cells that colonize all tissues including the germ line. In the progeny of such animals it is possible to ensure that entire organs are populated with cells that both contain and express a potentially oncogenic transgene. For example, mice have been produced in which all breast epithelial cells express an activated *ras* oncogene, introduced in a progesterone-sensitive construct and hence specifically expressed in progesterone target tissues. Despite the universal expression of the oncogene in the target tissue, only a few cells there undergo malignant change. The remainder sometimes show hyperplasia. Hence there must be additional carcinogenic factors, which cooperate with the inserted oncogene but are specific to only a few cells in the gland. Rather similar results have been obtained from other experiments of this design, aimed at expressing different oncogenes in a variety of tissues. Only in the case of *erbB2* expression in mammary epithelium did widespread malignant transformation follow the expression of a single oncogene, and it is possible that this could be attributed to unusual, genetically determined susceptibility in the animals used. Direct evidence for gene interaction in carcinogenesis has also been provided by transgenic animals bearing double defects. Thus animals that express *myc* at high sustained levels within lymphoid cells and those that overexpress *bcl-2* both develop lymphoid tumours, but animals with both defects succumb to a more aggressive malignancy than those with either lesion alone. Similarly, animals bearing a transgene that expresses the oncogene *Wnt-1* constitutively in mammary tissue, develop mammary carcinomas, usually moderately well-differentiated. When deficiency in *p53* is also introduced into the germline, *Wnt-1*-expressing mammary tumours appear at an earlier age, and many adopt a more aggressive histology and behaviour. Similar gene cooperation in carcinogenesis is suggested by observations in human tumours such as the frequent association of c-*myc* dysregulation with overexpression of the oncogene *bcl-2*. c-*myc* gene expression appears to induce a 'high-turnover' state in which cells may either proceed around the replication cycle or die by apoptosis, depending on their circumstances at the time, whereas *bcl-2* facilitates survival and inhibits induction of apoptosis. Hence the coexistence of these two oncogene abnormalities may have a ready explanation.

In tumour progression, loss of heterozygosity at the *p53* locus frequently coexists with mutation of the residual *p53* allele. This can be interpreted to mean that residual wild-type *p53* still exerts some tumour-suppressing effect, despite the 'dominant negative' nature of the mutation (Chapter 9). For rather less obvious reasons, loss of the residual wild-type *ras* alleles appears to be associated with tumour progression in cells of animal tumours that bear activated *ras* oncogenes.

Some tumours bear specific genetic defects at unexpectedly low frequency. These may indicate that one genetic event can supply the tumorigenic stimulus more usually associated with another. Attention to such absences of involvement may therefore lead towards the definition of critical pathways in carcinogenesis, dependent on multiple gene products, only one of which need be abnormal to subvert the whole (Figure 12.4). Thus sarcomas seldom show mutations in *p53*, but commonly amplify the *MDM2* gene, which codes for a protein that binds, and perhaps limits the effectiveness of cellular p53. Similarly *p53* mutations are seldom seen in cervical carcinoma cells in which there is infection with HPV16 or 18, but do appear in the minority in which HPV is not involved. Presumably altered p53 function is critical for the malignant transformation of these cells, but may arise either through mutation of the *p53* gene itself or sequestration of its product through binding to the viral E6 oncoprotein. Mutation of p53 protein already

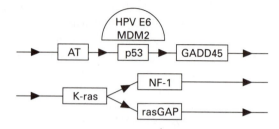

Figure 12.4 Diagrammatic summary of two pathways that may cooperate in carcinogenesis. The upper line shows the DNA repair pathway involving *p53*, and the lower line a portion of the *ras* signalling pathway. Abnormality in any one element in either pathway may influence the efficiency of the pathway as a whole. If carcinogenesis requires both pathways to be defective together, it is clear that a wide variety of possible combinations of abnormality could be responsible. In more detail, AT represents the products of the ataxia telangiectasia genes. These may be responsible for detecting DNA damage, and are abnormal in the recessive disorder of ataxia telangiectasia, a cancer-prone condition in which chromosome breaks and other manifestations of karyotypic instability are seen. The half-moon shape represents proteins that bind p53 protein and prevent it from interacting with its downstream effector sites. Two such proteins are known: the papillomavirus oncoprotein E6 and the endogenous proto-oncogene product, MDM2. GADD45 is a protein induced by p53 after DNA damage. The activity of the K-ras protein is regulated by at least two other proteins that activate its GTPase and so eventually terminate the Ras signal, rasGAP (GTPase activating protein and NF-1). Oncogenic mutations in *ras* genes, such as occur in many tumours, produce proteins in which the GTPase activity is sustained. NF-1 is abnormal in some tumours, and is inherited in defective form in the tumour-susceptibility syndrome neurofibromatosis.

inactivated through binding to HPV E6 would not be expected, as it would convey no further growth advantage to the cell in which it had occurred.

These observations invite further speculation. Although many different combinations of genetic events appear in tumours at different sites, it is possible that the essence of multistage carcinogenesis is to alter a rather small number of critical pathways within the cell, by affecting one element within each. Clear knowledge of these pathways, and how to reinstate them by pharmacological means, might then lead to new means of treatment of rather widespread applicability. Some encouragement for this hope comes from experiments in which single oncogenic mutations were corrected within cancer cells, with resulting loss of tumorigenicity. Thus, despite the fact that many steps may be needed to lead from normal to cancer, reversal of even one of them may have profound effects on cellular behaviour.

12.7 FURTHER READING

Armitage, P. and Doll, R. (1954) The age distribution of cancer and a multistage theory of carcinogenesis. *Br. J. Cancer*, **8**, 1–12.

Fishel, R., Ewel, A., Lee, S. *et al.* (1994) Binding of mismatched microsatellite DNA sequences by the human MSH2 protein. *Science*, **266**, 1403–1405.

Hamilton, S. R. (1992) Molecular genetics of colorectal cancer. *Cancer*, **70**, 1216–1221.

Hastie, N. D., Dempster, M., Dunlop, M. G. *et al.* (1990) Telomere reduction in human colorectal carcinoma and with ageing. *Nature*, **346**, 866–868.

Livingstone, L. R., White, A., Sprouse, J. *et al.* (1992) Altered cell cycle arrest and gene amplification potential accompany loss of wild type p53. *Cell*, **70**, 923–936.

Sinn, E., Muller, W., Pattengale, P. *et al.* (1987) Coexpression of MMTV/v-Ha-*ras* and MMTV/c-*myc* genes in transgenic mice: synergistic actions of oncogenes *in vivo*. *Cell*, **49**, 465–475.

Varmus, H. E., Godley, L. A., Roy, S. *et al.* (1994) Defining the steps in a multistep mouse model for mammary carcinogenesis. *Cold Spring Harbor Symp. Quant. Biol.*, **59**, 491–499.

13	Genomic instability and cancer

Stanley Venitt

13.1 INTRODUCTION

Molecular analysis of human cancers and their histopathological precursors shows that mutations in at least five genes can be identified during development of cancer (Chapter 12). Moreover, malignant tumours often display signs of genomic instability, including chromosomal anomalies, gene amplification and aneuploidy. As methods for detecting mutant genes in tumours become more sensitive and the number of candidate genes increases it is likely that most human cancers will be found to harbour more mutant genes than those already known to occur.

13.2 THE BACKGROUND MUTATION RATE OF NORMAL HUMAN CELLS CAN ACCOUNT FOR ONLY TWO OR THREE OF THE MUTATIONS FOUND IN HUMAN CANCERS

Each cell division of a somatic diploid mammalian cell requires the accurate and timely distribution of 6×10^9 base pairs of DNA (2.2 m in length) to each daughter cell. In long-lived species like *H. sapiens*, this process must operate accurately and unremittingly over many decades. Errors can arise if the replication machine selects the wrong nucleotides during polymerization of a new daughter strand using the template provided by the parental strand. Such misincorporation can result in mismatches. Several editing mechanisms prevent accumulation of coding errors in during DNA replication, which is remarkably accurate, with error frequencies ranging between 10^{-9} and 10^{-11} per base pair replicated. The background mutation rate is estimated to be about 1.4×10^{-10} mutations per base pair per cell generation, based on measurement of forward mutations in the hypoxan-

Molecular Biology for Oncologists. Edited by J. R. Yarnold, M. R. Stratton and T. J. McMillan. Published in 1996 by Chapman & Hall, London. ISBN 0 412 71270 9

thine-guanine phosphoribosyl-transferase (*HPRT*) gene in cultured human lymphoid cells, or electrophoretic protein variants at unselected loci. This background mutation rate cannot account for more than two or three mutations, bearing in mind that cancer is a clonal disease originating from a single mutant cell. It cannot explain the numbers of mutant genes already known to occur in cancers, let alone for the larger numbers that may be found in the future, even if clonal evolution is assumed to increase the selective pressure on mutations that confer a large growth advantage.

13.3 IS A MUTATOR PHENOTYPE THAT INCREASES THE RATE OF MUTATION NECESSARY FOR PROGRESSION OF CARCINO-GENESIS?

Loss or impairment of mechanisms that prevent mutations arising during and after DNA replication should increase the rate of mutation. Cells thus impaired would display a **mutator phenotype** and suffer genomic instability. They and their offspring would be more than usually prone to further mutagenesis. Were this to arise early in carcinogenesis it would explain the genomic instability that is characteristic of human cancers and the number of mutant genes observed in them. Mutator phenotypes, and their corresponding genotypes, have been discovered in the human germline, and in human cancers, suggesting indeed that genomic instability provides a powerful drive to carcinogenesis.

13.4 WHAT IS MUTATION AND WHAT CAUSES IT?

Changes in the content or arrangement of information in DNA can occur in a variety of ways, and at several levels, including replacement of one nucleotide by another within a codon, loss or addition of nucleotides, or of chromosomal segments, recombination and changes in the number of complete chromosomes in a cell (Chapter 2). Mutation occurs spontaneously or may be induced by physical and chemical agents.

13.5 THE RAW MATERIAL FOR MUTATION IS DNA DAMAGE

DNA is a relatively simple chemical polymer containing a variety of chemically reactive sites. It undergoes slow chemical degradation at physiological pH and temperature. It is subject to attack by physical, chemical and biological agents including the aqueous medium in which it is bathed, background ionizing radiation, solar radiation, endogenous and exogenous mutagens and viruses. Such agents can damage DNA and, unless the damage is repaired or the damaged cells stop dividing or die, coding errors may occur that, when replicated, will produce

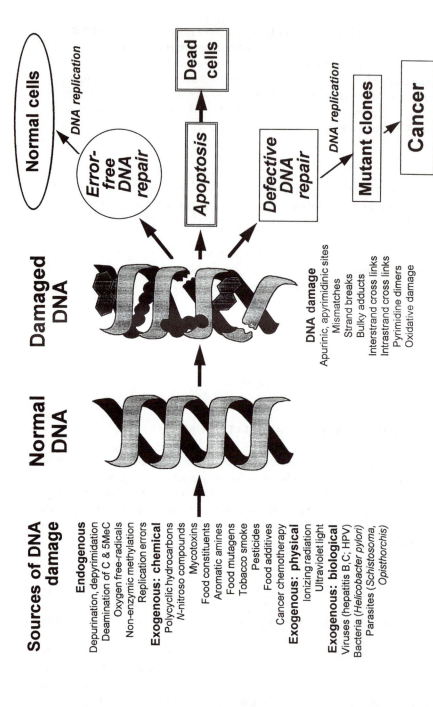

Figure 13.1 Sources of DNA damage that can lead to mutation and cancer, and fate of cells that sustain such damage.

mutant clones. Figure 13.1 summarizes different sources of DNA damage and the fate of damaged cells.

13.6 ENDOGENOUS MECHANISMS OF MUTAGENESIS INCLUDE DEAMINATION OF 5-METHYLCYTOSINE

The DNA of vertebrates is methylated at the 5 position of cytosine by a methyltransferase that conveys a methyl group from *S*-adenosylmethionine to form 5-methylcytosine (5mC). In mammals about 3% of cytosines in DNA are methylated and about 70% of 5mC occurs in 5'CpG3' dinucleotides. This global methylation of DNA is peculiar to vertebrates and is believed to compartmentalize the genome into active and inactive regions for the regulation of gene expression. Methylation occurs in both DNA strands, is symmetrical and the pattern of methylation is stably inherited during cell division. In the mammalian genome CpG dinucleotides are under-represented by about one-quarter of the proportion expected from the observed frequency of C and G in the DNA. There is a corresponding excess of TpG and CpA dinucleotides. This evolutionary depletion of CpG dinucleotides probably results from the readiness of 5mC to undergo deamination to form thymine, resulting in 5mC→T transition mutations. Despite the rarity of CpG dinucleotides and the existence of a G-T mismatch repair system, this transition is about 10 times more frequent than other transitions and is common in inherited diseases. Over 35% of germ-line point mutations that lead to human genetic disease are CpG→TpG transitions. These are also found at high frequency in cancers, e.g. in the *p53* tumour suppressor gene (Chapter 2). Thus, 5mCpG sites are particularly prone to mutation and 5mC is an endogenous mutagen. Cytosine itself can also undergo deamination to uracil, an 'erroneous' base. However, this is removed at high efficiency by uracil DNA glycosylase.

Biomethylation of DNA brings with it further dangers to genomic stability. *S*-adenosylmethionine is a weak alkylating agent capable of methylating DNA. The major products are 7-methylguanine and 3-methyladenine. The latter interferes with Watson–Crick hydrogen bonding and is rapidly removed by a specific enzyme, whereas 7-methylguanine is poorly repaired but is chemically unstable and is lost by chemical depurination.

13.7 REACTIVE OXYGEN SPECIES, LIPID PEROXIDATION AND NITRIC OXIDE ALSO GENERATE ENDOGENOUS MUTATIONS

Reactive oxygen species such as hydroxyl radicals, generated during normal aerobic metabolism, cause DNA damage and mutation. The major product of attack by hydroxyl radicals is 8-hydroxyguanine, which erroneously base-pairs with adenine, thus generating transversion mutations after DNA replication. 8-Hydroxyguanine is removed from DNA by a specific glycosylase, which also removes

other oxidized purine products. There are other mechanisms that generate mutagens endogenously; for example lipid peroxidation produces the mutagen malon-
dialdehyde. Chronic infection with certain viruses, bacteria and parasites causes
cancer in man or predisposes to it (Figure 13.1). Nitric oxide (NO), a short-lived
chemical messenger produced by many different cell types for a variety of physiological functions, is found at elevated concentrations in chronic infections, and
could be crucial to the mechanism by which such infections cause cancer.
Formation of mutagenic N-nitroso compounds by reaction of NO with secondary
amines, direct deamination of DNA bases and direct oxidation of DNA after formation of peroxynitrite or hydroxy radicals are possible mechanisms.

13.8 MANY EXOGENOUS CARCINOGENS CHEMICALLY MODIFY THE STRUCTURE OF DNA TO FORM ADDUCTS

A characteristic of most exogenous chemical carcinogens is their covalent binding to DNA to form **adducts**. Some carcinogens are intrinsically reactive and form
DNA adducts directly. These carcinogens are electrophilic – they acquire electrons
during chemical reactions. DNA contains many nucleophilic centres – atoms that
donate electrons (e.g. N-2, N-7, O-6 atoms of guanine). DNA adducts form when
electrophiles react with nucleophilic centres in DNA. Electrophilic carcinogens
include alkylating agents such as nitrogen mustard, melphalan and chlorambucil
(used in cancer chemotherapy), aromatic N-oxides, and aromatic nitro compounds.

Other carcinogens are not electrophilic *per se*, but are converted to electrophiles by eukaryotic oxidative metabolism. They include polycyclic aromatic
hydrocarbons, aromatic amines, alkyl and arylnitrosamines and many natural
products such as mycotoxins and plant products. Cyclophosphamide is an example of an indirectly acting cancer chemotherapy agent.

Adducts range in size and complexity from simple alkyl groups to bulky multi-
ring residues from chemicals such as polycyclic aromatic hydrocarbons and
aflatoxins. Some chemicals form intrastrand crosslinks between adjacent bases on
the same strand and/or interstrand crosslinks between each strand of the duplex.
In most cases, a single chemical will produce several different DNA adducts. This
may be due to the production of several different metabolites from the same chemical, to reaction of a single reactive species with atoms of different nucleophilici-
ties in the DNA molecule, or to a combination of both. Stereochemical and
physicochemical constraints and base-sequence context also play a part in determining the spectrum of DNA adducts formed by a given compound.

13.9 DNA ADDUCTS LEAD TO MUTATION

The biological consequences of adduct formation depends to a large extent on the
nature of the adduct and its precise location in the DNA molecule. For example,

a methylation at N7-G is much less mutagenic than at O6-G, since the latter participates in hydrogen-bonding during complementary base-pairing while the former does not. However, a bulky adduct such as that formed by aflatoxin B_1 at N7-G is highly mutagenic, because it causes gross distortion of the DNA structure. Adduct formation can lead to base-substitution, deletion and addition, and secondary damage such as depurination, and therefore to point and chromosomal mutation.

13.10 HOW IS REPLICATION ACCURACY MAINTAINED AGAINST SUCH AN ONSLAUGHT?

The fidelity of DNA replication is maintained by three processes.

- **Base selection**: DNA polymerase discriminates between correct and incorrect nucleotides during the insertion step of strand polymerization.
- **Exonucleolytic proofreading**: if the wrong nucleotide is incorporated, it is excised by a $3' \rightarrow 5'$ exonuclease before the nascent daughter chain is elongated.
- **Mismatch repair**: the mismatch repair system consists of at least 10 proteins whose structure is highly conserved in organisms ranging from prokaryotes such as bacteria, to lower and upper eukaryotes such as yeasts and mammals. These proteins combine in a molecular machine which operates as follows. All possible mismatches (G:T, A:C, G:G, A:A, A:G, T:T, C:T, C:C) provoke a strand-specific incision that stimulates exonucleolytic excision (chipping away of nucleotides) of a stretch of single-stranded DNA that includes the mismatched base and the incision. This leaves a single-stranded region with the correct base sequence that provides a template for reconstruction, by specific enzymes, of a new and corrected opposite strand.

13.11 OTHER FORMS OF DNA REPAIR HAVE EVOLVED TO MAINTAIN GENOMIC STABILITY

Error-free repair

Error-free repair is the best understood mode of DNA repair and operates at several levels of complexity. The simplest is enzymatic reversal of damage and includes photoreactivation, whereby pyrimidine dimers induced by short-wavelength UV are snipped apart by enzymes which require absorption of light of longer wavelength for their activity. Another example is the methyltransferase which transfers alkyl groups from O6-MeG to an acceptor protein, thus restoring base-pairing.

Nucleotide excision-repair is more complicated. The first step may involve glycosylases referred to earlier. These recognize and cut out modified bases (e.g. alkylpurines and other DNA adducts) from the sugar–phosphate DNA chain, leav-

Table 13.1 Recessively inherited disorders of DNA processing and repair that predispose to cancer

Disease	Cancer predisposition	Main clinical features	Spontaneous chromosomal damage
Xeroderma pigmentosum	Skin cancers	Photosensitivity	None
Fanconi's anaemia	Acute myelogous leukaemia	Short stature, skeletal abnormalities, pancytopenia	Chromatid breaks and non-homologous recombination
Bloom's syndrome	Leukaemias, lymphomas and carcinomas	Small birth size, growth retardation, facial telangiectasia, immunodeficiency	Breaks and non-homologous recombination; sister-chromatid exchange
Cockayne's syndrome	None known	Photosensitivity, dwarfism, mental retardation, microcephaly, retinal and skeletal abnormalities	None
Ataxia telangiectasia	Lymphoid leukaemias; lymphomas	Radiosensitivity, cerebellar ataxia, telangiectasia, immunodeficiency	Breaks and non-random recombination

ing behind apurinic sites (AP), which are substrates for another enzyme (AP-endonuclease), which cuts the phosphodiester chain at the AP. These cuts are substrates for an exonuclease that removes the deoxyribosephosphate and one or a few nucleotides. The damaged strand now contains a gap opposite its undamaged partner strand, whose base sequence is intact. This gap is filled, by base-pairing, by polymerases, and the patch of nucleotides is reconnected by ligases.

Error-prone repair

This is a strategy that allows cell survival at all costs, even at the risk of incurring a high level of mutation. It is particularly well understood in bacteria although similar mechanisms are known to occur in eukaryotes. The essential feature of error-prone repair systems is that the normal rules of complementary base-pairing are suspended in order to allow DNA replication to proceed even on a damaged template containing premutational lesions. As a result, many wrong bases are inserted in the newly synthesized DNA strand. Recombination of DNA strands during replication of damaged DNA may also contribute to the increased levels of mutagenicity.

Agents that induce chromosomal damage in cells in vitro	Complementation groups	Gene location (gene)	Molecular defect
Ultraviolet light (sister chromatid exchange)	7 + 1 variant	19q13.2 (*ERCC2*) 2q21 (*ERCC3*) 13q32–33 (*ERCC5*)	Defects in whole-genome excision repair
Bifunctional akylating agents	4	9q22.3 (*FACC*)	Defects in DNA–interstrand crosslink repair (?)
Ultraviolet light, various DNA-reactive agents	?	?	Defects in DNA ligase (?)
Ultraviolet light (sisters chromatid exchange)	2	10q11–21 (*ERCC6*)	Defects in excision repair of actively transcribed genes
Ionizing radiation; radiomimetics such as bleomycin	4	11q22–23 (*ATM*) (mutated in all four complementation groups)	Defects in a putative phosphatidylinositol-3-kinase involved in mitogenic signal transduction and cell-cycle control

13.12 DNA REPAIR IS HETEROGENEOUS

DNA repair occurs at different rates and to varying extents in different DNA sequences. It occurs preferentially at active genes, such that DNA adducts and other lesions are removed from the transcribed strand, but not the untranscribed strand. Repair is often linked to the rate of gene transcription, more repair occurring at high rates of transcription than at low rates. Such strand bias in DNA repair suggests links between cellular factors that signal DNA damage, gene transcription and DNA repair.

13.13 RARE RECESSIVE GERM-LINE MUTATIONS IN DNA-PROCESSING AND REPAIR GENES PREDISPOSE TO CANCER

There are several rare, recessively inherited single-gene defects that confer genomic instability (Table 13.1). Each confers an excess risk of cancer, and it is reasonable to assume that the molecular defects described contribute directly to the increased risk of cancer. In some cases (e.g. xeroderma pigmentosum) an exogenous mutagen (UV light) is required to trigger the sequence of events leading to cancer, while in others (e.g. Bloom's syndrome) it is unclear as to whether exogenous or endogenous mutagens are involved.

13.14 FEMALE AT HETEROZYGOTES ARE AT ELEVATED RISK OF BREAST CANCER

It is clear that AT homozygotes suffer a high relative risk of cancer, particularly lymphoma and leukaemia. But is there an excess cancer risk in AT heterozygotes? Several studies show that female relatives of AT patients suffer about a fourfold excess risk of breast cancer, although there is no consistent evidence of an excess risk for any other cancer. Taking into account the likely population frequency of the *AT* gene, AT heterozygotes would account for between 1% and 13% of breast cancer cases, with 3.8% being the best estimate. The *AT* gene codes for a putative kinase involved in mitogenic signal transduction and cell-cycle control in response to DNA damage, and cells from AT heterozygotes are more sensitive to ionizing radiation than are normal cells. Thus, genomic instability in response to DNA damage may explain a small but significant fraction of breast cancer. Apart from its other possible uses, detection of AT carriers by DNA testing may also make it possible to determine whether the use of even low doses of X-rays in mammographic screening poses a breast-cancer risk to this group.

13.15 DOMINANTLY INHERITED MUTATOR GENES ARE MISMATCH-REPAIR GENES

Mismatch-repair genes are dominantly inherited DNA-repair genes that, when mutated, confer susceptibility to cancer. They are expressed in many types of normal cells and their function and conservation during evolution suggests that they play a crucial role in maintaining genomic stability.

13.16 MISMATCH-REPAIR GENES WERE DISCOVERED BY STUDYING MICROSATELLITE INSTABILITY

A striking feature of the human genome is the occurrence of stretches of mainly intronic DNA in which certain base sequences are tandemly repeated. One such family of repetitive DNA elements is the microsatellite which consists of simple oligonucleotide repeats (Chapter 27). For example, about 50 000-100 000 $(CA)_n$ repeats are scattered throughout the human genome and many are polymorphic for length, i.e. the number of units ('n') within the microsatellite varies between individuals. Such polymorphic alleles show stable Mendelian inheritance and are indispensable markers for linkage analysis. Sometimes amplification of the number of repeats in a microsatellite can lead to loss of function of the gene in which it occurs – this a **mutation** rather than a **polymorphism**. Certain genetic disorders (e.g. fragile-X syndrome and Huntington's disease) are caused by unstable trinucleotide repeats in germ-line DNA.

13.17 HOW DOES MICROSATELLITE INSTABILITY ARISE?

DNA replication errors may result from misalignments between the parental strand and the daughter strand. If misalignment occurs at repetitive sequences a premutational intermediate can form that is stabilized by neighbouring base pairs. Frameshift mutations (deletion or insertion of one base) may occur depending on whether the unpaired nucleotide is in the parental strand or the daughter strand. Another possibility is slippage, followed by correct incorporation, then realignment to yield a base substitution. Di- and trinucleotide repeats are particularly prone to replicative slippage or stuttering. Studies in yeast show that DNA polymerase has a very high slippage rate on templates that contain repeated sequences, but that most of the errors are corrected by mismatch repair. Mismatch errors may also arise in the germ line during meiosis, for example by the production of heteroduplex DNA sequences generated by recombination between DNA strands from each parent.

13.18 DEFECTS IN MISMATCH REPAIR PRODUCE THE 'REPLICATION ERROR' (RER⁺) PHENOTYPE

Microsatellite instability is readily detected by PCR of appropriate markers followed by polyacrylamide gel electrophoresis. Tumours showing microsatellite instability are said to display the RER⁺ ('replication error') phenotype, implying that defective mismatch repair is the underlying mechanism. The RER⁺ phenotype is seen in about 15% of sporadic colorectal cancer, and in a variety of other sporadic tumours. The location and function of genes for the RER⁺ phenotype were discovered in families with hereditary non-polyposis colorectal cancer (HNPCC). In this autosomal-dominant trait affected individuals develop tumours of the colon (mainly right-sided), endometrium, ovary, stomach and other organs, often before age 50. Unlike the recessive DNA-repair disorders mentioned earlier (with frequencies of homozygotes less than 1 in 20 000), HNPCC is one of the commonest of human genetic disorders, affecting as many as 1 in 200 people and accounting for between 5% and 10% of colorectal cancer.

13.19 *MSH2*, *hMLH1*, *hPMS1* AND *hPMS2* MISMATCH-REPAIR GENES

Microsatellite instability occurs at multiple unrelated loci throughout the genome in tumours in such patients, suggesting that it is a symptom of a defect rather than the defect itself. Studies of microsatellite instability in yeast suggested that a mutation in a human homologue of a yeast mismatch-repair gene would be a likely candidate for the predisposing gene in HNPCC. This proved to be correct – some HNPCC kindreds are heterozygous for germ-line mutations in a human homologue of a yeast mismatch-repair gene *MSH2* (itself a homologue of a bacterial mismatch-repair gene *MutS*). Similar mutations are seen in tumour tissue but not in normal tissue of patients with sporadic RER$^+$ colorectal cancer. Three more human mismatch-repair genes (*hMLH1*, *hPMS1* and *hPMS2*) – all closely homologous to bacterial genes – have been identified among different HNPCC kindreds. It is expected that additional genes that confer microsatellite instability will be discovered, since not all HNPCC patients whose tumours display the RER$^+$ phenotype carry mutations in the four genes already identified. Moreover, there are other human genes with homology to the bacterial versions of *hMLH1*, *hPMS1* and *hPMS2*, and to other components of the bacterial mismatch-repair system.

13.20 THE RER$^+$ PHENOTYPE MAKES CELLS MORE MUTABLE

Studies *in vitro* show that the mutation rate of $(CA)_n$ repeats (in a shuttle vector) in human RER$^+$ tumour cells is at least 100-fold greater than that in RER$^-$ cells and that this increased mutability is associated with a global defect in strand-specific mismatch repair that can be demonstrated biochemically. This deficiency is observed with microsatellite heteroduplexes and with heteroduplexes containing single base–base mismatches, suggesting that RER$^+$ cells may also suffer an increased rate of base pair substitution mutations. This suggestion is supported by studies of point mutation of a gene, *HPRT,* that is expressed in human cells. RER$^+$ cells have mutation rates 100-fold greater than those of RER$^-$ cells. Heterogeneity of mutation rates between different RER$^+$ cell lines suggest that there are different classes of RER$^+$ tumours.

13.21 DEVELOPMENT OF TUMOURS ASSOCIATED WITH MISMATCH-REPAIR GENES USUALLY REQUIRES INACTIVATION OF BOTH ALLELES

Both maternal and paternal mismatch-repair alleles carry somatic mutations in sporadic RER$^+$ colorectal tumours, as seems to be the case in HNPCC. In the latter, the first mutation is in the germ line and the second is somatic. This parallels the mechanism seen with tumour suppressor genes, where both copies are inactivated in tumours (Chapter 1). However, the second copy of a tumour suppressor gene is usually inactivated by loss of heterozygosity rather than by point mutation.

13.22 DO MUTANT MISMATCH-REPAIR GENES IN HNPCC INITIATE CANCER OR ACCELERATE ITS PROGRESS?

In HNPCC the RER$^+$ phenotype is restricted to neoplastic tissue (adenomas and carcinomas) and normal colonic mucosa appears to show no generalized increase in somatic mutation. Cancers in HNPCC originate within benign adenomas, which are clonal and may mark a single initiating mutation, probably in the *APC* gene. The progression of adenomas is marked by increasing size, dysplasia and villosity. A study conducted in New Zealand showed that the frequency and anatomical distribution of adenomas in at-risk members of HNPCC families was the same as in an autopsy population, suggesting that mutant mismatch-repair genes do not initiate the process of neoplastic transformation. On the other hand, adenomas in HNPCC families were more likely to show villosity, high-grade dysplasia and probably increased size. These findings are consistent with the observation that the RER$^+$ phenotype is seen in neoplastic but not normal tissues and that the effects of mutant mismatch-repair genes is to accelerate the progression of adenoma to carcinoma, but not to initiate adenoma development.

13.23 DOES THE RER$^+$ PHENOTYPE PREDICT BETTER SURVIVAL?

Examination at the molecular genetic level has revealed that colorectal cancers are of two distinct types.

- The most common are RER$^-$ tumours. These tend to be non-familial, aneuploid, show numerous allelic losses and are relatively aggressive.
- RER$^+$ tumours are less common, are often familial, tend to be diploid, with few allelic losses. They are often multiple and relatively less aggressive.

The use of DNA testing to establish RER status of tumours may well turn out to be of use in prognosis if it is confirmed, in much larger studies, using universally agreed criteria for defining the RER$^+$ phenotype, that patients with RER$^+$ tumours survive longer than those whose tumours are RER$^-$.

13.24 DOES THE RER$^+$ PHENOTYPE CONFER INCREASED RESISTANCE TO CHEMOTHERAPEUTIC ALKYLATING AGENTS?

There is preliminary evidence that the RER$^+$ phenotype might confer increased resistance to the cytotoxic effects of alkylating agents of the type used in cancer

chemotherapy. Should this be confirmed it would be another good reason to establish the RER phenotype of tumours and, if possible, to tailor the therapy to the particular phenotype.

13.25 DOES THE RER⁺ PHENOTYPE CONFER INCREASED RISK OF OTHER TUMOURS AND DOES IT IDENTIFY UNRECOGNIZED HNPCC KINDREDS?

Another reason for establishing the RER status of tumours by DNA testing is that
a patient with an RER⁺ tumour, but with no clear family history of HNPCC, could
have acquired a new germ-line mutation in one of the known mismatch-repair
genes, or be a member of a kindred carrying a germ-line mutation in a mismatch-
repair gene not yet characterized. DNA testing to establish RER status is itself a
considerable undertaking, and determining whether those individuals with RER⁺
tumours carry germ-line mutations in *hMLH1, hMSH2 hPMS1* or *hPMS2* is even
more burdensome. However, until such studies are undertaken, it will not be possible to estimate what proportion of patients with sporadic RER⁺ tumours represent familial syndromes of low penetrance, or late presentation.

13.26 THE *p53* TUMOUR SUPPRESSOR GENE: IS IT THE GUARDIAN OF THE GENOME?

The *p53* gene is thought to be crucial to the maintenance of genomic stability. It
is widely held that *p53* is 'the guardian of the genome' – it is a key component of
a system that detects DNA damage and stops the cell cycle in G1 to allow DNA
to repair, or provokes apoptosis, whereby genomically damaged cells die. In this
model, loss of *p53* allows cells to accumulate mutations. However recent work
suggests that *p53* also plays a direct role in controlling cell senescence and cell
death and that these are its primary functions. (The *p53* gene and its product are
described in detail in Chapter 9.)

13.27 EXPERIMENTS WITH p53 KNOCKOUT MICE TEST THE 'GUARDIAN OF THE GENOME' THEORY

Insights into what the *p53* gene does are being gained by using strains of knockout
mice in which both copies have been knocked out ('nullizygous' (-/-)) or are heterozygous (+/-) for a null mutation or for a specific point mutation. Such mice
develop normally but suffer very high spontaneous rates of cancer at various sites,
the tumours appearing much earlier than in normal mice. If the 'guardian of the
genome' theory is correct, such mice might be expected to exhibit genomic instability in their somatic cells. Fibroblasts from such mice do indeed show genomic

instability, manifested by increases in aneuploidy and other chromosomal anomalies when grown in culture. Moreover, fibroblasts from (-/-) mice do not exhibit the senescence characteristic of normal primary cultures. Gene amplification (selective multiplication of particular genes in response to selective pressure) is another manifestation of genomic instability, and cells from mice with mutant *p53* alleles exhibit this phenomenon in culture. Mice deficient in p53 given γ-irradiation develop more tumours more quickly than do normal mice and exhibit a twofold excess of micronuclei (diagnostic of double-strand chromosome breaks) in peripheral-blood cells. These experiments show that impairment or abrogation of p53 function increases spontaneous and radiation-induced chromosomal aberrations in mice.

13.28 p53 DOES NOT APPEAR TO GUARD THE GENOME AGAINST POINT MUTATIONS

However, the same does not appear to hold for point mutations. Spontaneous somatic point mutation, measured in liver, spleen and brain of mice carrying a transgenic bacterial shuttle vector system, was no higher in *p53* nullizygotes than in (+/+) mice and there was no difference in the mutational spectra. In a mouse skin carcinogenesis system, in which mice are given a carcinogen which induces point mutations followed by a tumour-promoting chemical, the number, size, and growth rate of benign papillomas were not raised in *p53* (+/-) mice compared with wild type. The yield of papillomas was reduced in *p53* (-/-) mice, but the papillomas progressed to malignancy much more rapidly. Progression rate was also greater in (+/-) than in (+/+) mice and was associated with loss of the remaining wild-type allele. Most tumours from all groups had activating point mutations in the H-*ras* gene. These experiments show that absence of p53 does not augment the frequency of initiation or the rate of promotion but greatly enhances malignant progression. These findings are supported by gene transfer experiments showing that mutant p53 expression does not confer growth advantage on colon adenomas or well differentiated thyroid tumour cell lines.

13.29 SENESCENCE, IMMORTALIZATION, TELOMERES AND TELOMERASE

It appears, therefore, that p53 limits further growth and development of tumours only after a particular and critical stage of tumour development has been reached. Such a critical stage is escape from senescence, a process also referred to as 'immortalization'. Senescence, the loss of a cell's ability to divide, is not to be confused with apoptosis, which is an active process of cell death. In culture, normal cells undergo a limited number of divisions and will then become senescent, whereas tumour cell lines grow indefinitely. One system that regulates senescence and prevents immortalization does so by maintaining the integrity of chromosome ends – 'telomeres'. Human telomeres consist of TTAGGG sequences tandemly repeated up to 1000 times. They appear to have two functions. The first is to protect chromosomes from recombination and loss. The second is to resolve the 'end-replication problem' which predicts the progressive and potentially catastrophic shortening of the 3' end of DNA molecules over many cycles of replication – a problem imposed by the fact that DNA polymerization is initiated from the site of a bound RNA primer and can proceed only in the $5' \rightarrow 3'$ direction. This problem is overcome by telomerase, a ribonucleoprotein that makes, by reverse transcription, a DNA copy of its own telomere antisense sequence and fuses it to the 3' terminus of the chromosome.

13.30 THE TELOMERE IS A MITOTIC CLOCK AND TELOMERASE IS THE SPRING THAT DRIVES IT

By virtue of this second function, the telomere has acquired a third function, namely as a mitotic clock.

Telomeres and telomerase control senescence in normal cells

- The number of divisions that normal primary human fibroblasts can undergo in culture is directly proportional to the initial length of their telomeres when placed in culture.
- The length of a telomere is determined by number of cell divisions and by the presence or absence of telomerase.
- 50–200 telomeric nucleotides are lost at each cell doubling.
- Germ cells express telomerase throughout life.
- Most somatic tissues do not express telomerase activity.
- Somatic tissues lose telomere length progressively and, in culture, stop dividing when telomeres have become critically short.

Malignant cells and tumours re-express telomerase and escape from senescence

- Immortal tumour cell-lines express telomerase.
- Telomeric length does not accurately reflect telomerase activity.
- Of 400 independent tumours analysed, 85% express telomerase activity.
- Benign proliferative lesions do not express telomerase.

Thus, telomerase expression appears to be stringently repressed in normal human somatic tissues but is reactivated in cancer, suggesting that re-expression of telomerase coincides with immortalization, which is a crucial step in progression to malignancy. A model of the role of telomeres and telomerase in cellular ageing and immortalization, based on *in-vitro* studies of human cells, is shown in Figure 13.2.

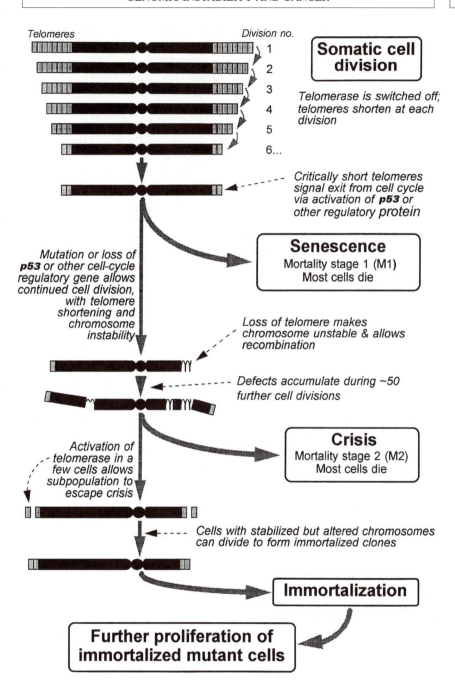

Figure 13.2 Model for the role of telomeres and telomerase in the control of senescence in mammalian somatic cells. This is based on studies of human cells grown in culture and is a modified version of the scheme presented in Rhyu, 1995.

13.31 HOW DOES p53 AND GENOMIC INSTABILITY FIT IN THE TELOMERE/TELOMERASE MECHANISM?

In this model, it is envisaged that senescence (M1) can be bypassed by loss of *p53* (or *Rb*) function. This allows proliferation of cells, which can become genomically unstable owing to erosion of telomeres. As genomic instability increases, most cells die (M2). However, in a few, telomerase is switched on again. This stabilizes the chromosomes and thereby immortalizes chromosomally mutant cells, which can then continue to proliferate. The fact that most tumours carry *p53* mutations and express telomerase suggests that there is powerful selective pressure for cancers to acquire these two properties. There are, however, some cancers that, as a class, do not carry *p53* mutations, notably germ-cell and embryonic tumours. Cells of such tumours might be expected to possess longer telomeres than normal somatic cells, and might therefore undertake many more mitoses before p53-mediated senescence intervened to prevent further cell division and risk of mutation.

The convergence of loss of *p53* with reacquisition of telomerase in most human cancers is another step in understanding the intimate link between genomic instability and cell-cycle control and suggests, in addition to the model described, a variety of hypotheses that are testable by molecular biological techniques.

13.32 FURTHER READING

Mutator phenotypes

Coleman, W. B. and Tsongalis, G. J. (1995) Multiple mechanisms account for genomic instability and molecular mutation in neoplastic transformation. *Clin. Chem.*, **41**, 644–657.

Loeb, L. A. (1991) Mutator phenotype may be required for multistage carcinogenesis. *Cancer Res.*, **51**, 3075–3079.

Methylation of CpG dinucleotides

Bird, A. (1992) The essentials of methylation. *Cell*, **70**, 5–8.

DNA damage and mutation

Lawley, P. D. (1994) Historical origins of current concepts of carcinogenesis. *Adv. Cancer Res.*, **65**, 17–111.

Lindahl, T. (1993) Instability and decay of the primary structure of DNA. *Nature*, **362**, 709–715.

Recessive DNA-repair mutations

Digweed, M. (1993) Human genetic instability syndromes: single gene defects with increased risk of cancer. *Toxicol. Lett.*, **67**, 259–281.

Mismatch repair and cancer

Eshleman, J. R. and Markowitz, S. D. (1995) Microsatellite instability in inherited and spo-radic neoplasms. *Curr. Opin. Oncol.*, **7**, 83–89.

Karran, P. and Bignami, M. (1994) DNA damage tolerance, mismatch repair and genome instability. *BioEssays*, **16**, 833–839.

Loeb, L. A. (1994) Microsatellite instability: marker of a mutator phenotype in cancer. *Cancer Res.*, **54**, 5059–5063.

Telomeres, telomerase and senescence

Rhyu, M. S. (1995) Telomeres, telomerase and immortality. *J. Nat. Cancer Inst.*, **87**, 884–894.

Wynford-Thomas, D., Bond, J. A., Wyllie, F. S. and Jones, C. J. (1995) Does telomere shortening drive selection for *p53* mutation in human cancer? *Mol. Carcinogenesis*, **12**, 119–123.

p53 and genomic instability

Smith, M. L. and Fornace, A. J. (1995) Genomic instability and the role of p53 mutations in cancer cells. *Curr. Opin. Oncol.*, **7**, 69–75.

PART TWO

Strategies for Cancer Prevention and Cure

Inherited predispositions to cancer: opportunities for genetic counselling

14

Victoria Murday

14.1 INTRODUCTION

Predisposition to cancer occurs in a number of different ways, and all modes of inheritance are possible, including dominant, recessive and X-linked. In many inherited cancer syndromes, the cancer susceptibility is the most obvious clinical feature of the condition. This is particularly true of the dominant susceptibilities to common cancers, such as cancers of the breast and bowel. However, there are a number of cancer susceptibility syndromes where other features of the phenotype bring the family to medical attention. In some of these susceptibilities the risk of cancer is lower, and clustering of cancer may not be evident in the family.

In addition to these single gene disorders of high penetrance, family studies frequently indicate an increased risk in the first-degree relatives of cancer patients. This is the case in relatives of colorectal cancer patients, for example, even after the high-risk dominantly inherited conditions have been excluded. The increase in risk is not sufficient to cause significant clustering of affected individuals in families, and can be due to less penetrant single gene variations or shared environmental factors.

14.2 GENETIC COUNSELLING

Genetic counselling is the term historically used to describe the interview which occurs when an individual attends a genetic clinic. Counselling is important in

Molecular Biology for Oncologists. Edited by J. R. Yarnold, M. R. Stratton and T. J. McMillan. Published in 1996 by Chapman & Hall, London. ISBN 0 412 71270 9

genetics and its non-directive nature, offering choices to patients, is the basis of the practice. However, much of the consultation, like any other outpatient appointment, is for diagnosis and management of disease, and this is carried out ordinarily using the history and examination of an affected individual. Most of the information that is given to patients about risk either to themselves or their relatives depends upon making an accurate diagnosis. With genetic disease, it may be the family history that holds the clue to diagnosis.

The family history

Establishing the pedigree is an important part of the interview. This is standardized to include the family history of cancer, other diseases, developmental and congenital abnormalities, and a history of miscarriages. Information about first- and second-degree relatives should be requested and, where appropriate, the family history should be extended as far as possible.

The age at which cancer was diagnosed, the site(s), the date of treatment and hospitals involved should be ascertained. This will allow assessment of risks to relatives and confirmation of diagnosis from hospital records. In addition, the diagnosis of a particular cancer syndrome may be possible from the pattern of cancers or associated non-malignant problems. It is important that the clinician has the necessary background knowledge to recognize any significant pattern, and be able to assess the risks from pedigree analysis.

Medical history and examination

It must be established from the history and examination whether the patient is an affected member or an at-risk member of the family, and the patient should be questioned on any symptoms indicative of cancer or congenital abnormalities. Some cancer syndromes have phenotypes that can be diagnosed in an individual. Frequently, it is the premalignant phenotype, such as adenomatous polyps in familial adenomatous polyposis (FAP), that enable a diagnosis to be made.

Initial clinical examination involves looking for any dysmorphic features and congenital anomalies. The skin should be carefully examined because many cancer syndromes are associated with dermatological features, such as pigmentary abnormalities, for example, freckles are seen in Peutz–Jeghers syndrome, *café au lait* patches in neurofibromatosis or Turcot's syndrome, basal cell naevi in Gorlin's syndrome (Table 14.1).

Assessing the risk

As a general rule, the occurrence of the same cancer in three close blood relatives is suggestive that there is a genetic susceptibility, particularly if individuals were affected at an early age.

If there are two close relatives with the same cancer, the general population risk

Table 14.1 Clinical features of cancer syndromes (MEN = multiple endocrine neoplasia)

Disease	Inheritance	Clinical features	Associated tumours
Familial polyposis coli	Autosomal dominant	Adenomatous polyposis; epidermoid cysts; supernumerary teeth; osteomas (especially jaw); congenital hypertrophy of retinal pigment epithelium (CHRPE)	Colorectal cancer; 10% upper gastrointestinal malignancy; thyroid cancer; benign fibrous desmoid tumours (10%)
Von Hippel–Lindau	Autosomal dominant	Retinal haemangiomas; cystic kidneys; pancreatic cysts	Cerebellar haemangioblastomas; cord haemangioblastomas; renal cell carcinoma; phaeochromocytoma
MEN type II	Autosomal dominant	Type II(a) no associated features; type II(b) gastrointestinal tract mucosal neuromata; prominent lips; thickened, anteverted eyelids; Marfanoid cubitus; nerve fibres in cornea; cutaneous neurofibromata; developmental delay	Medullary carcinoma thyroid; phaeochromocytoma; parathyroid tumours
Neurofibromatosis type II	Autosomal dominant	Often no café au lait spots of peripheral neurofibromata; cataracts	Bilateral acoustic neuromas; meningiomas; Schwannomas of dorsal roots; gliomas
Gorlin's	Autosomal dominant	Hypertelorism; macrocephaly; jaw cysts; basal cell naevae; pits in skin, especially palms and soles; linear calcification of falx; rib abnormalities; mild developmental delay; ovarian cysts	Medulloblastoma; basal cell carcinomas

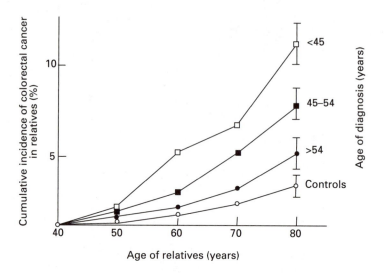

Figure 14.1 Cumulative risk of colorectal cancer in relatives of a patent diagnosed at various ages. (Source: adapted from St John, Dermott and Hopper, 1993.)

of that cancer is an important indicator of risk of genetic susceptibility, i.e. if a cancer is rare, then two cases in a family are less likely to have occurred by chance.

A single relative with a particular cancer often does not greatly increase the risk to relatives. The exception to this is if the relative is young, or had multiple primaries or a recognizable cancer syndrome. The risk of bowel cancer in the relatives of a single case illustrates the importance of age at diagnosis (Figure 14.1).

Occasionally, a malignancy may be known to occur frequently as a result of a germ-line mutation. Retinoblastoma is an example, a rare childhood malignancy of the eye, in which 40% of cases are due to a genetic susceptibility (Chapter 1). Some children have multifocal disease, which is almost invariably due to the presence of a germ-line mutation, with the risk for children of individuals with bilateral disease approaching 50%.

There is now published information on the risks for relatives of cancer patients, particularly for common cancers such as breast cancer and colorectal cancer. These are particularly useful for genetic counselling, permitting visual demonstration of risk assessment to the patient. The likelihood of a genetic susceptibility can be calculated, combining information on the number and age of affected individuals. The risk to the patient will depend upon their relationship to the affected family members, and their own age, since the risk will decrease the longer they remain free of disease.

When a specific diagnosis of cancer susceptibility is possible, then there may be more information available to impart, both in terms of the chances of developing cancer and possible non-malignant problems. For instance, if the *BRCA*1 mutation is the likely cause of breast cancer, then detailed information is available on the cumulative risk of both ovarian and breast cancer as well as the possibility of other cancers, such as bowel and prostate, for which there is an increased relative risk in affected individuals. A *BRCA*1 mutation may be suspected from the family structure, a dominant susceptibility to early breast cancer associated with ovarian cancer.

Mapping and cloning of genes for familial cancers has made it possible to confirm the clinical diagnosis with a molecular test in some instances. In many more families it allows identification of individuals at risk of disease within the family. The molecular tests therefore enable a more accurate assessment of risk (Table 14.2).

Discussion of cancer risk

This part of the interview involves communicating to the patient the results of the pedigree assessment, risk assessment and clinical examination. If a particular diagnosis is made, then information about the disease can be given, including the results of any molecular tests available for the confirmation of the diagnosis or for predictive testing.

Table 14.2 Genetic mutations in familial cancers (MEN = multiple endocrine neoplasia; NF = neurofibromatosis)

Disease	Location	Mutation analysis
Breast–ovarian	17q21	Not yet available
Von Hippel–Lindau syndrome	3p25	YES for *VHL*
Familial adenomatous polyposis	5q21	YES for *APC*
Gorlin's syndrome	9q	Not yet available
MEN Type II	10q11.2	YES for *ret*
MEN Type I	11Q	Not yet available
Wilms tumour	11p13	YES for *WT1*
Retinoblastoma	13q	YES for *Rb1*
Li–Fraumeni syndrome	17q	YES for *TP53*
NFI	17q11.2	YES for *NF1*
NFII	22q11.2 to q12.1	YES for *NF2*
Lynch syndrome	2p22	YES for *hMSH2*
Lynch syndrome	3p21	YES for *hMLH1*

Individuals attending genetic clinics may have a very rudimentary knowledge of genetics, and it is important that they have a simple explanation of Mendelian genetics and how risk is assessed. A simple explanation of how cancer develops as a result of somatic genetic events is also sometimes helpful. In this way patients can understand and come to trust the information they are given. If they are being given empirical risks, the method by which these figures are derived must be explained. If there are no data, this must also be discussed. If the geneticist has a clinical impression that there may be something unusual occurring, but this is no more than a hunch, this must be made clear. Having a risk figure is useful for the clinician because it may dictate what options for management are available, but these are only useful to patients if they are put into context, i.e. in relation to the population risk of that and other cancers. In particular, the age at which they are at greatest risk must be discussed, as this may affect the timing of prophylactic surgery and future cooperation in screening programmes.

Screening and prevention

The possibilities for screening and prevention should be discussed with the consulting individual. What is known about the value of any particular strategy including its rationale must be explored.

Since some individuals may wish to do nothing, it is important that this is also discussed as an acceptable option, and it may be the right decision for some people. In some instances prophylactic surgery needs to be discussed, but this must be approached with caution as some patients are frightened or even horrified at the suggestion. They may feel that this is confirmation from the doctor that their risk of cancer is unacceptably high, and may accentuate any fears they may have of the disease and its treatment.

14.3 THE USE OF MOLECULAR TESTS

Predictive testing

The number of cancer susceptibilities that have been mapped by genetic linkage is steadily increasing. This allows DNA analysis to be carried out, and individuals with a susceptibility gene to be identified, before the development of the disease. Tightly linked markers can be used diagnostically within families, but there are limitations to these DNA tests. Firstly, there must be the right family structure to enable the marker of high risk to be identified (phase known). Secondly, enough individuals from the family must be available to give samples. Thirdly, the affected parent of the individual being tested must be heterozygous at the marker to be informative, otherwise the chromosome (paternal or maternal) carrying the disease cannot be identified. The discovery of highly polymorphic tandem repeats and di- and trinucleotide repeats has greatly reduced the problem of uninformative markers.

Genetic heterogeneity

A further problem with predictive testing by genetic linkage arises if there is more than one locus responsible for the susceptibility. This is the case with familial cancers of both the breast and bowel. The number of families that can have predictive testing is limited by the degree of genetic heterogeneity, i.e. when more than one locus may cause the same condition. For instance, approximately half the breast–ovarian cancer families fail to show linkage to *BRCA*1; other gene loci must be involved. Each family becomes an experiment in itself, as linkage must be established in that particular family if predictive testing is to be carried out. The family also needs to be large so that there is at least a 95% likelihood of linkage. Most families are not suitable for testing, and patients are often disappointed that they are unable to have the test.

Identification and cloning of the gene involved in a cancer predisposition offers a solution to some of these problems. The mutation can then be sought by direct sequencing of the gene in a single affected individual, confirming if the gene is involved. The disease allele can then be looked for in other family members. Now the *BRCA*1 gene is identified, the number of families that can be tested will increase, since direct mutation analysis of an affected individual will confirm a *BRCA*1 family.

Mutation analysis originally required sequencing the gene in its entirety. Now, simpler techniques which screen for mutations will often be carried out first (Chapter 3). In this way the amount of sequencing is reduced. Mutation detection does however remain relatively labour intensive.

Genetic linkage analysis and direct mutation analysis are now possible for many different cancer susceptibilities (Table 14.2). In late-onset genetic disease susceptibilities, there is a consensus view that children should not be tested, unless there is to be a therapeutic intervention or change in management. Some cancer susceptibilities do require screening during childhood. For instance, screening for familial polyposis coli usually starts in early teenage years by sigmoidoscopy. DNA testing prior to this time will allow half the individuals to avoid having this invasive procedure. Testing seems entirely reasonable in this instance, particularly because prophylactic surgery is successful.

The value of testing for other cancer susceptibilities, where the value of screening and prevention is unknown, is less clear-cut. Many of the issues are relevant that have been discussed at length in relation to testing for other adult-onset genetic diseases, such as Huntington's chorea, where prevention is not possible. It has been demonstrated that using a set protocol for individuals having predictive testing for Huntington's chorea helps to minimize the problems experienced. It allows individuals to have time to decide if they really want the test and for what reason. There may be many reasons why individuals may wish to have a predictive test. They may want to know if they have the gene before starting a family, or to make plans for their own future. In other situations, it may be that they want to make choices concerning having prophylactic surgery or entering into screening or chemoprevention studies.

Facing a high risk of breast cancer is particularly difficult for some women. Often there have been several deaths from the disease in the family. As this is often a mother who has died when the patient was in her teenage years, the memories can be particularly painful. Since there may already be a great deal of anxiety about the disease, it may be very traumatic to find that the chance of having the gene for early breast cancer is high. It is therefore recommended that a formal protocol is followed when offering predictive testing for either *BRCA*1 or *p53*. It is probably a good idea to follow these protocols for some of the other more worrying conditions, such as von Hippel–Lindau disease. Initially, the pros and cons and accuracy of the test are explained to the patient. There is a compulsory psychological assessment. Patients are then left to consider for a while whether or not to have the test, and if they decide to proceed, they are seen again to discuss their reasons for wishing to do so. It is only then that the blood sample for testing is collected. The disclosure session is carefully planned so that the patient knows how long s/he will have to wait for the result. Following this, s/he is seen at suitable intervals to ensure that s/he has accepted the result and is not having any problems.

14.4 THE USE OF MUTATION ANALYSIS IN PATIENT MANAGEMENT

Retinoblastoma as an example

The child of an individual who once had bilateral retinoblastoma has almost a 50% risk of developing retinoblastoma. Individuals presenting with bilateral disease must therefore have germ-line mutations. In the affected children with a positive family history, 68% have bilateral tumours and 32% have unilateral disease, whereas in the general population, 25–30% of children present with bilateral tumours. It is therefore assumed that 10–15% of patients with unilateral tumours and no family history carry a germ-line mutation.

Penetrance of the gene is 90%, so single cases, bilateral and unilateral, may receive the mutation from an unaffected parent. The risks to siblings are therefore 5% for the siblings of bilateral disease and 1% for the siblings of unilaterally affected individuals. The siblings and offspring all therefore require intensive screening, although the majority are at a low risk. Mutation analysis in affected single cases identifies those cases with a germ-line mutation and reduces the number of individuals requiring screening. Reassurance can also be provided to the relatives of those individuals in whom no mutation is identified.

Use of radiotherapy

Individuals with a cancer susceptibility are frequently sensitive to carcinogens and to radiation. In addition there is a tendency for them to develop second tumours. Treatment by radiation may therefore be less than advantageous in these individuals. For example, treatment of a basal carcinoma by radiotherapy in an individual with Gorlin's syndrome will result in the development of multiple basal cell carcinomas in the radiation field.

Retinoblastoma patients provide another example. Of those with germ-line mutations 15–20% develop second tumours, especially osteosarcoma. There is thought to be no increase in risk to those without germ-line mutations. Radiotherapy will shorten the period between the malignancies. Following mutation analysis where a second cancer is likely, it may be possible to modify therapy to take this into account.

Prenatal diagnosis

Another useful application of the technology is prenatal diagnosis. A family's experience of the disease will often influence its decision to take up prenatal diagnosis. This is particularly true if the family has already lost a child with a malignancy. They may feel unable to take the risk with another child even if the tumour is only infrequently associated. This procedure can be carried out between 8 and 12 weeks by chorionic villous biopsy, allowing the option of termination in the first trimester.

14.5 POPULATION SURVEYS OF CANCER SUSCEPTIBILITY

The identification of cancer susceptibility genes makes it possible to examine populations of cancer patients and identify which individuals have a germ-line mutation. This is important in establishing the genetic contribution to the disease but is also important to the individuals identified and their families.

Li–Fraumeni syndrome as an example

A rare genetic form of breast cancer is seen in Li–Fraumeni syndrome. In these families, there is dominant inheritance of susceptibility to early-onset breast cancer and other cancers, including sarcomas, CNS tumours and leukaemias. Li–Fraumeni syndrome has been found to be due to an inherited mutation in the *p53* gene. In this condition, it is possible to have a genetic diagnosis by mutation analysis.

It is now possible to ask how many sarcomas, early breast cancers or second malignancies occur because of germ-line *p53* mutations. A number of population studies have now been done. For example, 181 random sarcoma patients (mostly osteosarcomas) were tested and two were found to have germ-line mutations. Both of these cases had a family history of cancer, in other words about 1% of the sarcoma patients. On the other hand, 15 sarcoma patients were selected because of a family history, or a history of multiple cancers, and six had germ-line mutations (three of these had no family history). This shows that the population rate is relatively low but that family history or a history of multiple tumours in an individual is a good indicator for testing the mutation.

Extending the phenotype using DNA testing

In the past, the phenotypes of cancer families have been defined clinically. The identification of cancer susceptibility genes enables families to be examined that do not fit the classical phenotype. This has been done using *p53* and the polyposis gene on chromosome 5, allowing the clinical phenotype to be expanded. For instance, a hereditary form of bowel cancer similar to polyposis, with a later age of onset and fewer adenomas, has now been linked to the locus on 5q.

14.6 CANCER GENES AND DEVELOPMENT (WILMS TUMOUR)

The possibility that genes expressed during development may also be involved in the development of cancer is perhaps best illustrated at the current time by the work done on the genetics of Wilms tumour. It may be inherited as an autosomal dominant but can also be associated with developmental abnormalities. About 2–3% of children with Wilms tumours have a chromosome deletion at 11p13. Beckwith–Wiederman syndrome, with a high risk of Wilms tumour, is sometimes associated with a duplication at 11p15. The deletion at 11p13 is associated with a chromosomal deletion syndrome with the acronym WAGR, standing for Wilms, aniridia, genital anomalies and gonadoblastoma, and mental retardation. Some males may be under-masculinized or have complete sex reversal. The Drash syndrome is a related syndrome but aniridia does not occur. In addition, children with Drash have an early onset neuropathy with mesangial sclerosis.

Molecular studies of the locus at 11p13 have identified the gene responsible, known as *WT1*. This is a zinc finger protein which functions as a transcription factor. This has enabled the identification of mutations in the Drash syndrome, all of which have so far been located in the important DNA-binding zinc finger motif. It is therefore very likely that the risk of Wilms tumour and the developmental abnormalities in the genital tract in this condition, and in WAGR, are a direct result of mutations or deletions of this gene.

The cloning of the aniridia gene, which is close but separate to the Wilms locus, explains the association of this abnormality with WAGR. It also explains the existence of families with dominant aniridia without evidence of Wilms susceptibility, and the lower risk of Wilms with sporadic aniridia than is seen in children with WAGR. In the past, all children with aniridia needed screening for Wilms. Now identification in these children of a mutation limited to the aniridia gene means that they are not at increased risk of Wilms tumour and do not require screening.

14.7 FURTHER READING

Claus, E. B., Risch, N. J. and Thompson, W. D. (1990) Age at onset as an indicator of familial risk of breast cancer. *Am. J. Epidemiol.*, **131**, 961–972.

Goodrich, D. W. and Lee, W. H. (1990) The molecular genetics of retinoblastoma. *Cancer Surv.*, **9**, 529–533.

Pritchard-Jones, K. and Hastie, N. D. (1990) Wilms' tumour as a paradigm for the relationship of cancer to development. *Cancer Surv.*, **9**, 555–578.

St John, D. V. B., McDermott, F. T., Hopper, V. L. *et al.* (1993) Cancer risk in relatives of patients with common colorectal cancer. *Ann. Intern. Med.*, **118**, 785–190.

Tyler, A., Ball, D. and Crauford, D. (1992) Huntington's disease in the United Kingdom. *Br. Med. J.*, **304**, 1593–1596.

15	Molecular basis of radiation sensitivity

Anthony T. Gordon and Trevor J. McMillan

15.1 MOLECULAR RADIOBIOLOGY IS THE UNDERSTANDING OF THE CELLULAR RESPONSE TO RADIATION AT THE MOLECULAR LEVEL

Molecular radiobiology has established its place in modern biology. The fundamental nature of the subcellular processes involved in the response to either ionizing radiation or UV light means that studies are relevant not only because of the importance of radiation in the environment and medicine, but also because radiation is a useful probe in basic cell and molecular biology. The aims of this field of study include the understanding of the biochemical pathways that are involved in the cellular response to radiation in order to determine the critical processes in its cytotoxic and mutagenic effects. There are therefore two principal approaches: the study of the postirradiation response to DNA damage and the assessment of the relevance of these responses to cell killing and mutation. The understanding of the molecular and genetic basis of the cellular response to UV damage is more advanced than for ionizing radiation. However, due to the therapeutic relevance in oncology the primary emphasis in this chapter will be placed on ionizing radiation.

15.2 CRITICAL DNA DAMAGE INDUCED BY IONIZING RADIATION IS RARE AND PROBABLY INVOLVES DNA DOUBLE-STRAND BREAKS

Ionizing radiation damages DNA directly by the ionization of atoms within DNA or indirectly by the ionization of water molecules which in turn react with DNA. A radiation dose of 1 Gy will produce 150 000 ionization events in the nucleus but

Molecular Biology for Oncologists. Edited by J. R. Yarnold, M. R. Stratton and T. J. McMillan. Published in 1996 by Chapman & Hall, London. ISBN 0 412 71270 9

only about 1000 strand breaks. In addition to single- and double-strand breaks in the DNA low LET radiation also produces a wide range of types of base and sugar damage and cross links between different parts of the DNA or between DNA and proteins (Chapter 16). Of the various types of radiation-induced damage there is evidence to suggest that it is the DNA double-strand break that is the primary lethal lesion. Therefore most emphasis has been placed on the induction and repair of double-strand breaks in the analysis of the basis of radiosensitivity. Double-strand breaks are induced at a rate of approximately 1 per chromosome per gray so these are very rare indeed, which makes them difficult to study. In addition, not all double-strand breaks have the same chemical structure, when the nature of the ends of DNA or the incidence of adjacent DNA damage is considered, so that it may only be a subset of these that are biologically relevant.

15.3 NATURAL RADIOPROTECTORS CAN LIMIT THE AMOUNT OF RADIATION-INDUCED DNA DAMAGE

The postirradiation processes that occur in a cell are shown in Figure 15.1.
Following ionization events a series of chemical processes acts on free radicals

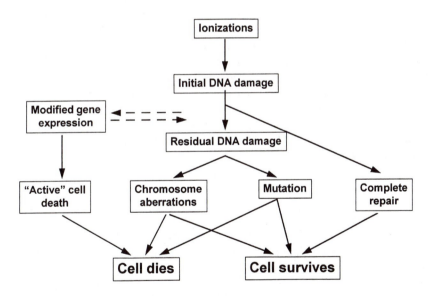

Figure 15.1 The physical deposition of energy by ionizing radiation is followed by a series of events that can modify the impact on the cell. Initial DNA damage is acted on by repair processes which, while efficient, still leave residual damage. Some of the resulting mutations and chromosome aberrations can be lethal while others do not affect proliferative capacity. DNA damage modifies the expression of some genes which can lead to active cell death or initiate pathways that may facilitate cell survival.

to directly reverse DNA damage or scavenge the products of water ionization before they damage DNA. Of particular importance is oxygen, which can 'fix' radiation-induced damage leading to a much greater level of double-strand breaks compared with cells irradiated in the absence of oxygen. Various enzymes influence damage induction by scavenging free radicals (e.g. superoxide dismutase) or by the catalysis of neutralization of radicals by the transfer of protons. An example of the latter is the action of glutathione-S-transferase in the reaction of glutathione with free radicals. Modification of this activity in cells has a significant effect on the cytotoxicity of radiation, especially in the absence of oxygen. In addition to these diffusible chemical protectors it is now clear that the proteins associated with DNA in chromatin (Chapter 28) have a significant radioprotective effect. Histone H1 in particular can protect DNA and if H1 is experimentally removed from isolated chromatin there is a twofold increase in the number of double-strand breaks induced by radiation.

15.4 MOST DNA DOUBLE-STRAND BREAKS ARE REJOINED WITHIN 2 HOURS AFTER IRRADIATION

DNA double-strand breaks are generally rejoined efficiently by cells. Breaks disappear exponentially after treatment with a half time of 20–40 min, so that there is little evidence of further rejoining of double-strand breaks 2–4 hours after irradiation. Most cells manage to rejoin > 90% of the double-strand breaks even when very large doses (> 100 Gy) are given, although assays are not yet available to allow us to assess how accurate this rejoining is. The importance of the rejoining process is demonstrated by the marked sensitivity exhibited by mutant cells isolated from hamster cell lines which are known to have a severe defect in double-strand break rejoining (Figure 15.2).

Recent progress has been made in identifying some of the processes that are influential in double-strand breaks rejoining (see 15.5 and Chapter 16) but our knowledge is still far from complete.

15.5 DNA DAMAGE LEADS TO CELL DEATH

While it is apparent that DNA damage is the primary cause of cell death after irradiation it is not always clear how death occurs. There appear to be three patterns of cell behaviour: cells either progress through one or two divisions before proliferation ceases, they do not survive the first mitosis after irradiation or they do not reach the first mitosis. In the first two cases it is clear that the DNA damage results in gross damage to chromosomes and since the amount of chromosome damage correlates very well with cell death it is believed that it is this damage that is lethal. This lethality is mediated through an inability to undergo a regular mitosis because of the presence of aberrant chromosomes (e.g. those with ring structures that can

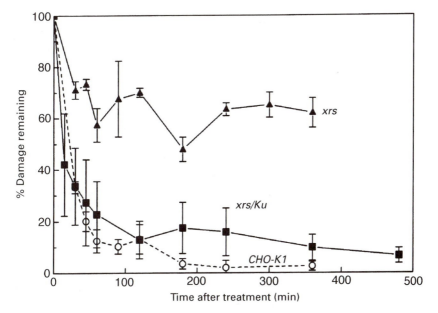

Figure 15.2 Conferring radioresistance on *xrs* cells. The *xrs* radiosensitive mutants isolated from Chinese hamster ovary cell have a severe defect in their ability to rejoin DNA
double-strand breaks such that 4 hours after treatment there are still 50–60% of breaks
remaining. This contrasts with < 10% remaining in the parental CHO cell line. By trans-
fecting the *Ku80* gene into the *xrs* cells the ability of the cells to rejoin double-strand
breaks is restored.

not undergo an even division) or by metabolic disruption due to the loss of genetic
material on chromosome fragments.

Cells that do not reach mitosis after irradiation either stick in the G1 phase of
the cell cycle and become increasingly abnormal in terms of size and appearance
or they undergo apoptosis (see below and Chapter 21).

15.6 THE RELATIVE RADIOSENSITIVITY OF CELLS IS LARGELY DETERMINED BY THE ABILITY TO LIMIT DAMAGE INDUCTION AND REPAIR DNA DAMAGE

A major experimental approach in this field has been the analysis of DNA damage induction and repair in cells that differ in their radiosensitivity. In rare cases a
significant defect has been identified that has been demonstrated to be directly
responsible for a change in sensitivity. The best example of this is the defect in the
radiosensitive *xrs* mutants that were isolated from Chinese hamster cells. These
were recognized as having a marked defect in double-strand breaks rejoining
(Figure 15.2) and now that the defective gene has been isolated (see below) it has

been shown that correction of this defect brings about the restoration of radio-resistance. In most other cases the molecular basis of variations in sensitivity have not been identified although correlations between radiosensitivity and damage induction, rate of repair, fidelity of repair and final residual damage have been reported. There have also been suggestions that the extent of radiation-induced cell cycle delays and apoptosis can contribute to the relative sensitivity of some cells.

15.7 CELL CYCLE REGULATION AND THE MAMMALIAN CELL RESPONSE TO IONIZING RADIATION

After exposure to ionizing radiation mammalian cells can delay G1, S and G2 phases of the cell cycle. Early theories suggested that these cell cycle delays allowed cells time to repair DNA damage, thus making them more resistant, and while this still may be partially true the relationship between cell cycle delay and radiosensitivity is turning out to be more complex as we learn more about the control of the cell cycle (Chapter 7).

The ionizing-radiation-induced mammalian G1 cell cycle delay seems to be highly dependent on the normal function of the tumour suppressor gene *p53*. This has been most clearly demonstrated with *p53*-null mice, which lack a G1 delay and are more radioresistant than mice with normal *p53* status. These null mice are more highly prone to radiation-induced carcinogenesis, possibly due to the abrogation of the *p53* dependent apoptosis (see below). At the molecular level cells appear to block in G1 because of the elevated expression of normal *p53*, leading to the activation of downstream cell cycle control proteins, principally the cyclin-dependent kinase inhibitor (CDKI) p21 (alternatively known as CDI1, SDI1, WAF1). CDKI p21 then interacts with other proteins such PCNA to block replication and cyclin-kinase complexes to block the cell in G1 (Chapter 7). The major limitation of our current understanding of the ionizing radiation G1 cell cycle delay is our lack of knowledge of the signals upstream of p53 which presumably detect DNA damage and stimulate the p53 response.

In mammalian cells much less is known about the precise control mechanisms of the ionizing-radiation–induced G2 delay. Although it has been shown that levels of the cyclins A and B1 vary before and during G2 delays, it is unknown whether these cyclins vary in response to or in control of the G2 delay. It has been shown that the addition of low levels of caffeine to cells can reduce the G2 delay, sensitize cells to radiation and increase the levels of cyclin B1 mRNA but again the direct relationship between these endpoints is unclear. Some insight into the mechanisms of the ionizing radiation G2 delay has come from the transformation of cells with the oncogenes H-*ras* and v-*myc*. Transformation with both oncogenes produces cells that are significantly more radioresistant than their parental cell lines and they show a greater G2 delay after ionizing radiation. The protein products of these oncogenes are important components of signal transduction pathways that control cell growth. How such pathways are initiated in response to DNA damage remains to be determined. In yeast cells the picture of the G2 delay is clearer. As yeasts can be easily mutated and transformed, a large number of mutants have been isolated, such as *rad9*, that are defective in their ionizing radiation induced G2 delay. Such mutants are radiosensitive and have been shown to be defective in gene products involved in the monitoring of DNA prior to mitosis.

15.8 APOPTOSIS OCCURS AFTER IONIZING RADIATION BUT THE GENERALITY OF ITS ROLE AS A DETERMINANT OF RADIOSENSITIVITY IS UNCLEAR

The importance of apoptosis in ionizing radiation cell death is highly cell-type-dependent, with certain radiosensitive cells such as those from the haemopoietic and lymphatic systems being more prone to rapid radiation-induced apoptotic cell death. In tumour cells the mechanisms of ionizing-radiation-induced cell death vary, with mitotic cell death and necrosis being observed as often, or more often, than apoptosis. It is also unclear whether the mode of ionizing radiation cell death determines the inherent sensitivity of each cell type, although certainly for many sensitive cells apoptosis appears to be a major mechanism of cell death. For example, stem cells at the base of the crypts of the small intestine commonly undergo apoptosis as part of their homeostatic control mechanism and they are very sensitive to ionizing radiation. However, differentiated cells higher in the crypt appear to suppress this ionizing-radiation-induced apoptosis. Thus the high sensitivity of crypt stem cells to apoptosis may be part of an efficient self-screening mechanism against carcinogenesis which could explain the lower than expected incidence of primary tumours in the small intestine.

While it is clear that apoptosis is an important mode of cell death after irradiation of some cell types there is still debate as to whether in other systems it can be simply a description of the 'death process' rather than a significant determinant of **whether** a cell dies. Evidence from murine tumour systems suggests that in contrast to other forms of rapid apoptosis, apoptosis after ionizing radiation can be somewhat delayed (i.e. > 24–48 h). In some cases apoptosis can even be a post-mitotic event, which makes the distinction between apoptotic and mitotic cell death very difficult. In a group of experimental systems it has been suggested that it is the timing of apoptosis that is a critical indicator of radiosensitivity (resistant cells show evidence of apoptosis later than sensitive cells) and that genetic manipulation of the susceptibility of cells to undergo apoptosis does not necessarily affect the final clonogenic potential of the cells. Thus the signalling pathways leading to apoptosis and in particular the initial apoptotic trigger after irradiation require further investigation.

Table 15.1 Some isolated genes involved in the control of ionizing radiation sensitivity of mammalian cells

Gene	Method of isolation	Gene function
XRCC1	Complementation of hamster mutant EM9	Associated with DNA ligase III, involved in SSB rejoining
XRCC5 (Ku80)	Complementation of hamster mutant xrs and genomic mapping	DNA endbinding activity, part of DNA-dependent protein kinase, involved in DSB rejoining
XRCC6 (Ku70)	Complementation of hamster mutant sxi-1	DNA endbinding activity, part of DNA-dependent protein kinase, involved in DSB rejoining
XRCC7 (SCID, p350 DDPKcs)	Genomic mapping	Part of DNA-dependent protein kinase, involved in DSB rejoining
ATM	Mapping of gene responsible for human ataxia telangiectasia	Part of signal transduction pathways?
DNA ligase 1	Protein isolation followed by identification of coding gene	Ligation of DNA ends

15.9 THERE HAS BEEN GREAT PROGRESS IN THE IDENTIFI-CATION OF GENES THAT CAN AFFECT RADIOSENSITIVITY

The understanding of the genetics of sensitivity to ionizing radiation is developing extremely rapidly, in a manner very similar to those that produced our detailed knowledge of human UV DNA repair mechanisms (Table 15.1).

Central in both cases has been the use of laboratory-generated DNA-damage-sensitive mutant cells and cells from inherited human DNA repair disorders. The genes mutated in some of these cell lines that produce the ionizing radiation DNA repair defects have been identified using powerful molecular biological techniques, thus allowing their role in DNA repair to be characterized.

A recent important discovery has been the cloning and partial characterization of the gene responsible for the human ionizing-radiation sensitive syndrome ataxia telangiectasia (AT), previously localized to chromosome 11q22/23. Patients with this syndrome have developmental disorders, show an increased predisposition to certain cancers and are characteristically sensitive to ionizing radiation. Furthermore, 0.5–1.4% of the population are thought to have one mutated AT allele (heterozygous), which may make these individuals more cancer-prone, particularly female heterozygotes developing breast cancer. Epidemiological studies first determined the chromosomal location of the AT gene locus, then the gene itself has been recently identified using a positional cloning technique. This technique relies on the use of specific DNA sequences (usually microsatellite markers), which have been shown to be closely linked to the gene locus of interest, by the analysis of the products of natural recombination in affected families. These markers were then used to identify DNA fragments in DNA libraries, first in artificial chromosome libraries, then in cosmids and finally in cDNA clones. The sequences in the genome corresponding to these cDNA clones were screened for mutations in cells from AT individuals until a gene was found that was mutated in all AT cell lines tested. This *ATM* (AT-mutated) gene has been shown to produce a primary 12 kbp cDNA, although many smaller transcripts are also produced. Interestingly, this gene is mutated in all complementation groups, suggesting that the AT phenotype is a product of mutations in just one gene rather than in any one of four genes, as previously thought. The predicted ATM protein shows some homology to proteins involved in important signal transduction pathways and yeast DNA repair proteins. With this discovery exact molecular phenotypes of this disorder may now be determined, including the defect responsible for the ionizing radiation sensitivity.

Laboratory-generated radiosensitive mutant cells have provided much of the information that we presently have about the mechanisms of DNA repair after ionizing radiation. Included in this group of cells are the *xrs* mutants mentioned previously. Using chromosome transfer and positional cloning techniques, the mutated gene responsible for the radiosensitivity of the *xrs6* cell line was localized down to a small area of human chromosome 2 and a known candidate gene, *Ku80*, identified as the mutated gene. The Ku80 protein was first identified as an autoantigen from patients with autoimmune disease. Its relevance to ionizing radiation was first suspected through its ability to bind to the free ends of DNA, and the reduction in DNA end-binding activity in *xrs* cells. A further important feature of *xrs* cells is that they are deficient in V(D)J recombination (Chapter 16). The linking of ionizing radiation DNA repair proteins and V(D)J recombination suggests a close relationship between these two processes. This linkage has been further confirmed by the finding that mice with the SCID (severe combined immune deficiency) disorder not only have a suppressed immune system due to a defect in V(D)J recombination but are also radiosensitive. The mutated gene in these mice codes for the catalytic subunit of a DNA-dependent protein kinase (DDPKcs) which also contains the two Ku proteins (70 and 80 kDa). On binding to free ends of DNA the Ku proteins complex with the DDPKcs, activating the complex, which then phosphorylates a number of cellular proteins.

15.10 THERE MAY BE A RELATIONSHIP BETWEEN CELLULAR RESPONSE TO RADIATION AND PREDISPOSITION TO CANCER

It has been recognized for some time that some human syndromes that exhibit abnormal sensitivity to cytotoxic agents have a predisposition to certain types of cancer. Among these is the ionizing-radiation-sensitive ataxia telangiectasia, where a high frequency of lymphomas are observed in the homozygotes. AT heterozygotes are also believed to have a high cancer susceptibility. There is a little evidence that exposure to ionizing radiation can be the cause of the tumours in these individuals but it is unlikely that this is commonly the case. It is more likely that the biochemical defect leading to the increased cancer incidence also alters the sensitivity to the cytotoxic agent. Some intriguing data relevant to this have come from the study of ionizing-radiation-induced chromosome aberrations in cells irradiated in the G2 phase of the cell cycle. It has been reported that even when no abnormal sensitivity to the killing effects of ionizing radiation is known, a large number of cancer prone syndromes can be detected in this so-called 'G2 assay'. These syndromes include xeroderma pigmentosum, Gardner's syndrome and ataxia telangiectasia. Thus the possibility of a close relationship between chromosomal sensitivity to ionizing radiation and cancer predisposition is raised and deserves considerable experimental attention.

15.11 WHAT CAN THE ONCOLOGIST DO WITH THIS INFORMATION?

There are two key areas of application for this information. The first, which should be applicable in the short term, is the pretreatment prediction of tumour and normal tissue response to therapy. The radiosensitivity of the cells within a tumour is one of the primary determinants of tumour radioresponsiveness and biochemical and molecular indicators of the sensitivity will be a useful aid to treatment decisions. There is increasing evidence that the degree of normal tissue reaction after radiotherapy can be indicated by the cellular sensitivity of fibroblasts isolated from the patient. In this case the aim will be to use the molecular information to identify sensitive individuals in order to avoid severe reactions. We would also wish to identify resistant patients who may be able to tolerate increased doses of radiation in order that dose escalation might lead to better cure rates.

As well as response prediction, information on the molecular basis of radiosensitivity should provide a basis on which the modification of radiation response can be developed. The identification of critical biochemical steps and the genes that can control these steps will allow the rational design of agents that can sensitize or protect cells to radiation. Both chemical modifiers and agents that use the informational technology outline in Chapter 19 are possibilities for this approach.

15.12 FURTHER READING

Aldridge, D. R., Arends, M. J. and Radford, I. R. (1995) Increasing the susceptibility of the rat 208F fibroblast cell line to radiation-induced apoptosis does not alter its clonogenic survival dose-response. *Br. J. Cancer*, **71**, 571–577.

Iliakis, G. (1991) The role of DNA double strand breaks in ionising radiation-induced cell killing. *Bioessays*, **13**, 641–648.

McMillan, T. J. and Peacock, J. H. (1995) Molecular determinants of radiosensitivity in mammalian cells. *Int. J. Radiat. Biol.*, **66**, 639–642.

Parshad, R., Sanford, K. K. and Jones, G. M. (1983) Chromatid damage after G2 phase x-irradiation of cells from cancer prone individuals implicates deficiency in DNA repair. *Proc. Nat. Acad. Sci. USA*, **80**, 5612–5616.

Savitsky, K., Bar-Shira, A., Gilad, S. *et al.* (1995) A single ataxia telangiectasia gene with a product similar to PI-3 kinase. *Science*, **268**, 1749–1753.

Taccioli, G. E., Gottlieb, T. M., Blunt, T. *et al.* (1994) Ku80: product of the XRCC5 gene and its role in DNA repair and V(D)J recombination. *Science*, **265**, 1442–1445.

<table>
<tr><td>16</td><td># How do cells repair DNA damage caused by ionizing radiation?</td></tr>
</table>

16 | How do cells repair DNA damage caused by ionizing radiation?

Simon N. Powell, Lisa A. Kachnic and P. Rani Anné

16.1 CELLS HAVE EVOLVED AN ABILITY TO REPAIR DNA DAMAGE

For a given radiation dose, a specific amount of energy is deposited in cells, tissues or water, which results in ionization damage to DNA. Differences in cell sensitivity to radiation are due to differences in the chemical and biological processing of that damage, collectively called 'repair'. Repair in this wide definition covers a broad range of subjects: the physics of energy deposition by ionizing radiation; the initial measurable damage inflicted in the DNA; the chemical and biological modification of this damage; and finally the biological consequences of residual radiation-induced damage.

16.2 RAPID CHEMICAL EVENTS ARE THE FIRST LINE OF DEFENCE

Energy is deposited in a variety of ionization densities. For 1 Gy absorbed dose of cobalt-60 irradiation, approximately 1000 sparse ionizations, 20–100 moderate clusters, 4–40 large clusters and 0–4 very large clusters occur. Different radiation qualities produce different relative amounts of ionization densities. The production of sparse ionizations is found to be in inverse proportion to the relative bio-

Molecular Biology for Oncologists. Edited by J. R. Yarnold, M. R. Stratton and T. J. McMillan. Published in 1996 by Chapman & Hall, London. ISBN 0 412 71270 9

logical effectiveness (RBE) and therefore is thought to be biologically less important. The relative number of ionization clusters tends to match the RBE, and is thus thought to be biologically more relevant.

There are rapid biochemical mechanisms that can modify the conversion of ionizations into biologically relevant damage. For example, one immediate effect of ionization is loss of hydrogen atoms from DNA, and this could be reversed rapidly by hydrogen donor molecules such as those containing sulphydryl groups. These molecules would need access to damaged sites and necessarily would be either small or already present adjacent to DNA. Reduced glutathione (the most abundant sulphydryl hydrogen donor molecule) and cysteamine are possible candidates. Artificial manipulation of thiol levels has a clear effect on initial DNA damage.

Another mechanism of 'immediate' biochemical modification of DNA damage is rapid enzymatic metabolism of the products of water radiolysis. Hydrogen peroxide is generated from hydroxyl radicals, and is converted to water and oxygen by catalase, or to water and dehydrogenated co-substrate by peroxidases. Superoxide radicals are catalysed by superoxide dismutase. Thus levels of these enzymes may determine the amount of DNA damage sustained.

Initial damage can be measured in the DNA of a cell after immediate biochemical modification has occurred but before biological processing (i.e. enzymatic reactions of relatively long half-time) has taken place. A variety of types of damage can be detected and are summarized in Figure 16.1.

The measurement of single- or double-strand breaks is based on DNA fragmentation in alkaline (denaturing) or neutral conditions respectively. DNA fragments can be separated according to size by a variety of means. Methods include velocity sedimentation, filter elution, or pulsed-field gel electrophoresis. DNA–protein crosslinks are quantified by the reduction in strand breaks detected when no protein digestion is allowed in the preparation of the test DNA. By contrast, base damage can be measured and characterized by chemical methods and

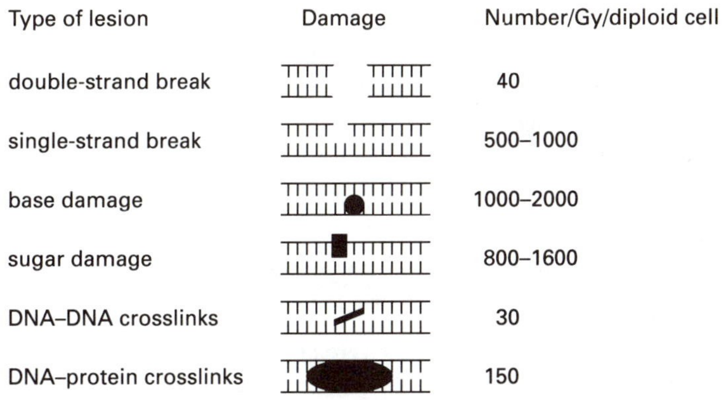

Type of lesion	Damage	Number/Gy/diploid cell
double-strand break		40
single-strand break		500–1000
base damage		1000–2000
sugar damage		800–1600
DNA–DNA crosslinks		30
DNA–protein crosslinks		150

Figure 16.1 Types of DNA damage detected after exposure to ionizing radiation.

these have, for example, detected more than 20 radiation products of thymine. Specific antibodies have been made to recognize certain types of base damage, and this can provide an alternative means of measurement.

Of the radiation-induced lesions listed, the double-strand break (DSB) appears to be associated most closely with cell lethality. The word association should be stressed: correlation exists between the number of DSBs and lethal events, but this does not show which DSBs are lethal or exclude the possibility that some other lesion produced in proportion to DSB is the critical lesion. By altering the damaging agent the ratio of DSBs to SSBs can be varied. With hydrogen peroxide the DSB:SSB ratio is reduced and this causes less cell lethality per lesion, while with bleomycin this ratio is increased and more cell deaths per lesion are found. Hence, it is concluded that the DSB seems to be the more important lesion biologically.

The majority of DNA DSBs (i.e. DNA fragments) are rejoined, but it is not clear whether the residual lesions remain because of saturation of repair processes leading to a failure to rejoin or a misrepair. Specific subsets of DSB, either by nature or position, may have a higher probability of repair failure. It has been proposed that the ability to repair a DNA DSB may be influenced by the presence of other minor types of damage in the immediate vicinity. Locally multiply damaged sites (LMDS), where a cluster of ionizations has caused a group of lesions close together, may result in a greater probability of loss of genetic material.

16.3 CELLS HAVE DEVELOPED A WIDE VARIETY OF REPAIR MECHANISMS FOR DEALING WITH THE MANY DIFFERENT TYPES OF DNA DAMAGE

Repair pathways appear to operate a repair hierarchy:

1. reversal of damage;
2. excision repair, for damage confined to a single strand;
3. repair for double-strand damage by recombination or end-joining.

These three mechanisms repair the diversity of lesions from small simple lesions to the restoration of large complex lesions. If the function of one component is defective, repair defaults to the next stage in the hierarchy.

The reversal of damage in DNA is the most direct mode of DNA repair. Examples of this mechanism of repair include:

- enzymatic photo-reactivation of pyrimidine dimers;
- repair of guanine alkylation;
- repair of single-strand breaks by direct rejoining;
- repair of sites of base loss by direct insertion.

For most of these examples a single gene is required for a specific function. At first sight, this seems an economical use of genetic information compared with the multiple gene products required for the more complex repair processes. However, since a large variety of DNA lesions are induced, this type of specific direct reversal of damage is likely to apply to those lesions that are produced frequently and spontaneously. These repair mechanisms maintain the integrity of the DNA sequence: a so-called 'housekeeping' function.

Excision repair

Excision repair involves cleavage of the sugar–phosphate backbone to remove the site of damage, which requires either single nucleotide excision or patch excision. The mechanism of damage recognition is unknown. Many have assumed that it is detected by the distortion to the double helix caused by the damage. Recognition of damage is followed by local unwinding of the DNA in the region of damage, which allows access of repair enzymes. Single nucleotide excision can occur by the following sequence:

• a glycosylase, which removes the base damage and leaves an apurinic or apyrimidinic (a.p.) site;
• an a.p. endonuclease which cuts 5' to the a.p. site;
• the baseless deoxyribophosphate residue is removed by a specific 5' phosphatase enzyme. This enzyme is distinct from the 5' to 3' exonuclease activity of DNA polymerase I.

Patch excision involves a different set of enzymes: protein complexes are involved in damage binding and cleavage as well as DNA unwinding. The damaged patch is removed and the space is resynthesized by the action of DNA polymerase, and joined by ligase (Figure 16.2(a)). The exonuclease activity of DNA polymerase makes the length of the re-synthesized patch considerably larger than the size of the patch initially excised.

Excision repair is a multistep process, and is less rapid than the single-step actions involved in the direct reversal of damage. The precise kinetics of excision repair are difficult to define, because, for the removal of pyrimidine dimers resulting from ultraviolet light, the rate of removal depends on whether the sequence is actively transcribed or even whether it is the sense or the antisense strand. There is now a well established connection between transcription and excision repair demonstrated by the genetic syndromes of xeroderma pigmentosum groups B, D and G. These diseases demonstrate ultraviolet light sensitivity and general impairment of transcription: they are linked by impaired unwinding of the DNA (i.e. the genes encode helicases) required for both transcription and repair.

Mismatch repair

Mismatch repair has certain similarities with excision repair, in that there is single-strand cleavage around the site of the mismatch. However, the proteins involved are separate from excision repair, and the recognition enzymes are mismatch-specific (Chapter 13). A mismatch binding protein has recently been isolated,

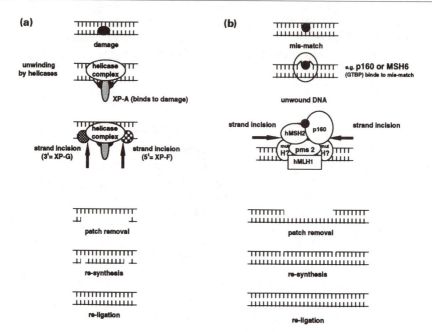

Figure 16.2 Removal of single-strand DNA damage has similarities to (**a**) nucleotide excision repair (NER) and (**b**) mismatch repair (MMR), illustrated here, in spite of using different proteins. The damage is detected by proteins which bind to the damage site. The protein is XP-A in NER, and a mismatch-specific binding protein (e.g. GTBP or p160 for G–T mismatches) in MMR. The DNA adjacent to the damage is unwound by helicases followed by strand incision in NER. In MMR the complex of associated proteins around the mismatch leads to strand incision. Single-strand patch removal, resynthesis and religation follow in both pathways.

which is a binding protein specific for G–T mismatches. The human homologues of the bacterial mismatch repair genes *MutS* and *MutL* (*hMSH* and *hMLH*) have shown remarkable evolutionary conservation of this repair pathway. Both the bacterial and human cells lacking functional *MutS* or *MutL* exhibit high rates of mutation. There are differences in the mechanism of cleavage between bacteria and humans: cleavage is directed opposite a methylated base in bacteria but is directed more by secondary structure of the DNA in eukaryotes (Figure 16.2(b)).

16.4 REPAIR OF DOUBLE-STRAND DAMAGE: REJOINING OR RECOMBINATION?

When double-stranded DNA damage is induced in cells, the first step in the removal of that damage is to 'clean up' the damaged region. As a minimum, this will involve the excision of at least one nucleotide from each strand, by the action of specific endonucleases. Double-stranded DNA damage is in close proximity on each strand (suggested by the clustering of ionizations) and excision repair of both strands will lead to a loss of sequence. Simple blunt-ended re-ligation of strand breaks is demonstrated *in vitro*, but this will not restore the sequence. Although ligase can rejoin breaks, the lack of this enzyme in yeast leads to hypermutability rather than clear radiosensitivity. Bloom's syndrome is a rare human disease characterized by ligase deficiency, early onset of cancer and hypermutability, but cell killing from radiation is only at the sensitive end of the normal spectrum. This suggests that simple rejoining may not be the dominant mechanism for closing double-strand breaks.

The association of radiation sensitivity with a lack of DNA recombination was initially made in the *rad52* mutant of the budding yeast, *S. cerevisiae*, but this association has now extended to mammalian cells with the observation of radiation sensitivity in *SCID* mice (severe combined immune deficiency as a result of impaired V(D)J recombination). Fibroblasts from *SCID* mice were also found to be deficient in double-strand break repair using two damage assays, pulsed field gel electrophoresis and neutral filter elution, thus associating double-strand damage processing with recombination.

Recombination in yeast is dominated by homologous recombination, i.e. DNA exchange occurs at sites of sequence similarity, which is often the homologous chromosome in diploid cells. By contrast, non-homologous recombination is the more frequent process in mammalian cells. Non-homologous recombination means that little or no apparent homology exists at the site of DNA strand exchange. It is suspected that non-homologous recombination involves short-sequence homology, perhaps as few as 2–6 base pairs (Figure 16.3).

Any recombination event will result in at least short sequences of heteroduplex DNA at the site of strand-exchange, and thus mismatch repair. Long tracts of heteroduplex are therefore likely to result in non-viable cells.

16.5 REPAIR-DEFICIENT MUTANTS: IDENTIFICATION OF GENES AND UNDERSTANDING FUNCTION

Progress in resolving rejoining or recombination will come from understanding the function of repair proteins. The study of mammalian repair-deficient mutants has led to substantial progress in the understanding of repair of DNA damaged by ionizing radiation. It is thought that radiosensitive mutants lack one or more key components of the repair process. If two different mutants lack different components of repair, a cell formed by their fusion will lack neither repair component. They will have 'complemented' each other and the cells will be radioresistant. If the same component is missing from both mutants, complementation with cell fusion will not occur.

Ionizing-radiation-sensitive mutants of rodent cells form at least nine complementation groups. The human DNA repair gene that complements the EM9 mutant of AA8 CHO cells was the first to be isolated and called *XRCC1* (**X**-ray sensitivity **c**ross-**c**omplementing). It also corrects the high frequency of sister-

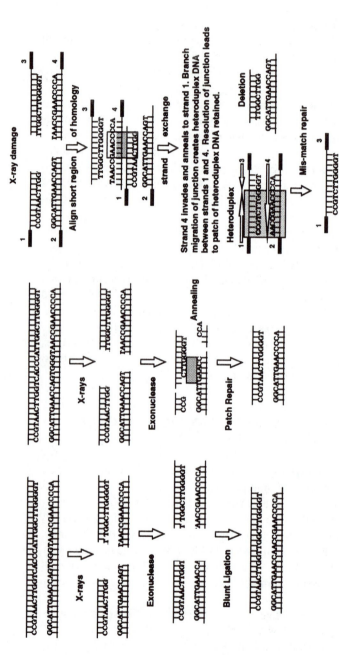

Figure 16.3 Possible mechanisms in double-strand break repair. In (a), exonuclease and blunt-end ligation is shown. This is unlikely to be the dominant method of repair, since ligase deficiency does not lead to extreme radiosensitivity. (b) Exonuclease digestion may reveal short sequences available for strand annealing (GAACC to CTTGG). (c) Recombination between the same short sequence homology will also allow double-strand break closure. This generates a small deletion and a retained short region of heteroduplex DNA which is followed by mis-match repair. The end result is only 2 bp different from strand annealing. This method is favoured by models in which the exposed termini are protected by DNA end-binding proteins.

chromatid exchanges and single-strand break rejoining defect in EM9. *XRCC1* appears to act as a co-factor for DNA ligase III, but how this corrects radiation sensitivity to ionizing radiation is not yet clear.

The double-strand-break-rejoining-deficient mutants *xrs*, XR-1 and V3 were found to be V(D)J recombination deficient, like SCID cells. When *xrs* was found to lack a protein that bound to double-strand DNA ends using a gel-mobility assay, and mobility was further reduced by an antibody to Ku antigen, the gene (*XRCC5*) was cloned as the 80 kDa protein subunit of Ku antigen. This protein forms part of a protein complex: the DNA-dependent protein kinase (DNA-PK or SP-1 kinase). It seemed highly likely that other components of the complex may well be other *XRCC* genes (Figure 16.4).

The p350 subunit (now commonly referred to as DNA–PK$_{CS}$, the catalytic subunit of the protein complex) is deficient in the V3 hamster mutant (*XRCC7*) and mouse SCID cells, and the other Ku protein subunit (70 kDa) is deficient in a previously unknown mutant, *sxi1* (*XRCC6*). Both Ku subunits have DNA end-binding activity; the complex of three proteins has kinase activity, but the *in-vivo* substrate of the kinase is not yet determined. The protein complementing the XR-1 mutant, which is also double-strand-break-rejoining and V(D)J-recombination-deficient, has not yet been isolated.

The four remaining *XRCC* proteins (2, 3, 8 and 9) have no identified sequence or functional defect established to date. Certain features of the *irs*1 mutant (*XRCC2*) are similar to ataxia telangiectasia (AT), but there is no established genetic link. AT cells are X-ray-sensitive, have impaired signalling of DNA damage, suggested by the lack of cell-cycle arrest and the impaired induction of p53 in response DNA damage, and also show abnormalities of DNA recombination. The pleiotropic AT phenotype is at the centre of many critical DNA damage and signalling pathways. Considerable interest was aroused recently when the AT gene

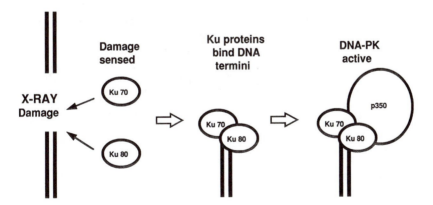

Figure 16.4 Double-strand break repair proteins. Three of the proteins involved in double-strand break rejoining have been identified as being part of a DNA-dependent protein kinase complex which is formed at the site of the DNA break. The important substrates for DNA-PK and other associated proteins are as yet unknown.

was cloned. It was found to have features in common with a phosphatidyl-inositol 3'-kinase (a known signal transduction intermediate) and with the fission yeast *S. pombe rad3* and *mec1* genes (cell-cycle checkpoint mediators). To everyone's surprise, a single gene was found to be mutated in all the previously established AT complementation groups: the reasons for this remain unclear.

16.6 CONCLUSIONS

The most rapid progress in understanding the molecular biology of DNA repair has come from identifying functional groups of proteins. Transcription coupled repair of single-strand ultra-violet-light-induced damage was demonstrated by xeroderma pigmentosum (XP) cells types B, D and G. Defective helicase function leads to both excision repair defects and transcription failure. The identification of other proteins in the transcription/repair complex has led to the complete reconstitution *in vitro* of nucleotide excision repair by purified proteins. The proteins involved in the repair of DNA damage caused by ionizing radiation are involved in a complex mediating a certain type of DNA rejoining/recombination, and may not be specific to double-strand breaks in DNA. Identifying other proteins involved in this complex and the actions performed by the complex will be areas of intense interest over the next few years.

16.7 FURTHER READING

Friedberg, E. C., Walker, G. C. and Siede, W. (1995) *DNA Repair and Mutagenesis*, ASM Press, Washington, DC.

Haber, J. E. (1992) Exploring the pathways of homologous recombination. *Curr. Opin. Cell Biol.*, **4**, 401–412.

Savitsky, K., Bar-Shira, A., Gilad, S. *et al*. (1995) A single ataxia-telangiectasia gene with a product similar to PI-3 kinase. *Science*, **268**, 1749–1753.

Taccioli, G. E., Gottlieb, T. M., Blunt, T. *et al*. (1994) Ku80: product of the *XRCC5* gene and its role in DNA repair and V(D)J recombination. *Science*, **265**, 1442–1445.

Weinert, T. A. and Hartwell, L. H. (1988) The *RAD9* gene controls the cell cycle response to DNA damage in *Saccharomyces cerevisiae*. *Science*, **241**, 317–322.

Chemotherapeutic drug-induced DNA damage and repair

Robert Brown

17.1 DNA IS AN IMPORTANT TARGET FOR CHEMOTHERAPEUTIC DRUGS

The basis of chemotherapy is that drugs are selectively more toxic to the tumour than to the host. As our understanding of the biochemical differences between normal and tumour cells increases, the possibility of rationally designing drugs or therapies targeted to tumour-specific biochemical pathways becomes feasible. However, most existing clinically active drugs have been discovered by a combination of random screening and serendipity. In the National Cancer Institute (USA) screening programme for new antitumour drugs, it was estimated that over a 30-year period more than 370 000 compounds had been screened but only eight drugs had proved of clinical benefit. This leads to the questions – what are the cellular targets of the successful drugs that make them effective antitumour agents, and can this knowledge be used to aid the design of novel drugs or to improve therapeutic strategies with existing agents?

Many effective chemotherapeutic drugs act indirectly or directly on DNA and DNA-metabolizing enzymes. Drugs that cause direct damage to DNA include:

- platinum coordination complexes, such as *cis*-diamminedichlorplatinum(II) (cisplatin). These compounds form bidentate adducts, with the platinum moiety bridging adjacent bases on the same DNA strand (intrastrand) or bases on opposite strands (interstrand) (Figure 17.1).

Molecular Biology for Oncologists. Edited by J. R. Yarnold, M. R. Stratton and T. J. McMillan. Published in 1996 by Chapman & Hall, London. ISBN 0 412 71270 9

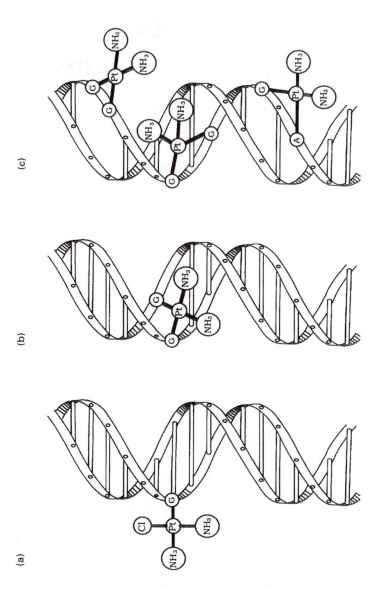

Figure 17.1 Schematic diagram of cisplatin–DNA adducts. **(a)** Monofunctionally bound; **(b)** interstrand crosslink; **(c)** intrastrand crosslinks.

- alkylating agents, such as nitrosoureas, mitomycin C, cyclophosphamide, melphalan, chlorambucil and busulfan. These compounds form monofunctional or bifunctional covalent bonds between the carbon of an alkyl moiety and a nucleophilic base of DNA. Bifunctional agents can give rise to intrastrand and interstrand crosslinks.

Indirect damage to DNA can occur by:

- blockage of DNA synthesis by nucleotide analogues such as 5-fluorouracil or cytosine arabinoside, or inhibition of dihydrofolate reductase by methotrexate;
- generation of free radicals which cause base damage and DNA strand breaks, e.g. bleomycin and quinones such as mitomycin C or doxorubicin;
- interference with proteins involved in DNA function, e.g. anthracycline-induced stabilization of the DNA topoisomerase II cleavable complex.

Many studies have correlated DNA damage induced by chemotherapeutic drugs with cellular toxicity. Tumour cells may be more susceptible than normal cells to the effects of these agents because of increased initial levels of DNA damage, decreased removal and repair of the drug–DNA lesions, or decreased tolerance of DNA damage. This chapter describes approaches to the measurement of drug-induced DNA damage and repair in tumour cells, and how this may influence future cancer chemotherapy.

17.2 MEASUREMENT OF DRUG-INDUCED DNA DAMAGE

Measurement of DNA damage at a cellular level is complicated by:

- the variety of lesions that result from exposure to any one agent;
- the heterogeneity of DNA damage throughout a cell population;
- the heterogeneity of damage throughout the genome.

Useful information regarding the amount of drug reaching the DNA can be obtained by measurement of the amount of drug bound to DNA following drug exposure. Assays exist for classical alkylating agents (gas chromatography), platinum coordination complexes (atomic absorption spectrometry, AAS) and intercalating agents (high performance liquid chromatography, HPLC). Although the levels of drug measured in this way often correlate with cytotoxicity, they provide no information about the specific DNA lesions that cause cell death or the repair of drug-induced DNA damage.

17.3 MANY CYTOTOXICS INDUCE DNA STRAND BREAKS

DNA filter elution methods have proved extremely useful in the identification and quantification of specific DNA lesions, including single- and double-strand breaks,

DNA–protein crosslinks and interstrand DNA crosslinks. In this technique, cells are lysed on a filter, most RNA and protein is removed by washing, and DNA is eluted by pumping buffer through the filter. The basic principle is that the rate of elution depends on the length of the DNA fragments and the nature of the lesion detected is determined by the characteristics of the lysis and elution buffers. DNA double-strand breaks are measured when the elution buffer has a neutral pH because the double-strandedness of DNA is maintained. Alkaline elution buffer denatures the DNA so that single-strand breaks are assessed.

Other assays frequently used for measuring double-strand DNA breaks are pulsed-field gel electrophoresis and the single cell gel electrophoresis (Comet) assay. Both assays depend on increased electrophoretic mobility of DNA due to changes in DNA fragment length which result from drug-induced double-strand DNA breaks. Pulsed-field gel electrophoresis measures changes in the entire cell population while the Comet assay has the advantage of analysing changes at a single cell level.

Double-strand breaks in DNA have a crucial role in the induction of cellular responses which lead to cell cycle arrest and apoptotic death. This has been demonstrated by specifically creating double-strand breaks by microinjection of restriction enzymes into cells. It has been suggested that the action of many chemotherapeutic drugs depends on the induction of double-strand breaks and drug-induced adducts may need to be converted into a double-strand break by DNA replication or aberrant repair.

17.4 MEASUREMENT OF SPECIFIC DRUG–DNA LESIONS IS SOMETIMES POSSIBLE

DNA damage induced by the platinum coordination complex cisplatin has been well characterized, and methods for the detection of specific cisplatin–DNA adducts have been developed. Drug-damaged DNA is enzymatically digested to produce a mixture of mononucleotides and cisplatin–dinucleotide adducts, which can be separated by HPLC. More recently, monoclonal antibodies have been raised against cisplatin–DNA adducts. These have been used to detect adducts at the cellular level by immunocytochemistry. This offers a simple means of examining both the frequency and the distribution of DNA lesions throughout the malignant and non-malignant cells in a solid tumour after treatment with cisplatin. Because of the practical difficulty of obtaining serial tumour biopsies in patients, cisplatin–DNA adducts have been measured in more accessible tissue samples (peripheral leucocytes and buccal smears) during and after chemotherapy. Although correlations have been made between the number of adducts per normal cell and tumour response, this is probably a rather complex method of measuring drug exposure and does not assess the DNA repair capacity of the tumour.

Several electrophoretic and chromatographic techniques provide sensitive methods for the detection and separation of DNA containing interstrand crosslinks.

These methods rely on the fact that crosslinks prevent the complete denaturation of the two complementing DNA strands by heat or alkali and the DNA will readily reanneal. Denatured single-stranded DNA can then be separated from double-stranded DNA that has reannealed as a result of an interstrand crosslink. DNA damage induced by crosslinking agents and its repair can be examined, at the level of specific genes in drug-treated cells, using Southern blot detection (Figure 17.2).

This general approach can also be used to detect certain types of DNA adducts and their repair. Enzymes are used that cleave DNA specifically at the site of the DNA adduct, converting it into a single-strand DNA break. Levels of single-strand breaks can then be detected by alkaline agarose gel electrophoresis and Southern blot detection (Figure 17.2).

The presence of damaged bases in DNA can inhibit the action of DNA polymerases which require correctly base-paired DNA as a substrate. Damage in DNA inhibits the activity *in vitro* of the polymerase used in polymerase chain reaction (PCR) assays. This inhibition of the PCR reaction is a highly sensitive method of measuring DNA damage *in vitro* and even in cells. Since the position in the DNA where the polymerase stops can be identified, this allows analysis of the specific bases in DNA that the drugs interact with.

17.5 CELLS VARY IN THEIR RESPONSE TO DRUG-INDUCED DNA DAMAGE

It has been postulated that the drug sensitivity of certain tumour types may be due to deficiencies in DNA repair in these cells, acquired during development of the tumour. Evidence for the importance of DNA repair pathways in carcinogenesis comes most directly from inherited tumour-prone human disorders, in which cellular hypersensitivity to radiation and DNA damaging drugs is associated with defects in DNA repair. Although these syndromes are themselves rare in the general population, heterozygous carriers have been estimated at frequencies of 0.1–1% and an increased risk of cancer has been suggested for some of these heterozygous carriers. Further evidence for a possible involvement of DNA repair in carcinogenesis comes from chemical carcinogenesis models in animals, where lack of repair of adducts may allow a premutagenic lesion to persist long enough to be fixed upon DNA replication. If defective DNA repair processes are involved in tumour development, identification of the particular defect in tumours and hence the DNA damage sensitivity could have important implications for the choice of chemotherapy treatment regimes.

Defective DNA repair is also implicated in microsatellite instability, which is a feature of some tumour cells (Chapter 13). The mammalian genome is punctuated with repetitive nucleotide sequences or microsatellites. These microsatellite sequences are normally replicated with high fidelity. However, DNA polymerases can increase or decrease the length of the microsatellite sequence by slippage of one DNA strand relative to the other, resulting in a frameshift mutation of the

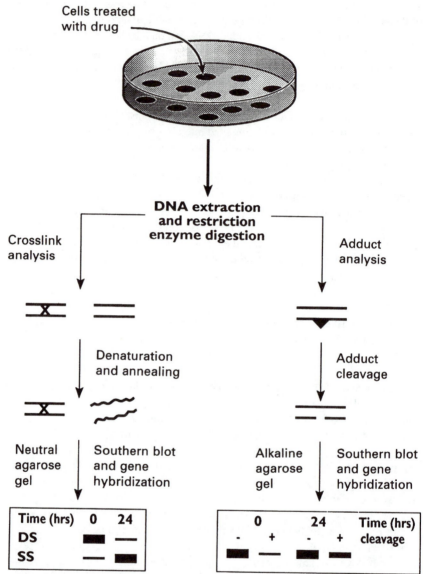

Figure 17.2 Detection of gene-specific DNA damage and repair. Cells are treated with DNA-damaging drug and total DNA is extracted and digested with restriction enzyme. For crosslink analysis (left) the DNA is denatured and allowed to reanneal rapidly. Only interstrand crosslinked DNA will reanneal correctly. The number of single-strand fragments present after reannealing will increase with repair of interstrand crosslinks. DS = double-stranded DNA fragment containing interstrand crosslinks; SS = single-stranded DNA fragment. The increased intensity of the SS DNA at 24 hours represents repair of crosslinks. For adduct analysis (right), lesion-specific single-strand breaks are created at the sites of damage. DNA fragments are resolved in an alkaline denaturing gel and

microsatellite. High levels of such mutations have been observed in hereditary non-polyposis colorectal cancer (HNPCC) and in many sporadic tumours. Since this microsatellite instability is believed to be due to replication errors, such cells have been termed RER$^+$. The genes responsible for this phenotype have now been identified and are homologous to mismatch repair genes in bacteria. A variety of such mismatch repair genes, including *MLH1* and *MSH2*, have now been shown to be mutated in tumours of RER$^+$ phenotype. Loss of activity of these genes in tumour cell lines has been shown to confer increased tolerance to DNA-damaging agents such as monofunctional alkylating agents, which may have important implications for resistance of tumour cells to such anticancer agents. At present it is not clear why decreased repair should lead to increased tolerance and resistance. One possibility is that defects in mismatch repair lead to a reduced fidelity of DNA replication. This in turn allows the replication complex to bypass the lesion in the DNA, leading to tolerance of DNA damage.

As techniques for separating genomic DNA into functionally distinct subfractions have been developed, it has become apparent that DNA damage and repair are not homogeneous throughout the genome. Active chromatin has certain structural features, such as a lower level of DNA condensation, which appear to render it more susceptible to cytotoxic damage. It has also recently been shown that repair of DNA damage is non-homogeneous; for example regions of the genome containing transcriptionally active genes are repaired more efficiently than the genome as a whole. The importance of repair of transcriptionally active genes is reflected in Cockayne's syndrome, where failure of preferential repair leads to neurodegeneration and photosensitivity but not, interestingly, to an increase frequency of malignancy.

Analysis of DNA repair in tumours is not straightforward. Many of the assays used to examine repair and cellular sensitivity of cell growth in tissue culture are not applicable to tumours. Even when cell lines from tumours are available, it is impossible to compare the tumour with its normal cell of origin. Therefore much of the work on DNA repair and response of tumours to chemotherapy is based on the analysis of alterations in cell lines that have been selected for acquired resistance to chemotherapeutic drugs.

17.6 DRUG RESISTANCE MAY BE DUE TO ENHANCED DNA REPAIR

Evidence for increased DNA repair as a mechanism of chemotherapeutic drug resistance has been shown in cisplatin-resistant cells. Resistant cells selected by chronic exposure to cisplatin have been shown to have increased unscheduled

---

probed for the gene of interest by Southern analysis. The reduction in intensity of the band observed at 0 hours after treatment (+) represents the presence of adducts in the DNA and the increase in intensity of the band at 24 hours after treatment (+) represents the repair of damage.

DNA synthesis, i.e. DNA synthesis that is part of patch-filling during repair after exposure to drug. Some resistant cells also remove the major cisplatin-induced lesion (the Pt-GG intrastrand crosslink) more rapidly than do sensitive cells. Similarly, cisplatin-sensitive testicular tumour cells appear less proficient in the removal of adducts than resistant bladder tumour cells. One disadvantage of those types of repair assay that treat cells with cisplatin is the unknown effects of the drug on cellular components other than DNA, which could affect the cellular repair machinery. An assay system that provides information concerning the cellular responses to damaged DNA, without the confounding effects of other damaged cellular molecules, is the plasmid reactivation assay (Figure 17.3).

In this assay plasmid DNA is treated *in vitro* with cisplatin and then introduced into the cells of interest. The plasmid DNA contains a 'reporter' gene whose expression can easily be quantified. Platination of the DNA inhibits expression of this reporter gene, but after transfection into repair-proficient cells, removal of the platinum adducts increases reporter gene expression. Increased reactivation of a

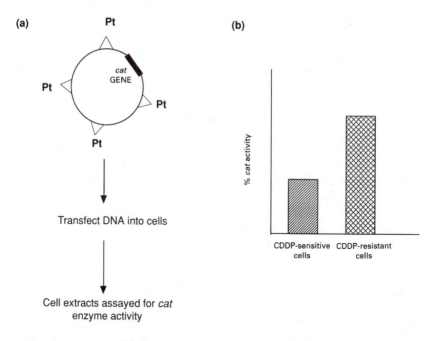

Figure 17.3 The plasmid reactivation assay. **(a)** A plasmid containing a reporter gene such as the *cat* gene, which encodes the enzymes chloramphenicol acetyl transferase (CAT), is treated with cisplatin. After transfection of undamaged or cisplatin-treated plasmid into cells, the cells are grown to allow the expression and repair of the *cat* gene. **(b)** CAT activity can then be measured in extracts from the cells and can be compared between cisplatin-sensitive and -resistant lines.

cisplatin-treated reporter gene has been observed in cisplatin-resistant cells, again supporting the involvement of enhanced DNA repair in cisplatin resistance.

While the types of assay described above can be used to show increased removal of damaged DNA in cell lines, they cannot readily be used to examine repair levels in tumour biopsies. The relevance of DNA repair mechanisms to clinical drug resistance can only be assessed when suitable means of assaying repair in tumour biopsies become available. Possible approaches include the identification of repair genes and measurement of their expression in tumours, or the measurement of specific biochemical activities in tumour cell extracts.

17.7 SOME PROTEINS INVOLVED IN THE RESPONSE TO DNA DAMAGE HAVE BEEN IDENTIFIED

DNA repair synthesis assay

DNA excision repair by cell extracts can be measured *in vitro* by the incorporation of radiolabelled nucleotides into damaged DNA (Figure 17.4).

This assay has been used to measure repair of ultraviolet-light-, acetylaminofluorine- and cisplatin-induced DNA damage by human cell extracts. Extracts from xeroderma pigmentosum (XP) cells, known to be defective in this type of DNA repair, have been shown to be deficient in repair synthesis activity. The assay has been used to identify and purify some of the proteins required for excision repair and aided the understanding of the molecular mechanisms of nucleotide excision repair in human cells. This information may be useful for designing drugs that can target specific proteins and inhibit DNA excision repair and there is obvious potential to use this assay to analyse the repair activity of resistant tumour cells.

Damage-recognition proteins

Protein–DNA complexes have decreased mobility compared with either DNA or protein alone under appropriate electrophoresis conditions. Damaged oligonucleotides are retarded in a gel when proteins are attached (Figure 17.5(a)). Western blots of proteins can be probed with damaged oligonucleotides to allow the formation of DNA–protein complexes (Figure 17.5(b)).

Damage recognition proteins (DRPs) binding to UV- and cisplatin-damaged DNA have been identified in cell extracts. A cisplatin-resistant cell line that has increased levels of damaged plasmid DNA reactivation has overexpression of a UV-DRP. The cisplatin-DRP have been shown to bind only to the Pt-GG intrastrand crosslink, but not to interstrand crosslinks or monofunctional forms characteristic of the inactive *trans*-DDP. This specificity of interaction may implicate cisplatin-DRP in the antitumour activity of the platinum coordination complexes.

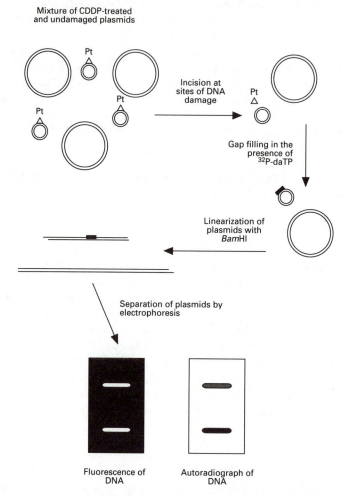

Figure 17.4 The repair synthesis assay. Undamaged and cisplatin-treated plasmids of different sizes are incubated with cell extract in the presence of radiolabelled dATP. Following the repair reaction, plasmids are separated by agarose gel electrophoresis and monitored for incorporation of labelled dATP to detect damage-dependent repair synthesis.

Alkyltransferases

Alkylation of DNA by chemotherapeutic drugs such as CCNU and BCNU can be directly reversed by the transfer of the alkyl moiety via an alkyltransferase from the alkylated base to a cysteine residue in the active site of the enzyme. Alkyltransferases are thus suicide proteins, irreversibly inactivated after binding an alkyl group. Cells which are deficient in O^6-alkylguanine alkyltransferase are

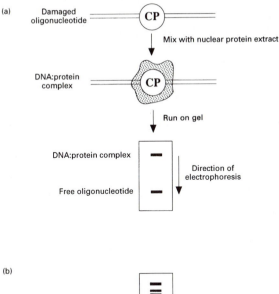

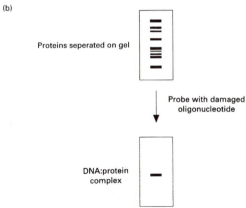

Figure 17.5 Assays of DNA-damage-recognition proteins. **(a)** Mobility shift assay for cisplatin-DNA binding proteins. Oligonucleotide DNA is treated *in vitro* with cisplatin to produce cisplatin–DNA adducts. After labelling with ^{32}P-, the oligonucleotide is incubated with cell extract and the ability of proteins to bind to the damaged DNA can be assessed by gel electrophoresis and autoradiography. Proteins binding to the DNA are observed as a slow migrating retardation complex. If they are damage specific, they bind to the damaged DNA but not to undamaged DNA. **(b)** Southwestern assay of cells. Cell extracts are separated by polyacrylamide gel electrophoresis and transferred to nitrocellulose. The nitrocellulose blot is incubated with labelled oligonucleotide, either undamaged or containing cisplatin–DNA adducts. Proteins binding to the DNA can be assessed by gel electrophoresis and autoradiography. Damage recognition proteins should only be observed on the blot incubated with the cisplatin-treated DNA.

unable to repair methylation of the O^6 position of guanine. These are termed mer⁻ (**m**ethyl **e**xcision **r**epair deficient), while cells that are methyl excision repair proficient are termed mer⁺. Mer⁻ cells are hypersensitive to simple methylating, ethylating and chloroethylating agents, in comparison to mer⁺ cells. However, human tumours have generally been found to express the mer⁺ phenotype and evidence correlating enhanced alkylation repair in human tumours and alkylating agent treatment failure is scanty. It is considered likely that resistance to the methylating agent dacarbazine and the chloroethylnitrosoureas (BCNU, CCNU and methyl-CCNU) is due to enhanced repair by alkyltransferase.

DNA polymerases

There are at least five mammalian DNA polymerases. Since polymerase-β (pol-β) can be induced by some DNA damaging agents and its activity is independent of the replicative state, it has been assigned a DNA repair function. Pol-β activity has been shown to be elevated in cisplatin-resistant P388 leukaemia cells that have elevated levels of excision repair; Pol-α and -β levels are elevated in certain cisplatin-resistant colonic and ovarian carcinoma cell lines.

17.8 ARRESTING CELLS DURING PROLIFERATION MAY ALLOW MORE REPAIR TIME

DNA-damaging agents can induce arrest of proliferating cells at several points in the cell cycle. In particular DNA damage can induce a G1 arrest prior to S-phase and DNA replication, as well as a G2 arrest prior to mitosis. Studies of *RAD9* mutants of *Saccharomyces cerevisiae* have led to the concept that signal pathways are present in cells which will sense DNA damage and check progression through the cell cycle. This may allow the opportunity for repair of potentially lethal damage. Indeed, cells with mutations of the *RAD9* gene are defective in DNA-damage-induced arrests and are hypersensitive to DNA-damaging agents. Thus it has been hypothesized that tumours that lack such damage-induced arrest would be hypersensitive to DNA-damaging agents and that resistance could develop if cells regained or increased damage-induced arrest, allowing more repair time.

The G1 arrest induced in mammalian cells by anticancer agents such as ionizing radiation and topoisomerase II inhibitors has been shown to be dependent on functional activity of the tumour suppressor gene product p53. In certain tumour cell types loss of p53 function and loss of G1 arrest can cause increased sensitivity to chemotherapeutic agents; however, in other tumour cells loss of p53 function correlates with no change in sensitivity or increased resistance. Increased resistance associated with loss of p53 function is probably due to the requirement for p53 in DNA damage-induced apoptotic cell death. These observations exemplify how the effect of a given gene on drug resistance will be dependent on the expression of other genes in the cell, which in turn will depend on the cell or tumour type.

17.9 BLOCKING DNA REPAIR MAY SENSITIZE CELLS TO CYTOTOXICS

If increased DNA repair plays a role in drug resistance of tumours, inhibitors of DNA repair may have the potential to improve responses to chemotherapy. Methylxanthines such as caffeine have been shown to inhibit DNA excision repair, but also exert other effects that enhance the cytotoxicity of DNA-damaging agents: cells are prevented from arresting at the G2 phase of the cell cycle, possibly reducing the time available for DNA repair, and postreplication DNA repair is inhibited. These agents can enhance the cytotoxicity of DNA-damaging agents *in vitro*, but at present there is nothing to suggest they will improve the therapeutic ratio of drugs such as cisplatin.

Excision repair can also be inhibited by blocking the repair polymerization step. Two agents that may inhibit DNA repair polymerases are hydroxyurea and cytosine arabinoside (ara-c). Hydroxyurea inhibits ribonucleotide reductase, leading to depletion of dexoyribonucleotide triphosphates, the precursors required for DNA synthesis. Ara-C causes chain termination when incorporated into DNA. The combination of hydroxyurea and ara-C has been shown to enhance the cytotoxicity of cisplatin in a human colon cell line. Aphidocolin, an inhibitor of DNA polymerase-α and -δ, has been shown to inhibit repair of cisplatin-DNA adducts and to potentiate the toxicity of cisplatin in a human ovarian cell line.

Post-translational modification by ADP ribosylation is a ubiquitous mechanism to regulate enzyme activity. It has been shown that ADP ribosylation participates in excision repair. DNA ligase activity, involved in the final step in excision repair, is in part dependent upon post-translational modification by ADP ribosylation, and many other proteins involved in excision repair may also undergo ADP ribosylation. Inhibition of ADP-ribosylation, with 3-aminobenzaminidine or nicotinamide, can sensitize murine tumours to cisplatin.

While there remains uncertainty as to the role of DNA repair in determining the response of tumours to chemotherapy, it seems likely that an increased understanding of the biochemistry of DNA repair and its regulation will facilitate the design of novel and effective chemotherapeutic strategies.

17.10 FURTHER READING

Aboussekhra, A. *et al.* (1995) Mammalian DNA nucleotide excision repair reconstituted with purified protein components. *Cell*, **80**, 859–868.

Burt, R. K., Poirier, M. C., Luk, C. J. and Bolr, V. A. (1991) Antineoplastic drug resistance and DNA repair. *Ann. Oncol.*, **2**, 325–334.

Epstein, R. J. (1990) Drug-induced DNA damage and tumour chemosensitivity. *J. Clin. Oncol.*, **8**, 2062–2984.

Lowe, S. W., Riley, H. E., Jacks, T. and Honsman, D. E. (1993) p53-dependent apoptosis modulates the cytotoxicity of anticancer agents. *Cancer Res.*, **54**, 3500–3505.

Sancar, A. (1994) Mechanisms of DNA excision repair. *Science*, **266**, 1954–1956.

18 | Molecular basis of drug resistance

Sally L. Davies and Ian D. Hickson

18.1 VARIED MOLECULAR MECHANISMS ACCOUNT FOR INTRINSIC AND ACQUIRED DRUG RESISTANCE

Although there have been major advances in the treatment of certain cancers, most notably testicular teratomas, childhood leukaemias and Hodgkin's disease, the common adult solid tumours remain largely incurable. This failure of conventional radiation and cytotoxic chemotherapy reflects one or two major obstacles: *de novo* (also called intrinsic) resistance and acquired resistance. In many cases, the radio- and chemosensitivity of a particular tumour type reflects the cell type/tissue from which the tumour arose. Thus, pancreatic carcinoma and hypernephroma cells are, perhaps not surprisingly, as refractory to killing by chemotherapeutic agents as is the normal pancreatic and kidney tissue. However, this is not the complete story, since many tumours show a striking response to anticancer therapy, with the observation of a $\geq 90\%$ tumour cell kill not uncommon during a course of therapy. The problem under these circumstances is not one of *de novo* resistance, but of acquired resistance. Upon recurrence, these tumours frequently exhibit a completely different phenotype when challenged with drugs previously shown to elicit an antitumour effect. At this time, the tumour appears refractory not only to the previously used drug regimes, but also to agents that it has not previously been exposed to. This commonly observed clinical phenomenon reveals two aspects of acquired resistance to therapy. Firstly, the resistance mechanisms must be readily 'activated' by tumour cells during therapy. Secondly, cells possess the ability to acquire resistance to classes of drug not previously encountered. Although we shall use the term 'acquired resistance' to reflect this phenotypic change in the

Molecular Biology for Oncologists. Edited by J. R. Yarnold, M. R. Stratton and T. J. McMillan. Published in 1996 by Chapman & Hall, London. ISBN 0 412 71270 9

tumour population as a whole, it is possible, of course, that the change arises in the tumour as a result of a selection imposed by the therapy for those pre-existing clones that are intrinsically drug-resistant.

The molecular mechanisms of drug resistance are as varied as the agents used to attack tumour cells. In this chapter, we shall address three basic strategies used to prevent chemotherapeutic agents performing their prescribed role:

- an increased ability to extrude drug from cells;
- the detoxification of drugs by the overexpression of cytoplasmic drug metabolizing enzymes;
- the manipulation of nuclear proteins to prevent them acting as suitable targets for anticancer drugs.

These general strategies are depicted diagrammatically in Figure 18.1.

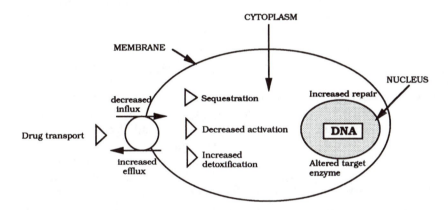

Figure 18.1 General strategies mediating drug resistance.

18.2 DRUG RESISTANCE CAN BE MEDIATED BY ALTERED DRUG EXTRUSION FROM CELLS

As outlined above, many drug-resistant tumours, or tumour cell lines *in vitro*, are resistant not only to the agents to which they have been exposed previously, but also to a wide variety of structurally and functionally distinct compounds. This phenomenon is termed multidrug resistance (MDR). The range of drugs that the MDR phenotype is relevant to is listed in Table 18.1. Most of these agents are natural compounds, or based on natural compounds, and are generally lipophilic.

As is frequently the case in research, several groups simultaneously identified, using different technologies, the molecular basis for the multidrug-resistant phenotype. It was known that cell lines exhibiting the MDR phenotype overexpress a membrane glycoprotein of molecular weight 170 000 termed the P-glycoprotein. The cloning and sequencing of cDNAs encoding this protein revealed that the P-

Table 18.1 Drugs associated with multidrug resistance – drugs transported by P-glycoprotein

- Anthracyclines, e.g. epirubicin, doxorubicin
- Vinca alkaloids, e.g. vincristine, vinblastine
- Etoposide
- Mitoxantrone
- Actinomycin D
- Colchicine
- Methotrexate
- Taxol

glycoprotein was closely related to a family of membrane proteins involved in the transport of a wide variety of substrates. A characteristic feature of this family is that they use the energy of ATP hydrolysis to transport molecules against a concentration gradient. Amongst the most notable members of this family are the Ring 4/11 peptide transporter of the human major histocompatibility complex, and the CFTR protein defective in cystic fibrosis patients. The P-glycoprotein is predicted to contain two transmembrane domains each consisting of six α-helical segments that span the membrane and an intracellular ATP binding domain (Figure 18.2). It is not clear at this stage whether the substrate for P-glycoprotein is a drug molecule located in the cytoplasm or in the membrane itself.

The MDR phenotype is generally associated with overexpression of the P-glycoprotein, although this is not a universal feature. As well as simple overexpression, it is possible for cells to acquire a form of MDR phenotype via mutation of the *MDR1* gene encoding P-glycoprotein. In many cases, these mutations confer resistance to only certain classes of drug. For example, one mutation confers selective resistance to colchicine.

A wide variety of agents have been identified that interfere in some way with drug transport mediated by the P-glycoprotein and therefore are able to reverse MDR. The most notable of these agents are listed in Table 18.2.

At least some of the MDR reversing agents are substrates themselves for the P-glycoprotein and can, therefore, be considered as direct inhibitors of the drug transport process. In other cases, however, the mechanism of action remains a mystery, although interference with post-translational modification of P-glycoprotein could be relevant in certain cases. The cAMP-dependent protein kinase (also called protein kinase A) has been implicated in the regulation of P-glycoprotein function.

In several MDR cell lines, there is no overexpression of P-glycoprotein: instead, a second member of the ATP-dependent membrane transporter family, which shows strong primary sequence homology to P-glycoprotein, is overexpressed. This protein is termed the multidrug-resistance associated protein or

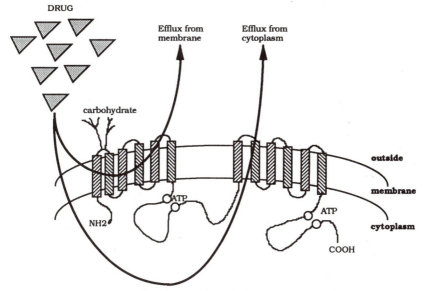

Figure 18.2 Structure of P-glycoprotein. Diagrammatic representation of two possible mechanisms of drug efflux by P-glycoprotein: extrusion from the cytoplasm or from the membrane. P-glycoprotein is depicted with two domains, each of six transmembrane segments (hatched), with an associated cytoplasmic ATP binding domain.

Table 18.2 MDR-reversing agents

- Calmodulin inhibitors
- Calcium channel blockers (e.g. verapamil)
- Local anaesthetics
- Cyclosporins
- Reserpine
- Quinidine and quinine
- Tamoxifen

MRP. Although the *MRP* gene was identified in multidrug-resistant lung cancer cell lines, recent evidence suggest that this putative drug transporter can generate an MDR phenotype in cells of several different tissue origins. Transfection of an expressing cDNA clone for MRP into cell lines confers an MDR phenotype. The characterization of MRP is not as well advanced as that of P-glycoprotein, but early indications suggest that MRP may be relevant to acquired drug resistance *in vivo*.

18.3 DRUG RESISTANCE CAN BE MEDIATED BY ALTERED DRUG DETOXIFICATION

The most important pathway for acquired drug resistance to many chemotherapeutic agents is an alteration to the cellular capacity to detoxify and/or sequester the drug and consequently render it harmless. The majority of the drug detoxification proteins are located in the cell cytoplasm, although nuclear detoxification can occur. The tripeptide glutathione (GSH; γ-glutamyl-cysteine-glycine) plays a central role in the detoxification of cisplatinum and many alkylating agents such as chlorambucil, melphalan and cyclophosphamide (Figure 18.3).

Many of these agents require some form of metabolism to generate the ultimate cytotoxic species. Indeed, it is the cytotoxic metabolites of the parent drugs which are generally thought to be the substrates for detoxification by the GSH pathway.

The cysteinyl group of GSH reacts spontaneously at a measurable rate at physiological pH with the electrophilic metabolites of alkylating agents and cisplatinum. GSH is present at millimolar concentrations in most human cells and therefore GSH alone could be a significant detoxification agent. However, it appears the GSH does not work alone. Instead, a family of cytosolic enzymes called the glutathione-S-transferases (GSTs) cooperate with GSH to detoxify many drugs. Different GST isozymes generally have different substrate specificities, although a degree of overlap in substrate range is evident amongst the different GSTs. For example, overexpression of the α-class GSTs is implicated in conferring resistance both to cyclophosphamide and to bifunctional nitrogen mustards

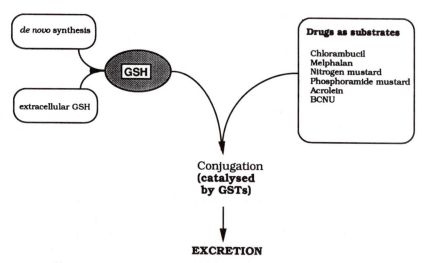

Figure 18.3 Cytoplasmic detoxification by GSH. GSH is formed from amino acid precursors, which can arise from normal cellular pools or from the breakdown of extracellular GSH. Electrophilic drugs bind to GSH in a reaction catalysed by GSTs. Following further metabolism, the water soluble metabolites are excreted.

like melphalan. This apparent versatility demonstrated by the α-class GSTs is probably the result of the fact that it is the phosphoramide mustard generated by the metabolism of cyclophosphamide that is the substate for detoxification. Biochemical studies have shown that purified α-class GST is able to catalyse the conjugation of melphalan to GSH.

Since GSH is consumed in the conjugation reaction, cells exposed to high concentrations of cytotoxic drugs show a marked depletion in GSH content. Under these circumstances, it is possible for cells to salvage GSH from the extracellular environment by breaking it down to its constituent parts and transporting the free amino acids into the cell to act as precursors for new GSH synthesis. The membrane associated enzyme γ-glutamyl-transpeptidase is involved in this process and many cell lines made resistant *in vitro* to alkylating agents overexpress this protein. Since the intracellular concentration of GSH is critical for drug detoxification, agents that interfere with GSH biosynthesis and/or activity have been sought for their potential as drug resistance modifiers. The prototypical agent is buthionine sulphoxamine (BSO), which inhibits the rate-limiting step in GSH biosynthesis. Coadministration of BSO and melphalan produces increased levels of cell kill compared to melphalan alone. Although BSO may not prove particularly useful as a chemotherapeutic agent, compounds that act via similar mechanisms could have some utility *in vivo*.

Detoxification of xenobiotics by GSH may require that the conjugates formed are either exported from the cell or transported into intracellular vesicles. A poorly characterized transporter called the GS-X pump is implicated in vesicle-mediated excretion of GSH-drug conjugates and a number of other compounds. The compounds transported into these vesicles are eventually eliminated from cells via exocytosis. The GS-X pump appears to be an ATP-dependent transporter which shows some similarities to the MRP protein described above. Whether MRP and/or a related ATP-dependent transporter is involved in GSH-drug conjugate excretion remains to be proved.

A second potential mediator of drug detoxification is the cysteine-rich protein metallothionein (MT). MT binds heavy metals and its expression is induced by exposure of cells to these metals. Overexpression of MT has been implicated in resistance not only to heavy metals, but also to many anticancer drugs, including cisplatinum and melphalan. Consistent with a protective role, cells deficient in MT expression, due to a targeted inactivation of the MT genes, are hypersensitive to a number of anticancer drugs, including cisplatinum and nitrogen mustards.

18.4 DRUG RESISTANCE CAN BE MEDIATED BY ALTERATIONS IN NUCLEAR DRUG TARGETS

DNA topoisomerase II (topo II) is a nuclear enzyme that regulates chromosome structure via an ability to catalyse the interconversion of different topological forms of DNA. Consequently, this enzyme has many roles in DNA metabolism, including replication, recombination and chromosome condensation and segregation. Topo II is also the primary target for many commonly used anticancer drugs, including mitoxantrone, doxorubicin, etoposide (VP-16) and epirubicin (Table 18.3). These topo II-targeting agents exert their cytotoxicity via the stabilization of a reaction intermediate called the 'cleavable complex', formed when the enzyme becomes covalently bound to the 5' end of cleaved DNA (Figure 18.4). Collision of the replicating machinery is thought to initiate a series of events which culminate in cell death.

A second class of topo II-directed agents exert their effect without the formation of cleavable complexes. These compounds, which include the dioxopiperazine derivatives, ICRF-159 and -193, inhibit the normal catalytic cycle of topo II and prevent sister-chromatid separation at anaphase. In the absence of active topo II, intertwined sister chromatids may be broken during attempted mitotic chromosome segregation.

A number of factors that influence the cytotoxicity of topo II-directed agents have been identified. The most common observation is that topo II content correlates with drug sensitivity. Cells expressing low levels of topo II accumulate fewer topo-II-mediated strand breaks and are therefore less sensitive to topo II-directed agents. This is a commonly observed mechanism of acquired drug resistance in cell lines exposed to topo II-targeting drugs. In contrast, cells expressing high levels of topo II accumulate more topo II-mediated strand breaks and are therefore drug-hypersensitive. This mechanism of resistance/hypersensitivity only operates in the case of cleavable complex-forming drugs. Drugs that exert their cytotoxicity without the formation of DNA damage are likely to be less cytotoxic under conditions where topo II levels are elevated.

It is likely that several factors serve to regulate topo II activity and therefore susceptibility to inhibition by antineoplastic drugs. For example, a number of point mutations within conserved domains of topo II have been shown to reduce the extent to which topo II-targeting drugs induce cytotoxicity. It seems likely that this may be one mechanism by which tumour cells evade the toxic effects of chemotherapy. In addition, topo II is a phosphoprotein *in vivo* and modulation of its phosphorylation status influences not only catalytic activity but also the degree of inhibition by topo II-targeting drugs. Finally, topo II must be localized to the nucleus during interphase to mediate drug-induced cell killing, and some drug-resistant cell lines contain a truncated topo II enzyme which is located primarily in the cytoplasm. Thus, cell lines and tumours have at their disposal a wide range of possible mechanisms for manipulating the structure/activity of topo II in such a way as to retain sufficient activity for viability but to provide less scope for antineoplastic drugs to exert their cytotoxicity.

In conclusion, human cells have evolved a myriad of strategies for protection against environmental toxins. When challenged by antineoplastic drugs, tumours can utilize these protective systems to resist the toxic effects of drugs. It may be naive to assume that cytotoxic agents can be identified that are not subject to at least one of the common protective mechanisms. However, the challenge is to

Table 18.3 Topo-II-targeting drugs

Drug class	Examples	Intercalator	Method of action
Acridines	Amsacrine (m-AMSA)	Yes	Cleavable complex stabilizers
Anthracyclines	Doxorubicin Epirubicin	Yes	Cleavable complex stabilizers
Anthracenediones	Mitoxantrone	Yes	Cleavable complex stabilizers
Ellipticines	Ellipticine	Yes	Cleavable complex stabilizers
Actinomycins	Actinomycin D	Yes	Cleavable complex stabilizers
Epipodophyllotoxins	Etoposide (VP-16) Teniposide (VM-26)	No	Cleavable complex stabilizers
Bis-piperazinediones	ICRF-159, 193	No	Non-cleavable complex stabilizers, inhibit enzyme in closed clamp conformation
Isoflavanoids	Genestein	No	Cleavable complex stabilizers
Anthracenyl peptides	Merbarone	No	Antagonizes cleavable complex formation by intercalators
Hexasulphated naphthylurea	Suramin	No	Unknown; binds to topo II

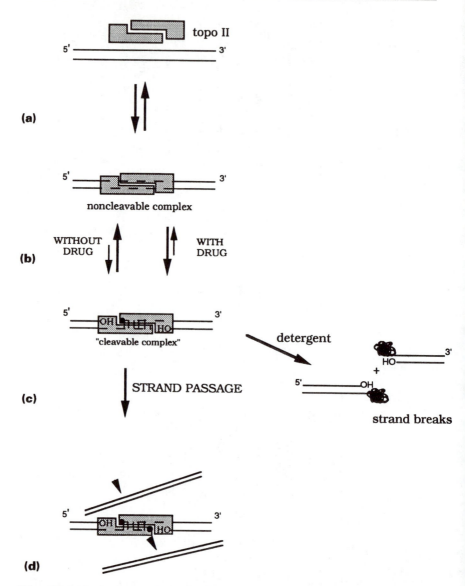

Figure 18.4 The catalytic cycle of topo II and the action of topo II-targeting drugs.
(a) Homodimeric topo II (indicated by stippled blocks) interacts non-covalently with
duplex DNA to form a non-cleavable complex (b). The enzyme cleaves both strands of
the DNA, binding covalently to the 5′ end of the nicks via an O^4 phosphotyrosyl linkage.
These covalent structures are termed the 'cleavable complexes' (c). The strand breaks
can be revealed at this stage by denaturation of the topo II, for example by detergents.
Normally the topo II passes a second intact duplex through the break (d). Treatment with
topo II poisons leads to an accumulation of cleavable complexes in cellular DNA.

understand fully the molecular mechanisms of drug resistance and to design strategies to overcome them.

18.5 FURTHER READING

Bradley, G. and Ling, V. (1994) P-glycoprotein, multidrug resistance and tumour progression. *Cancer Metastasis Rev.*, **13,** 223–233.

Gottesman, M. M. and Pastan, I. (1993) Biochemistry of multidrug resistance mediated by the multidrug transporter. *Annu. Rev. Biochem.*, **62**, 385–427.

Pommier, Y. (1993) DNA topoisomerases I and II in cancer chemotherapy: update and perspectives. *Cancer Chemother. Pharmacol.*, **32**, 103–108.

Tew, K. D. (1994) Glutathione-associated enzymes in anticancer drug resistance. *Cancer Res.*, **54**, 4313–4320.

Watt, P. M. and Hickson, I. D. (1994) Structure and function of type II DNA topoisomerases. *Biochem. J.*, **303**, 681–695.

19 | Drug discovery and cancer genetics

Mark R. Crompton

The successful treatment of human tumours with radiotherapy or chemotherapy is hampered by problems associated with non-responsiveness, development of resistance and toxic side effects. It is now a commonly held belief that significant advances in the success rates of treatment will require the design of novel therapeutic strategies that are radically different from those that are employed at present.

19.1 MUTANT ONCOPROTEINS OFFER NEW THERAPEUTIC TARGETS

Tumour development follows the accumulation of multiple cooperating mutations. These mutations cause enhanced, reduced or altered/novel functions of the proteins encoded by the affected genes (Chapter 1). The outcome is changes in the behaviour of clones of cells such that they are capable of survival, proliferation, invasion and the recruitment of enhanced blood supply (angiogenesis) under circumstances where normal cells would be kept in check.

The identification and isolation of the genes that are mutated in tumours allows the development of assay systems for drugs that are targeted at the gene products. This strategy of directing therapies at specific molecules in cancer cells should have several advantages, particularly in relation to selectivity. Several potential targets are only expressed at certain developmental stages in specific normal cell types, and only a subset of these expressing cells may actually require the function of the molecule at any time. Thus, targeted drugs that modify the function of the affected molecule may show exquisite specificity of action *in vivo*. This could,

Molecular Biology for Oncologists. Edited by J. R. Yarnold, M. R. Stratton and T. J. McMillan. Published in 1996 by Chapman & Hall, London. ISBN 0 412 71270 9

in theory, allow the administration of doses that are lethal to cancer cells without unacceptable side-effects. Further advantages of directed therapies might be the ability to treat tumours that respond poorly to existing approaches and the design of combination therapies that act synergistically on tumour cells and reduce the likelihood of clones evolving simultaneous multiple resistance.

A variety of molecules are being studied in the hope of developing new therapies, with the intention of inhibiting deregulated functions, 'replacing' lost activities or using tumour-associated molecules as selective 'receptors' for therapeutic agents. The diversity of biological processes being targeted extends beyond the traditional search for cytostatic/cytotoxic activity on cultured tumour cells; the abilities of many tumour types to evade the immune system, circumvent normal differentiation signals and programmes, metastasize and induce angiogenesis are all the subjects of extensive investigation.

This chapter concentrates on a few related approaches that have shown some promise to date. In particular, the focus will be on attempts to modify intracellular signalling pathways that are deregulated in tumour cells. Other chapters will discuss complementary approaches.

19.2 INTRACELLULAR SIGNALS ARE SUITABLE TARGETS FOR DRUGS

Figure 19.1 shows a schematic and greatly simplified outline of some intracellular signalling pathways that may be relevant to the aberrant behaviour of tumour cells (Chapter 5).

In this depiction, regulatory interactions between molecules are represented by double-headed arrows and serve to demonstrate that molecules involved in the interpretation of intercellular regulatory communication, e.g. growth factor receptors, interact with and influence various types of cytoplasmic signalling molecule. These in turn are involved in multiple intracellular signalling pathways that often involve the generation of 'second messenger' molecules or the initiation of protein phosphorylation cascades. By means that are mostly poorly understood, these pathways often modify transcription factors and thus alter the expression of panels of genes that determine the behaviour or fate of a cell. In the context of cell proliferation, research is beginning to clarify how growth factors might initiate and regulate progression through the cell cycle, either by inducing modifications to components of the cell cycle machinery or by regulating their expression. Some important nuclear cell cycle regulatory molecules are shown in Figure 19.1; they are described in greater detail in Chapters 7 and 9.

The molecules in Figure 19.1 that have been highlighted by shading are those whose genes are often found to be altered in some way in human tumours. In fact, a high proportion of the mutations involved in tumour development affect intracellular signalling molecules or components of the cell cycle machinery (genes for transcription factors are also often mutated in tumours). Although the products of mutated genes would be the most obvious targets for the development of tumour

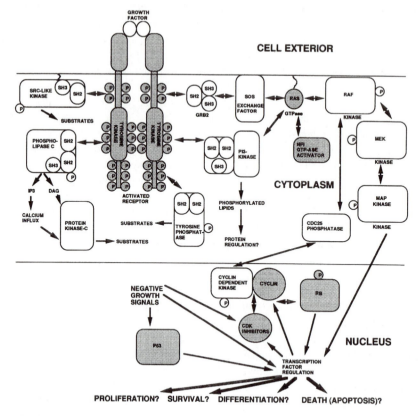

Figure 19.1 Regulatory interactions between some intracellular signalling molecules. Double-headed arrows = regulatory interactions; single-headed arrows = consequences of activities; shading = molecules genetically altered in human tumours; SH2/SH3 = protein domains involved in protein–protein interactions; DAG = diacylglycerol; IP3 = inositol trisphosphate; P = regulatory protein phosphorylation.

therapies, current research is directed at all aspects of intracellular signalling. Two of the research directions that are showing particular promise at present will now be described in more detail.

19.3 PROTEIN-TYROSINE KINASES (PTKs) ARE TARGETS FOR NEW DRUGS

Protein kinases are enzymes that catalyse the transfer of phosphate groups (usually from ATP) on to specific amino acid side chains in polypeptides. They are central components of intracellular signalling pathways. Phosphorylation of tyrosine is rare in cells compared to phosphorylation of serine and threonine, but the diverse enzymes that mediate this modification are extremely important and regulate many aspects of normal growth and development. They are classified into two broad groups: receptor protein-tyrosine kinases (PTKs) that span the cell membrane with an extracellular ligand-binding domain and an intracellular catalytic domain (e.g. many growth factor receptors), and non-receptor intracellular PTKs (e.g. Src family kinases).

There is evidence for a causal role of PTK deregulation in human tumours. Mutated PTK-encoding genes have been described in tumours such as gliomas, leukaemias, lymphomas and thyroid carcinomas, while many other tumour types such as breast tumours express high levels of active kinases that are apparently unaltered in their structure. In many cases, the overexpression or activation of these enzymes in cell or animal models results in experimental tumorigenic transformation. Conversely, inhibition of their expression or activity in tumour cells often causes reversion of transformation. Further roles for PTKs in tumour development are apparent from the discovery that many angiogenic factors secreted within tumours signal to vascular endothelial cells through receptor PTKs.

The likely central role of PTKs in tumour development makes the approach of inhibiting their catalytic function a particularly promising one. The fact that these enzymes perform crucial roles in normal physiology (e.g. the insulin receptor is a PTK) has to be taken into account when designing such strategies, but there are good reasons to think that a useful degree of selectivity is attainable. Many PTKs are expressed in a restricted range of normal tissues; in such cases drugs designed to be specific or selective for the inhibition of a particular enzyme are likely to have fewer side-effects than general PTK inhibitors. Further selectivity may be possible through exploiting the degeneracy that may be inherent to many signalling pathways. Evidence for this degeneracy can be seen in transgenic mice with homozygous disruptions of genes encoding signalling molecules. These often result in defects in far fewer cell types than predicted from a survey of the molecules' tissue expression patterns. Such observations can be explained as reflecting either highly plastic developmental processes in which cells have the ability to respond to losses of gene function by compensatory changes in the expression of other genes, or the fact that many cells express multiple signalling molecules that may possess substantially overlapping or redundant functions. Striking examples of redundancy have come from the study of the Src-like PTKs; mice lacking any one of three widely expressed members of this enzyme family develop with only relatively minor defects, whereas double mutants bred to lack pairs of them fail to reach adulthood. Since several human tumour types, including carcinomas of the colon and breast, have elevated PTK activity associated with Src-like kinases, it may be possible to inhibit a specific Src family member and achieve a therapeutic response without serious side-effects in the normal tissues.

19.4 PROTEIN-TYROSINE KINASE (PTK) INHIBITOR DISCOVERY AND DESIGN

Several starting points are being used in the search for selective PTK inhibitors. Past experience suggests that natural products will provide a rich source of novel chemical structures, some of which may possess useful biological properties.

Many investigators are, therefore, screening extracts of organisms such as plants, fungi and microorganisms for activities that score in the biochemical assays (e.g. inhibition of PTK activity) that they have developed. The hope that natural products will provide the basis for new drug development is one of the economic and pragmatic arguments for ecological conservation strategies aimed at preserving biodiversity. Other potentially valuable bioactive structures may be found in the extensive catalogues of compounds generated by chemical and pharmaceutical research in other fields.

Another approach to drug discovery is to take known chemical structures with desirable properties, and to use these as templates for the development of related structures with increased potency or selectivity in the particular system under study. There are many reported examples of interesting molecules that have been identified by a combination of the above approaches, erbstatin and its derivatives being good examples. Erbstatin is a compound that was isolated from cultures of *Streptomyces*, and found to inhibit the PTK activity of the receptor for epidermal growth factor. The structure of erbstatin has similarities with that of tyrosine, the substrate for phosphorylation catalysed by PTKs (Figure 19.2), and an extensive series of variations on the erbstatin theme has been synthesized. Some of these have been reported to be selective inhibitors of specific PTKs, and to be able to inhibit the growth of tumour cells transformed with the cognate activated PTK.

It is the hope of many molecular biologists working on cancer that their discoveries may indicate rational routes to the development of novel therapies. In the field of PTK research, the recent description of the three-dimensional structure of the catalytic domain of the insulin receptor has given us a close up of the structural basis behind the enzyme's interactions with its substrates, ATP and protein-tyrosine. Information relating to PTKs involved in tumour development may allow the design of drugs that interfere with these interactions and thus inhibit the enzymatic activity. Different PTKs exhibit selectivity for the proteins that they phosphorylate, presumably because they recognize structural determinants in the polypeptide sequences surrounding the tyrosines to be modified. It is, therefore, hoped that inhibitors of particular kinases can be designed based on the tyrosine peptide bind-

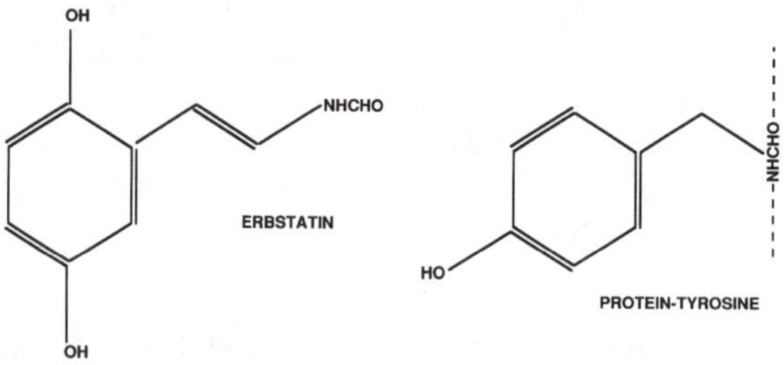

Figure 19.2 Structure of a PTK inhibitor and the substrate with which it competes. The structural similarities between the natural product erbstatin and protein-tyrosine may partly explain the tyrosine kinase inhibitory properties of the former.

ing site. Studies on known selective kinase inhibitors that interfere with ATP binding suggest that the ATP binding site may also be an appropriate target for the rational design of selective PTK inhibitors.

19.5 MUTANT Ras AND THE DESIGN OF FARNESYLTRANSFERASE INHIBITORS

Ras proteins are plasma-membrane-anchored molecules that perform important 'switching' roles in signal transduction pathways in many normal and neoplastic cells (Figures 19.1 and 19.3, and Chapter 6).

They bind guanine nucleotides and catalyse the hydrolysis of GTP to GDP; at any one time the ratio of Ras-GTP to Ras-GDP in cells is regulated by multiple signalling pathways through several Ras regulatory proteins. Ras-GTP is the active form of Ras that contributes to the regulation of several 'downstream' signalling molecules. In the best understood example, Ras-GTP binds the Raf protein kinase, thus relocating Raf at the plasma membrane. In this location, Raf is able to activate a cascade of protein kinases (the MAP kinase cascade) that regulate numerous intracellular processes (Figure 19.3). There is very good evidence that Ras proteins are important to the development of many human tumour types. In particular, genes encoding Ras proteins are mutated in approximately 30% of human tumours to produce products that are preferentially in the GTP-bound form in cells; the consequence of this is deregulation of signalling that contributes to transformation. Since normal Ras proteins have also been shown to be necessary for cell transformation induced by oncogenic protein tyrosine kinases, Ras activity is likely to be important in many human tumours.

One of the routes by which Ras proteins are regulated is depicted in Figures 19.1 and 19.3. The 'adaptor' protein Grb2 connects plasma membrane tyrosine kinase receptors (PTKs) to the cytoplasmic Ras-signalling pathway. Grb2 has specific amino acid sequences called SH2 and SH3 domains which bind specific phosphorylated tyrosine residues on activated PTKs and the Sos exchange factor, respectively. The association of the Grb2/Sos complex with activated PTKs allows Sos to promote exchange of nucleotides on Ras proteins, thereby favouring accumulation of the active GTP-Ras form. The interactions regulated via Grb2 are exemplary of many signalling mechanisms, in that they are mediated by SH2 domains binding peptides containing phosphorylated tyrosine residues, and by SH3 domains binding peptide sequences rich in proline residues. The selectivity of individual SH2 and SH3 domains for the sequences that they bind suggests that they may be suitable targets for the development of modulators of signalling processes, and current research is aimed at identifying therapeutically useful specific inhibitors of such interactions. The altered proteins encoded by mutated *ras* genes in tumours also represent very attractive targets for the development of specific drugs; however such compounds have not been reported to date.

Membrane association is required for signalling by both normal and mutant Ras proteins, and significant progress has been made in designing drugs that take

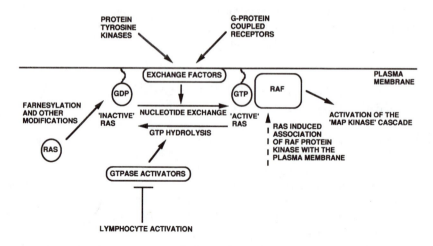

Figure 19.3 Regulation and function of Ras proteins. Ras proteins 'switch' between active and inactive states in a manner that is regulated by a variety of signalling 'inputs', some of which employ tyrosine kinases and interactions mediated by SH2 and SH3 domains. One of the functions of active Ras is to recruit Raf to the plasma membrane, where the latter can initiate further signalling processes that contribute to tumorigenesis. The signalling function (and possibly regulation) of Ras requires its farnesylation-dependent membrane association. Drugs designed to modulate 'Ras pathways' could, therefore, be directed at a number of different types of molecule.

advantage of this fact. Ras proteins are among a small set of proteins that are modified with a particular lipid structure, the farnesyl group. This modification is a prerequisite for further processing of Ras proteins, and is responsible for their association with the plasma membrane. Protein farnesylation is mediated by the cellular enzyme farnesyltransferase, which catalyses the transfer of farnesyl groups from farnesyl pyrophosphate (FPP) to protein cysteine residues present in specific sequence motifs (the 'CAAX' box, where C = cysteine, A = aliphatic, X = any). Employing the known structures of the substrates of farnesyltransferase, structural analogues of FPP and the CAAX box have been synthesized alone or in combination to produce farnesyltransferase inhibitors. Figure 19.4 shows one 'peptidomimetic' approach that has been employed in the design of substrate analogue farnesyltransferase inhibitors, and points out two of the factors that have to be taken into consideration when designing drugs that need to act *in vivo*, i.e. cell permeability and intracellular metabolism. When the effects of these drugs on cells have been examined, very encouraging results have been obtained.

Despite Ras function being thought to be required for normal cell growth, farnesyltransferase inhibitors block the functions of mutant Ras in transformed (neoplastic) cells without affecting normal cells. This paradox may be explained in several ways. The doses of the drugs used may inhibit the enhanced Ras activity associated with transformation while leaving sufficient Ras activity intact to mediate normal cellular responses. An alternative explanation, for which there is now experimental evidence, is that cells contain molecules that can compensate for loss

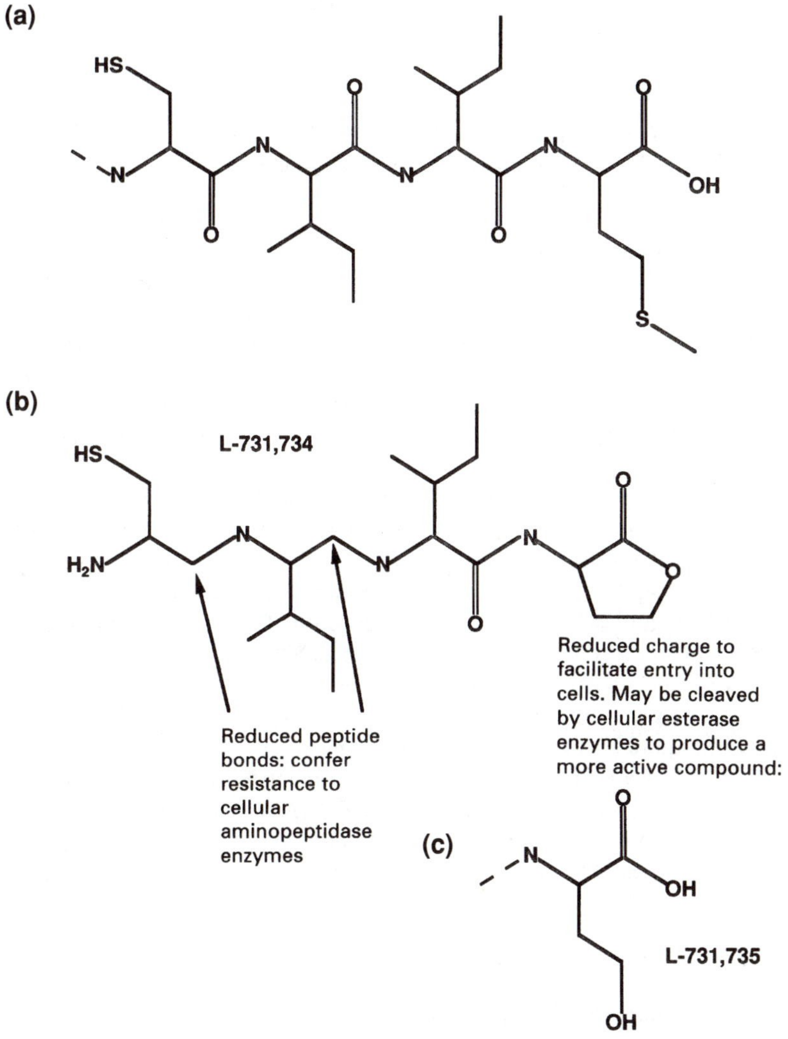
(a)
HS
N
O
N
O
N
N
O
OH
O
S
(b)
HS
L-731,734
H2N
N
N
N
O
O
O
Reduced charge to facilitate entry into cells. May be cleaved by cellular esterase enzymes to produce a more active compound:
Reduced peptide bonds: confer resistance to cellular aminopeptidase enzymes
(c)
N
O
OH
L-731,735
OH

Figure 19.4 A strategy for the design of farnesyltransferase inhibitors. The compound L-731,734 (**b**) was designed to mimic the 'CAAX' region of a Ras protein (**a**) that is farnesylated by farnesyltransferase. Resistance to degradation and cellular permeability were important factors that were taken into account. The compound L-731,735 (**c**) differs from L-731,734 at one end and is more like the CAAX box; it is a more potent inhibitor of farnesyltransferase *in vitro*. Intracellular enzymes may convert the cell-permeable L-731,734 to the more active L-731,735. (Source: modified from Kohl *et al.*, 1993.)

of Ras function when farnesyltransferase is completely inhibited. A molecule similar to Ras proteins in structure and function, but employing a different modification for membrane targeting, has been identified, and its expression in a variety of normal cells may contribute to their resistance to the effects of farnesyltransferase inhibitors. Whatever the mechanisms involved, farnesyltransferase inhibitors may prove to be selective inhibitors of the development of tumours with deregulated Ras signalling.

19.6 DRUG SCREENING STRATEGIES INCREASINGLY EXPLOIT MOLECULAR BIOLOGICAL TECHNIQUES

The 'non-rational' strategies of drug discovery described above, e.g. those involving the screening of natural products for protein-tyrosine kinase (PTK) inhibitors, will only be effective if extremely high throughput assays are available. The molecular cloning of molecules such as PTKs has allowed them to be expressed at high levels in various biological systems that facilitate their purification in active forms. The combination of these molecular biological techniques with advanced sample handling technology has led to the development of biochemical screens for enzyme inhibitors. In the 'hierarchy' of screens that are employed in drug discovery programmes of the type described here, the biochemical screens are often followed by assays using cultured cell lines. Appropriate cell lines are chosen to have specific properties related to transformation (such as altered morphology or deregulated growth) that are attributable to the activities of the signalling molecule (e.g. PTK) under investigation. Such cells may already be available and characterized, or they may be specifically generated for the purpose of research or drug screening by gene transfection. Issues such as cell permeability, drug stability/metabolism and likely toxicity can begin to be addressed in such assays. Animal studies of drug efficacy, metabolism and toxicity would be the final development stage, and would be required before trials on human subjects could begin. Murine models of neoplastic disease can be produced by transplanting human tumour cells into immunocompromised animals (xenograft models), or by generating genetically modified animals that develop frequent tumours as a result of oncogene expression and/or disruption of tumour suppressor genes.

19.7 ACKNOWLEDGEMENT

Many thanks to Mike Fry for helpful comments.

19.8 FURTHER READING

Brunton, V. G. and Workman, P. (1993) Cell-signalling targets for antitumour drug development. *Cancer Chemother. Pharmacol.*, **32**, 1–19.

Carboni, J. M., Yan, N., Cox, A. D. *et al.* (1995) Farnesyltransferase inhibitors are inhibitors of *ras*, but not r-*ras2/TC21*, transformation. *Oncogene*, **10**,1905–1913.

Fry, D. W., Kraker, A. J., McMichael, A. *et al.* (1994) A specific inhibitor of the epidermal growth factor receptor tyrosine kinase. *Science*, **265**, 1093–1095.

Kohl, N. E., Mosser, S. D., deSolms, S. J. *et al.* (1993) Selective inhibition of ras-dependent transformation by a farnesyltransferase inhibitor. *Science*, **260**, 1934–1937.

Levitzki, A. and Gazit, A. (1995) Tyrosine kinase inhibition: an approach to drug development. *Science*, **267**, 1782–1788.

The molecular biology of endocrine responsiveness in breast cancer

20

Elizabeth Anderson and Anthony Howell

20.1 OESTROGEN RECEPTOR PROTEIN IS THE KEY MEDIATOR OF OESTROGEN ACTION AND THE RELATIONSHIP BETWEEN ITS STRUCTURE AND FUNCTION IS NOW UNDERSTOOD

It is 100 years since a response to endocrine manipulation in advanced breast cancer was first observed and over 25 years since the first anti-oestrogen, tamoxifen, was introduced into clinical practice. Tamoxifen has now become the single most important endocrine therapy for early and advanced breast cancer and is in clinical trial as a preventative agent for women at increased risk of the disease. A substantial number of women will not show a response to endocrine treatment. Moreover, in advanced breast cancer, all responders to therapy eventually relapse and die of their disease. In this chapter we will explore what is known concerning the mechanisms of endocrine responsiveness and resistance at the molecular level.

When radiolabelled oestrogens of high specific activity were synthesized in the 1960s, it was discovered that endocrine-responsive breast tumours and other organs such as the endometrium contain a protein called the oestrogen receptor (ER). This protein binds oestradiol (E_2) with high affinity and is the key mediator of oestrogen action and response to endocrine therapy. In recent years, characterization of both the ER protein and its mechanism of action has progressed rapidly to reveal a strong relationship between the structure of the receptor and its function. It is clear that this relationship determines to a large extent the clinical response to anti-oestrogen therapy. Furthermore, elucidation of the interactions

Molecular Biology for Oncologists. Edited by J. R. Yarnold, M. R. Stratton and T. J. McMillan. Published in 1996 by Chapman & Hall, London. ISBN 0 412 71270 9

between the receptor, its agonists and antagonists has led to the development of potentially more effective anti-oestrogens and to strategies that may avoid the inevitable resistance and relapse seen with tamoxifen treatment.

20.2 THE STRUCTURE OF THE OESTROGEN RECEPTOR (ER) CAN BE CLEARLY RELATED TO ITS FUNCTIONS

The *ER* gene isolated from a cultured breast cancer cell line (MCF-7) was one of the first steroid receptors to be cloned and characterized. The MCF-7 *ER* mRNA is 6.3 kb in length, encodes a protein that contains 595 amino acids and has a molecular weight of 66 000. The *ER* gene is situated on chromosome 6, is more than 140 kb in length and contains eight exons. Analysis of the primary amino acid sequence reveals that the ER is a member of a superfamily of nuclear transcription factors that encompasses all the steroid receptors as well as those binding tri-iodothyronine, retinoic acid and vitamin D_3. Also included in this family are a large number of 'orphan' receptors whose ligands, the molecules that bind to receptors, have yet to be identified. The nuclear receptors contain several regions or domains in which the amino acid sequences are conserved between family members, suggesting common functions. Some of these domains form distinctive structures that are essential for receptor function. The ER is a representative nuclear receptor in that it contains all the conserved domains characteristic of the family.

The six functional domains of the ER are shown in Figure 20.1 and are conventionally labelled A–F. Region C (the DNA-binding-domain or DBD) in the centre of the molecule confers the ability to bind DNA and is highly conserved between species and between members of the nuclear receptor family. Region E (the hormone-binding-domain or HBD) contains the sequence to which the ligand binds and is also highly conserved between species and between different nuclear receptors. The hormone-binding domain (Region E) also contains the amino acid sequence that initiates gene transcription after oestradiol binds to the ER. This is known as the ligand-inducible activating function (or AF-2) of the ER molecule. There is another transcriptional activating function (AF-1) situated in the A/B region of the ER, the action of which appears to be ligand-independent. Three other domains have been identified within the ER; one mediates binding to heat shock protein (HSP), the second is a sequence that directs localization of the receptor to the nucleus whereas the third is involved in receptor dimerization (see below). There is some functional overlap between the different domains: for example, region E appears to participate in receptor dimerization, nuclear localization and binding to HSP in addition to containing the hormone-binding domain (Figure 20.1).

Many oestrogen target tissues express variant ER molecules as well as the wild-type ER. These variants are thought to arise when the introns are removed from the hnRNA and the exons are spliced together to create mRNA. During this

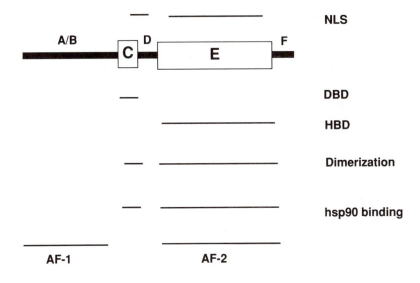

Figure 20.1 Functional domains of the human oestrogen receptor. The human oestrogen receptor (ER) is divided into six functional domains (A–F). The hormone-binding domain (HBD) resides in region E and region C contains the DNA-binding domain (DBD). Sequences that direct nuclear localization (NLS) have been found on the boundary between regions C and D and also in region E. The sequences that participate in receptor dimerization are in regions C and E, which also contain the heat shock protein (HSP) binding domains. Finally, there are two regions that mediate activation of transcription: transactivating function 1 (AF-1), which is constitutively active, is contained within regions A and B. The oestrogen-inducible transactivating function (AF-2) is in region E.

process, exons may be removed, creating receptors lacking specific functional domains, which are known as splice variants. For example, deletion of exon 5 through this alternative splicing process results in a receptor protein where part of the hormone-binding domain is missing. Several splice variants have been described in malignant breast tumours and their presence has been proposed as a cause of anti-oestrogen resistance. However, the variants have also been described in normal breast tissue, which suggests that their role in breast tumours may be of limited significance, although this has yet to be confirmed.

Although early studies suggested that the ER might be a cytoplasmic protein that is translocated to the nucleus after hormone-binding, it is now clear that the receptor resides in the nucleus whether or not its ligand is present. Recent studies have shown that there is an active, energy-dependent mechanism for transporting ER to the nucleus called the **nucleocytoplasmic shuttle**. Although the receptor is constantly being lost from the nucleus through passive diffusion, the shuttle transports it back, ensuring that ER is always in the nucleus. Once the ER has been transported to the nucleus it 'docks' with an HSP via two separate regions of the receptor (Figure 20.1). One of these regions is situated at the carboxy end of the DNA-binding domain, the other is in region E and both are required for efficient ER–HSP complex formation. At least one other protein has been shown to participate in the ER–HSP heterocomplex. Likewise, with the receptors for progesterone and glucocorticoids, it appears likely that other heat shock proteins assist in heterocomplex formation. The function of the heterocomplex with heat shock proteins is still unclear. Although studies on other steroid receptors suggest that binding to HSP maintains the receptor in the conformation required for ligand binding, this does not appear to be the case for the ER. The most likely explanation is that formation of the complex with heat shock proteins prevents the free receptor from binding to DNA.

20.3 THE OESTRADIOL–ER COMPLEX BINDS TO A SPECIFIC DNA SEQUENCE INVOLVED IN GENE REGULATION CALLED THE OESTROGEN RESPONSE ELEMENT (ERE)

Oestradiol (E_2), the endogenous ligand for the ER, diffuses freely into the cell and its nucleus where it binds to a specific domain in region E of the receptor. This domain is an independently folded structure that forms a pocket of hydrophobic amino acids containing the ligand binding site.

Binding of E_2 to the ER initiates the series of events shown in Figure 20.2. The first step is dissociation of the ER from its heterocomplex with the heat shock proteins, which allows two ER molecules to bind together to form a dimer. This is termed the **ER homodimer** as both the components are identical, in contrast to the receptors for progesterone (PR) and retinoic acid (RAR), which are composed of dissimilar monomers. Dimerization is mediated by sequences found within both the hormone-binding domain and the DNA-binding domain. The dimerization and the ligand-binding functions of the hormone-binding domain can be separated, suggesting that dimerization can take place in the absence of ligand. However, it appears that this does not occur because of the interaction of the ER with HSPs. The second dimerization domain situated within the DNA-binding domain may serve to further stabilize binding of ER to DNA.

Binding of E_2 and subsequent ER dimerization enables the receptor to bind specific DNA sequences called **oestrogen response elements** (**EREs**) found close to oestrogen-regulated genes. Their role is to regulate gene transcription. The first ERE was identified in studies on the *Xenopus* vitellogenin gene and is 13 base pairs (bp) in length, containing two 5 bp sequences, separated by a 3 bp spacer (see Figure 20.3).

As the two 5 bp repeats are exactly complementary to each other, the vitellogenin ERE is a perfect palindrome. However, it is clear that other EREs are not perfect palindromes and that half sites, containing just one half of the ERE, are also active as enhancers of ER-activated gene transcription. Several EREs may be present in the promoter regions of oestrogen-responsive genes and these may act synergistically to enhance gene transcription. The situation is complicated still

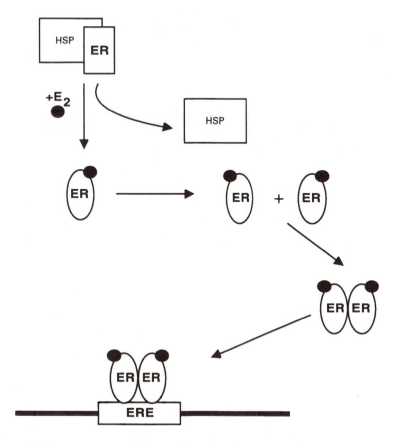

Figure 20.2 Dimerization of the ER and binding to the response element. Binding of oestradiol (E_2) to the receptor causes a conformational change in the receptor protein and release from its complex with the heat shock proteins (HSP). Two ER molecules bind together to form an ER homodimer, which then binds to specific DNA sequences called oestrogen response elements (EREs) in the vicinity of genes that are oestrogen-regulated. The ER–ERE complex acts to stimulate gene transcription.

further by the fact that EREs are very similar to the response elements for other nuclear receptors such as the progesterone and glucocorticoid receptors. Some specificity is conferred by differences in the length of the spacer and by the orientation of the repeats in the palindromes. However, it is clear that tissue and hormone specificity is also determined by other sequences flanking the response elements that have yet to be completely defined.

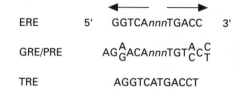

ERE	5' GGTCA*nnn*TGACC	3'
GRE/PRE	AG^A_GACA*nnn*TGT^A_CC^C_T	
TRE	AGGTCATGACCT	

Figure 20.3 The oestrogen response element prototype. The prototype oestrogen response element (ERE) isolated from the *Xenopus* vitellogenin gene is a perfect 13 bp palindromic sequence consisting of two 5 bp inverted repeats separated by a 3 bp spacer. The direction of these nucleotide repeats is indicated by the arrows above the sequence and '*n*' indicates any choice of nucleotide. The palindromic nature of the sequence is revealed in the complementary strand (not shown), which contains the same nucleotides in reverse order. The response elements recognized by the glucocorticoid and progesterone receptors (GRE/PRE) and the thyroid hormone receptor (TRE) are also illustrated. These demonstrate that response elements for the different nuclear receptors contain similar nucleotide sequences. This suggests that specificity must be conferred by other sequences that flank the response element.

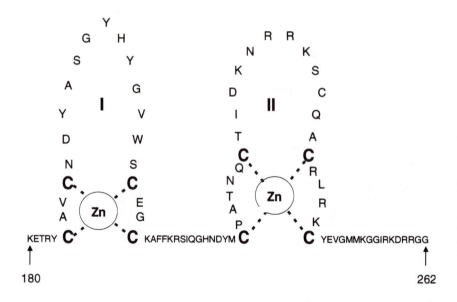

Figure 20.4 The zinc fingers of the DNA binding domain (DBD). The DNA-binding domain of the oestrogen receptor is rich in cysteine (C) residues and the position of these cysteines does not vary between steroid receptors. The cysteine residues are probably covalently linked to a zinc atom to create two loops or 'zinc fingers', labelled I and II. These structures are reminiscent of the zinc fingers in other factors that initiate gene transcription.

20.4 THE REGION OF OESTROGEN RECEPTOR PROTEIN THAT BINDS DNA FORMS FINGER-LIKE PROJECTIONS THAT INTERACT SPECIFICALLY WITH THE OESTROGEN-RESPONSE ELEMENT

The ER interacts with the ERE through the DNA-binding domain situated in region C of the molecule. This region is the most highly conserved between species and other members of the nuclear receptor family, and investigation has yielded important new information on its structure and the nature of its interaction with DNA. Sequencing of the DNA-binding domain reveals that it is rich in cysteine residues, the positions of which are invariant between steroid receptors. It is thought that these cysteine residues are 'cross-linked' by zinc to form two loops or 'zinc fingers' (Figure 20.4), which are reminiscent of the zinc fingers in other factors that initiate gene transcription.

The points of contact between the fingers of the ER and DNA have now been mapped; this shows that the first or N-terminal finger determines the specificity of the ER–ERE interaction and makes direct contact with the DNA bases. It is possible that the second zinc finger interacts with the phosphate backbone of DNA and mediates interactions between the ER and other proteins.

Exactly how the ER enhances gene transcription is, as yet, unclear.

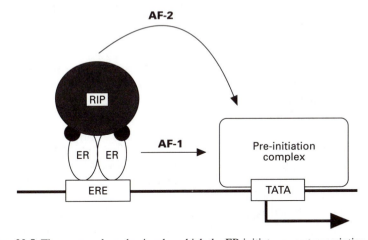

Figure 20.5 The proposed mechanism by which the ER initiates gene transcription. The schema illustrates an oestrogen response element (ERE) upstream of an oestrogen-regulated gene. The nucleotide sequence TATA (called the TATA box) indicates the starting point for transcription of the gene, which proceeds in the direction of the heavy arrow. Receptor dimerization and binding to the ERE is thought to stabilize the preinitiation complex that assembles around RNA polymerase at the TATA box before transcription begins. In this model, the two transcriptional activation domains (AF-1 and AF-2) co-operate differently with the other factors required for initiation of gene transcription. It is thought that AF-2 binds, in a ligand-dependent manner, with intermediary proteins known as receptor interacting proteins (RIPs). How AF-1 interacts with the transcriptional machinery is not yet known.

Experiments *in vitro* suggest that the ER homodimer, once it has bound to the ERE, stabilizes the 'pre-initiation complex' (Figure 20.5), a group of proteins that form around RNA polymerase, the enzyme responsible for transcribing DNA into RNA, which binds to the core promoter of the gene (the DNA sequence at which transcription is started).

The pre-initiation complex is an assembly of a large number of proteins (probably more than 30) and is absolutely required for the initiation of RNA transcription. It is likely that the two transcriptional activation domains within the ER (AF-1 and AF-2) cooperate differently with the other factors required for initiation of gene transcription. AF-2, in particular, binds in a hormone-dependent manner to other proteins or coactivators that have variously been named receptor-interacting proteins (RIP) or oestrogen-receptor-associated proteins (ERAP). Binding to these proteins appears to be essential for the transcriptional activity of AF-2 *in vitro*. How AF-1 interacts with the transcriptional machinery is not known. Oestrogen-controlled genes that have been detected in breast tumours include those that encode the progesterone receptor (PR), growth factors such as transforming growth factor α and the protein pS2, and proteases such as cathepsin D, together with several of the genes that are up-regulated in the early stages of response to stimulators of cell division such as thymidine kinase. Continued stimulation with E_2 results in down-regulation of the ER protein in a negative feedback mechanism. The action of progesterone through the PR also results in inhibition of ER protein synthesis.

20.5 THE OESTROGEN AGONIST AND ANTAGONIST PROPERTIES OF TAMOXIFEN DEPEND ON WHICH ACTIVATION DOMAIN OF THE ER MOLECULE IS USED TO ENHANCE GENE TRANSCRIPTION

There are two major classes of anti-oestrogen: the first, typified by tamoxifen (Figure 20.6), are triphenylethylene derivatives whereas the second group of 'pure' or specific anti-oestrogens are based on the structure of oestradiol and are typified by ICI 164 384 or ICI 182 780.

Tamoxifen can be a complete oestrogen antagonist, a partial oestrogen agonist or a complete agonist, depending on the gene, the tissue or the species that is being studied. The oestrogen agonist properties of tamoxifen or its metabolites may contribute to the eventual failure of tamoxifen when used to treat advanced breast cancer.

An understanding of how tamoxifen exerts oestrogenic effects on breast tumours came when the functions of the different ER domains were elucidated. Like E_2, tamoxifen binds to the HBD of the ER and causes dimerization of two ER monomers (Figure 20.7).

This ER homodimer binds to the ERE, which is sufficient to allow AF-1 to operate in a constitutive fashion. Unlike E_2, however, tamoxifen does not induce

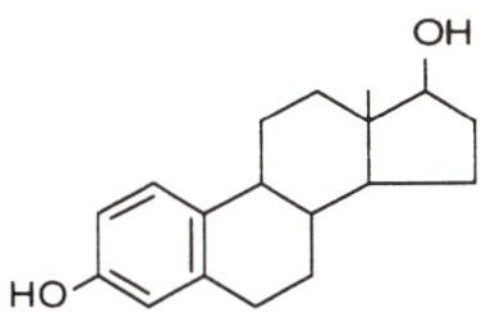

17 beta-OESTRADIOL

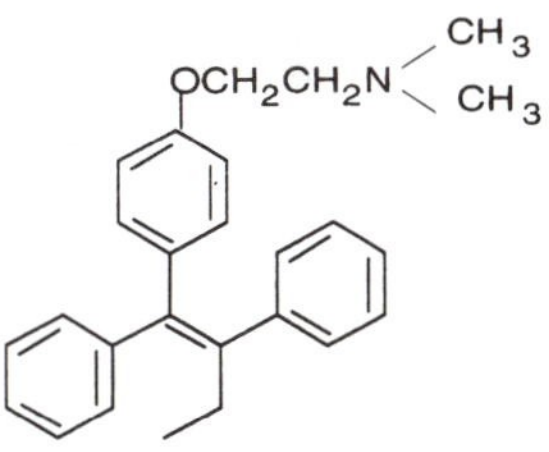

TAMOXIFEN

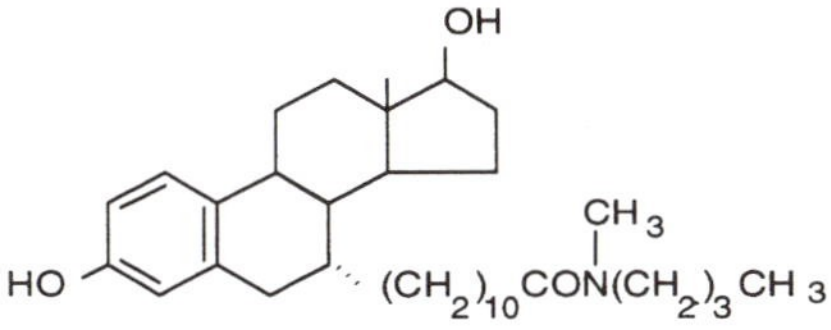

ICI 164384

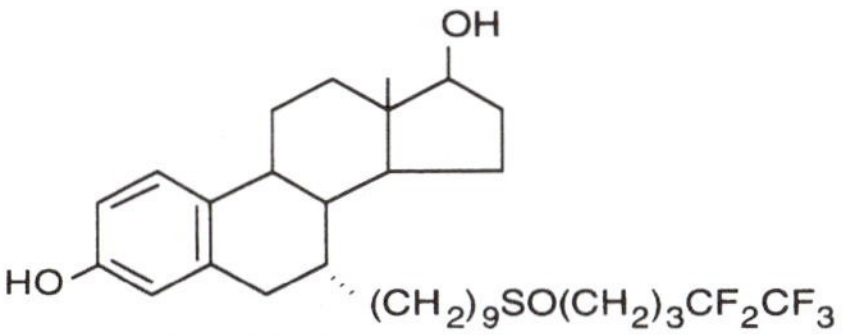

ICI 182780

Figure 20.6 The structure of oestradiol (E_2), tamoxifen and the specific anti-oestrogens. Tamoxifen is an non-steroidal antagonist as it is a triphenylethylene derivative. The two specific anti-oestrogens, ICI 164 384 and ICI 182 780, are based on the structure of E_2 but have a bulky side chain situated at the 7a position of the B ring of the steroid.

the changes in ER structure that are required for AF-2 to exert its ligand-dependent effects on gene transcription. Thus, where the activation of a gene is entirely dependent on AF-1, tamoxifen will act as a complete agonist. In contrast, tamoxifen will be a complete antagonist for genes whose transcription is controlled entirely by AF-2. Where the transcription of a gene is controlled by both AF-1 and AF-2, tamoxifen will be a mixed agonist/antagonist. The activity of tamoxifen can be changed from agonist to antagonist and *vice versa* by mutating the area of the ER that is involved in discriminating between ligands. However, there is very little evidence to suggest that these mutations occur naturally and one alternative explanation for tamoxifen agonism is that other signalling pathways within tumour cells could alter the transcriptional activity of the ER. Other explanations for tamoxifen resistance include the formation of oestrogenic tamoxifen metabolites or that the uptake and retention of tamoxifen within tumour tissue is impaired in some way. The presence of splice variants may also alter the type of response to tamoxifen although this has yet to be confirmed *in vivo*. However, it is possible that no single mechanism is responsible for the development of resistance to tamoxifen and that it is more likely to be due to a combination of the different factors.

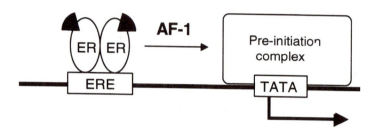

Figure 20.7 The proposed mechanism of tamoxifen action. The schema illustrates an oestrogen response element (ERE) upstream of an oestrogen-regulated gene. The nucleotide sequence TATA (called the TATA box) indicates the starting point for transcription of the gene, which proceeds in the direction of the heavy arrow. Like E_2, tamoxifen binds to and promotes dimerization of the ER and binding to the ERE. Binding of the tamoxifen–ER dimer allows AF-1 to operate in a constitutive fashion but the ER has not undergone the conformational change that is needed for AF-2 to become active. Where gene activation is entirely controlled by AF-1, tamoxifen will act as a complete agonist; conversely, tamoxifen will be a complete antagonist for genes controlled by AF-2 alone. In the case of genes that are under the control of both AF-1 and AF-2, tamoxifen will be a mixed agonist/antagonist.

The 'pure' anti-oestrogens are so called because they are devoid of oestrogenic activity and were developed in response to the need for an endocrine therapy to which patients would not become resistant. These compounds have a completely different mechanism of action to tamoxifen in that their major effect is to substantially reduce the cellular content of the ER by decreasing the half-life of the protein. It is thought that this reduction in half-life is caused by inhibition of the nucleocytoplasmic shuttle which, in turn, blocks re-entry of the receptor into the nucleus and promotes its degradation. Although ICI 182 780 binds to the ER, it disrupts dimerization, and consequently binding to the ERE, presumably because the bulky side chain physically interferes with the dimerization process (Figure 20.8). The early clinical studies with ICI 182 780 show that the ER is indeed, down-regulated in breast tumours *in vivo* and that the duration of response to the drug in advanced breast cancer patients who have failed tamoxifen treatment is longer than that for other second-line endocrine therapies.

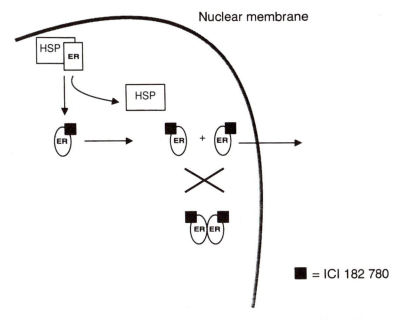

Figure 20.8 The proposed mechanism of ICI 182 780 action. The specific anti-oestrogens typified by ICI 182 780, bind to the ER and promote dissociation from its complex with the heat shock proteins (HSP). However, the ER bound to ICI 182 780 is unable to dimerize. As a consequence the receptor leaves the nucleus but, because it is bound to ICI 182 780, it cannot be 'shuttled' back. This leaves the receptor protein susceptible to degradation which, in turn, reduces its half-life and causes an apparent reduction in cellular content of the ER.

20.6 OTHER ENDOCRINE THERAPIES WORK BY REDUCING OESTROGEN PRODUCTION

Although tamoxifen is the most widely used first-line therapy for primary and advanced breast cancer, there are several other drugs with an endocrine mechanism of action that are frequently used as second-line agents. These may be broadly categorized into two groups: those that interfere with oestrogen synthesis and those that act by down-regulating the ER. The first category comprises agents that act by depriving the tumour of oestrogen. Some of these drugs, including aminoglutethimide, 4-hydroxyandrostenedione and Arimidex, inhibit the final step in oestrogen biosynthesis, which is the conversion of androgen into oestrogen and which is catalysed by an enzyme called **aromatase**. In premenopausal women, the major site of aromatase activity is the ovary whereas, after the menopause, the peripheral sites of aromatase activity such as adipose tissue, muscle and the tumour itself, predominate. The other type of drug that causes the tumour to be deprived of oestrogen is the gonadotrophin releasing hormone (GnRH) agonists, which inhibit pituitary secretion of the gonadotrophins, which, in turn, prevents secretion of oestrogen by the ovary. The second category of endocrine agents includes progestogens such as medroxyprogesterone acetate (MPA) and high doses of oestrogenic compounds such as diethylstilboestrol (DES). These appear to act by down-regulating the synthesis of the ER protein by the tumour cells, which then renders them insensitive to the actions of endogenous oestrogens. Why these second-line endocrine therapies should be effective after resistance to tamoxifen has developed is still not completely understood.

20.7 UNDERSTANDING THE MOLECULAR BASIS OF OESTROGEN RECEPTOR FUNCTION SHOULD PROVIDE MORE CLINICAL BENEFITS IN THE FUTURE

The characterization of the ER and its mechanism of action is an example of how molecular biological studies can directly affect the management of a disease. For several years, we have been able to identify those patients most likely to respond to endocrine therapy by measuring the ER content of their breast tumours. This process has been further refined by simultaneously measuring a protein, such as the PR, whose expression is dependent on the action of the ER. The results of recent studies into the mechanisms of anti-oestrogen resistance now offer the prospect of being able to identify those tumours on which tamoxifen will exert oestrogenic effects. Furthermore, these studies have provided a rational basis for the design of new pure anti-oestrogens. However as might be expected, the factors that affect ER expression and its mechanisms of action are rather more complicated than was first thought. Research into the nuclear receptor superfamily is a very rapidly progressing field from which radical new ideas regarding mechanisms of action are emerging. It is to be hoped that these new ideas will translate into future clinical benefits.

20.8 FURTHER READING

General reviews

Green, S. and Chambon, P. (1991) The oestrogen receptor, in *Nuclear Hormone Receptors*, (ed. M. G. Parker), Academic Press, London, pp. 15–38.

Rea, D. W. and Parker, M. G. (1995) Structure and function of the oestrogen receptor in relation to its altered sensitivity to oestradiol and tamoxifen. *Endocrine-Related Cancer*, **2**(1), 13–17.

Key works

Berry, M., Metzger, D. and Chambon, P. (1990) Role of the two activating domains of the oestrogen receptor in the cell-type and promoter-context agonistic activity of the anti-oestrogen 4-hydroxytamoxifen, *EMBO J.*, **9**, 2811–2818.

Cavailles, V., Dauvois, S., Danielan, P. S. and Parker, M. G. (1994) Interaction of proteins with transcriptionally active estrogen receptors, *Proc. Nat. Acad. Sci. USA*, **91**, 10009–10013.

Dauvois, S., Danielan, P. S., White, R. and Parker, M. G. (1992) Antiestrogen ICI 164,384 reduces cellular estrogen receptor content by increasing its turnover. *Proc. Nat. Acad. Sci. USA*, **89**, 4037–4041.

Fawell, S. E., White, R., Hoare, S. *et al.* (1990) Inhibition of estrogen receptor-DNA-binding by the 'pure' antiestrogen ICI 164,384 appears to be mediated by impaired receptor dimerization, *Proc. Nat. Acad. Sci. USA*, **87**, 6883–6887.

Kumar, V., Green, S., Stack, G. *et al.* (1987) Functional domains of the human estrogen receptor. *Cell*, **51**, 941–951.

21

Apoptosis as a therapeutic target

John A. Hickman

21.1 THE IMPOSITION OF DRUG-INDUCED CELL DEATH IS AN IMPORTANT STRATEGY FOR THE TREATMENT OF DISSEMINATED TUMOURS

The genetic instability of many human tumours (Chapter 13) and their tendency to progress to a more aggressive phenotype militates against treatments that inhibit growth but leave tumour cells viable. Escape from cytostasis exacts too high a price. With tumour cell death as the desired endpoint the obvious questions are: how can cells be made to die and how might tumour cell death be imposed selectively? This chapter discusses the opportunities to harness constitutive mechanisms of engaging cell death; these initiate a type of cell death described as **apoptosis**.

21.2 APOPTOSIS IS AN INTRINSIC MECHANISM OF CELL SUICIDE WHICH REMOVES CELLS WHICH ARE AGED, DAMAGED OR IN EXCESS OF REQUIREMENT

The sculpting of tissues and organs during development and the maintenance of appropriate cell numbers during homeostasis demands that controlled cell loss occurs alongside proliferative cell gain. The removal of cells which are in excess of appropriate number, or which are aged or are damaged, must occur in a manner that avoids the leakage of cellular contents. Without an orchestrated mode of cell loss and removal an inflammatory reaction would be initiated. Cell death by apoptosis provides the mechanism whereby cells are removed in a regulated way.

Molecular Biology for Oncologists. Edited by J. R. Yarnold, M. R. Stratton and T. J. McMillan. Published in 1996 by Chapman & Hall, London. ISBN 0 412 71270 9

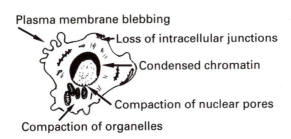

Plasma membrane blebbing

Loss of intracellular junctions

Condensed chromatin

Compaction of nuclear pores

Compaction of organelles

Figure 21.1 The apoptotic cell. The outer plasma membrane remains intact, although it begins to bleb. Cells lose cell–cell contact, and if attached to a substrate *in vitro* will detach from it. The nucleus contains condensed chromatin, and there is compaction of the organelles, such as mitochondria and nuclear pores, which are displaced from areas abutting the condensed chromatin. *In vivo*, apoptotic cells are rapidly phagocytosed either by viable neighbours or by macrophages; this rapid clearance may explain why so few apoptotic cells are observed in tissues. Failure to remove apoptotic cells within a reasonable time permits the breakdown of the plasma membrane, cell necrosis and an inflammatory reaction.

Apoptosis is a type of cell death defined by pathologists because of its distinct, conserved morphological features (Figure 21.1).

It is conceptualized as 'programmed' because it is a death driven from within the cell – essentially a cell suicide – with all cells poised on the precipice of self-destruction. The 'programme' to do this is considered to be intrinsic to all cells and it occurs without the loss of plasma membrane integrity. Self-destruction is prevented by the expression of genes that suppress apoptosis or by the provision of external signals promoting survival. These genes and their products are discussed below. Some of the genes that control cell death by apoptosis have been highly conserved in evolution. Indeed, genes essential for some developmental cell deaths in the nematode worm *Caenorhabditis elegans* have homologues in mammalian cells. Little is known about the biochemical events that commit a cell to an apoptotic cell death. Recent evidence suggests that there is an initiation of a proteolytic cascade in the cytoplasm, which ultimately results in the non-random cleavage of DNA to 50 kb fragments and, in some cells, to smaller fragments of integer weight 200 base pairs. These correspond to cleavage in between nucleosomes. The cleavage of DNA to 50 kb fragments is an irreversible event.

21.3 FAILURE TO UNDERGO APOPTOSIS AFTER DNA DAMAGE PROMOTES CARCINOGENESIS

In some cell types, particularly certain stem cells, DNA damage readily initiates apoptosis. The mechanism of this DNA damage-initiated apoptosis will be discussed below. The deletion of a cell with a damaged genome may be a preferable event for the organism to fixation of the DNA damage and its perpetuation in descendants as a mutation or chromosomal aberration. This might be especially important if these cells give rise to greatly amplified numbers of potentially damaged progeny. Thus, initiation of apoptosis prevents carcinogenesis by removing damaged cells with a potential to undergo malignant transformation. The ability of different cell types to initiate apoptosis after DNA damage depends on a number of factors: the ability of that cell to 'sense' DNA damage, the way in which the DNA damage sensor(s) are coupled to the pathways initiating apoptosis (or alternative responses such as cell cycle checkpoints and repair) and the menu of survival signals acting upon that cell. This idea has important implications not only for carcinogenesis but also for treatment. First, it suggests that cancer might more appropriately be conceived as a disease of cell survival, permissive for proliferation. Second, it suggests that cells that are intrinsically more capable of sustaining DNA damage (because they do not initiate apoptosis) will also be more difficult to kill by strategies that depend on the delivery of DNA damage. This idea is illustrated in Figure 21.2.

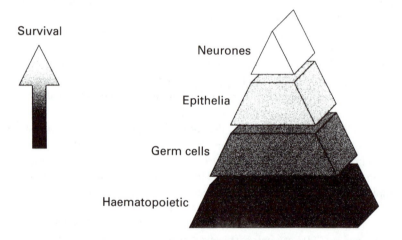

Figure 21.2 A hypothetical hierarchy of cell types according to likelihood of engaging apoptosis after damage. Cell types at the base have the lowest threshold for undergoing apoptosis. Within each stratum there are further hierarchies: small intestinal cells die more readily than colonic cells, for example (see Further Reading). Of these cell types, neurones are unique in being postmitotic; taking this property of neurones into consideration, there is an inverse relationship between the ease of apoptosis (low survival potential) and cancer incidence in a tissue type. Epithelial tumours form the major class in humans; haematopoietic tumours are of minor incidence. This hierarchy also suggests that haematopoietic tumours may be the most sensitive to therapy and epithelial tumours the most resistant. Additionally, it suggests that the side-effects of cytotoxic drugs will primarily be to the bone marrow, then the germ cells.

21.4 ANTICANCER DRUGS INITIATE APOPTOSIS

Although the past few years have seen a profound increase in our understanding of the mechanistic and molecular details of how cytotoxic antitumour drugs exert their damage on the cell (DNA damage, perturbation of nucleotide synthesis, interference with the function of microtubules, etc.) the precise reason why such perturbations result in cell death has been unclear. For example, topoisomerase poisons such as etoposide induce transient double-strand breaks in DNA. It has been considered that subsequent damage to specific genes may disable the cell because it runs out of some key gene product. There is no evidence to support or refute this. The observation that all classes of cytotoxic drugs are able to initiate cell death with the morphology of apoptosis in sensitive cells, irrespective of the type of perturbation they initiate, has changed perceptions as to how these drugs do, or often do not work. Because a variety of drugs, working to impose different types of damage, induce a conserved type of cell death, it is considered that the drug-induced damage *per se* is not directly responsible for cell death. Rather, the cell 'senses' the perturbation, or damage (such as DNA strand breaks), and responds to it according to phenotype (Figure 21.2). Thus, those cells which have a low threshold for the coupling of the damage 'signal' will engage apoptosis readily whilst those with a high threshold (and the ability to sustain damage) will not. A key question is now: what determines these thresholds?

21.5 THE TUMOUR SUPPRESSOR GENE *p53* IS A SENSOR OF CELL DAMAGE AND PLAYS A KEY ROLE IN DETERMINING THE FATE OF CELLS WHICH HAVE SUSTAINED DNA DAMAGE

The *p53* gene is the so-called 'guardian of the genome' (Chapter 9). Recent evidence suggests that p53 may be directly involved in the detection of DNA strand breaks by binding to the cleaved ends of a broken deoxynucleotide chain. By a mechanism which is not yet understood, but may involve phosphorylation, the p53 protein becomes stabilized after binding to DNA and then modifies gene transcription. p53 can both activate and suppress transcription (Figure 21.3).

Activation of transcription of the *WAF1* (also known as *CIP1*, *p21*) gene results in the inhibition of cell cycle progression as the WAF1 protein complexes with the cyclin-dependent protein kinases (CDK) required for the procession of cells from the G1 to S phase of the cell cycle (Chapter 7). This cell cycle checkpoint prevents the replication of a damaged DNA template and permits time for repair. The stoichiometry of WAF1 protein to the cyclin-dependent kinases is likely to be important in deciding the extent of the G1 checkpoint: leakage of damaged cells to S phase may initiate apoptosis. And here there is an apparently paradoxical role for p53, for not only is it essential for the imposition of a checkpoint but it is also required for apoptosis. Animals with inactivations of both *p53* alleles (homozygous-null or 'knockout' animals) provide thymocytes that fail to undergo apopto-

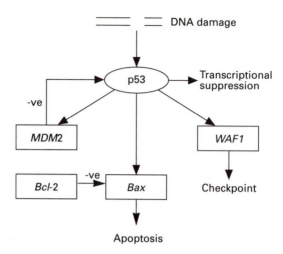

Figure 21.3 Detection of DNA damage by p53. Scheme to show the result of detection of DNA strand breaks by p53 and the pathways of gene transcription leading to the alternative fates of a cell cycle checkpoint (activation of *WAF1*) or cell death (increased expression of Bax). The activity of Bax can be counteracted by Bcl-2 expression and MDM2 activation can work in a self-regulatory feedback loop that suppresses p53 activity. The possibility of transcriptional suppression exists, which would provide a further control pathway.

sis after DNA damage inflicted by ionizing radiation or the drug etoposide. Similarly, the normal wave of cell death observed in intestinal epithelia 4–8 hours after irradiation is absent in *p53*-null mice. The loss of functional p53 therefore inhibits the onset of apoptosis. What is not clear is how different cells with functional p53 utilize it to 'check' their cell cycle or to die. How do cells 'decide' between these two outcomes? Thus far no clear relationship between the expression of p53 protein and fate exists (i.e. does more p53 protein mean more apoptosis?). This 'decision' making is discussed further, below.

Since it has been estimated that 60% of human tumours have non-functional p53 protein, and because enforced expression of p53 (by transfection of an artificial *p53* construct) in a number of tumour cell lines has resulted in their death by apoptosis, restoration of p53 function presents itself as a potentially important goal for drug hunters. One of the most encouraging results in support of such a strategy comes from the observation that the conformation of the non-functional mutant protein can be restored to that of functional protein by an antibody. Small molecules are being sought which might similarly restore p53 functionality through induction of conformational change. The question remains as to what exactly that function should be: should it restore the ability of p53 to activate transcription or to suppress transcription, or is apoptosis induced by some unknown feature of the p53 molecule? The evidence that suppression of *p53* transcription is critical for the onset of apoptosis is accumulating. One piece of evidence supporting this idea comes from studies where DNA-damage-induced cell death was not prevented by inhibitors of protein synthesis – implying that apoptosis can occur without the need to make new proteins (e.g. WAF1). This therefore presents operational problems for the drug screener – what should be the biochemical end-point of a screen looking for compounds that restore p53 function? Performing a whole cell assay, using cells with mutant p53 and measuring cell death after screening compounds is unsatisfactory because cell death can be induced in a p53-independent manner. There is a requirement for a biochemical end-point.

Although p53 protein clearly plays a pivotal role in the induction of cell death after genotoxic damage there are some caveats regarding strategies which aim to restore wild-type, functional p53 to tumour cells. The first is that the restoration may promote a cell cycle checkpoint, permitting repair of damage provided by cytotoxic drugs or radiation. The second is that it is becoming clear that p53-independent mechanisms of cell death exist that are initiated after DNA damage. These are delayed compared to the rapid, p53-dependent cell death observed in wild-type cells. The nature of this pathway and its importance to the modulation of a response to therapy in human tumours is not yet clear. In addition, the expression of p53 will induce apoptosis as long as certain members of a family of death-suppressing genes are not present. This is the *bcl-2* gene family (Figure 21.3) and their proteins, in common parlance, act 'downstream' of p53.

21.6 SURVIVAL AND DEATH ARE ALSO REGULATED BY THE RATIO OF MEMBERS OF THE *bcl-2* FAMILY OF GENES, IMPORTANT TARGETS FOR DRUG HUNTERS AND EXPRESSED IN MANY HUMAN TUMOURS

The survival of a cell depends upon receipt of signals from its partners and from the extracellular milieu, such as from the extracellular matrix and soluble growth/survival factors (e.g. nerve growth factor, insulin-like growth factor). In addition, a family of genes has been found encoding endogenous cellular proteins that form hetero- and homodimers together and which appear to contribute to the determination of the 'threshold' for survival. These are the *bcl-2* family. *Bcl-2* encodes a 26 kDa protein, of unknown activity, that suppresses cell death induced by a multitude of factors: growth/survival factor withdrawal, heat, X-rays and, very significantly, all classes of cytotoxic drugs. Bcl-2 protein contains domains that are conserved amongst other family members; these provide binding sites for hetero- and homodimerization. One of the homologues of *bcl-2*, called *bax*, accelerates cell death when expressed in cells deprived of growth/survival factors. It has been proposed that the formation of Bax–Bax protein homodimers promotes cell

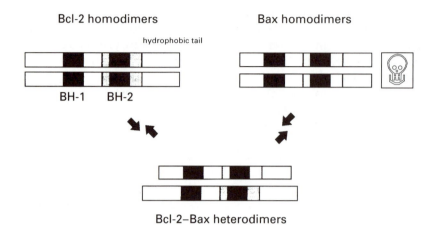

Figure 21.4 The survival 'rheostat' dependent on ratios of Bcl-2 to Bax protein. The two binding domains necessary for homo- and heterodimerization are shown (BH-1 and BH-2). An increase in Bax protein favours the formation of Bax homodimers, which is considered to accelerate apoptosis. An increase in Bcl-2 protein favours the formation of Bcl-2–Bax heterodimers, which suppresses apoptosis.

death whilst Bcl-2–Bax heterodimers suppresses death (Figure 21.4).

Although the activity of these proteins is unclear, their location in the cell is tantalizing: Bcl-2 has a hydrophobic tail that anchors it to the nuclear membrane and to the outer mitochondrial membrane. The ratio of expression of these proteins thus sets a 'rheostat' for survival or death. The control of *bax* gene expression has been suggested to be p53-regulated (Figure 21.3). Thus, after DNA damage the 'decision' to arrest at a cell cycle 'checkpoint' or to die may depend upon the ratio of Bcl-2–Bax and upon the stoichiometry of WAF1 to cyclin-dependent kinases. For example, in testes tumour cell lines (exquisitely sensitive to DNA damaging cytotoxic drugs) there is wild-type *p53*, modest levels of *bax* but no *bcl-2* expression. DNA-damaging agents cause modest increases in *WAF1* (*p21*) expression. The cells do not arrest at a G1 checkpoint but instead engage apoptosis, probably in S-phase.

Bcl-2 protein has been reported to be expressed in a whole variety of human tumours, including breast and colon carcinomas, androgen-refractive prostate cancer and neuroblastomas. A variety of studies have shown that expression of Bcl-2 protein suppresses drug- and radiation-induced apoptosis. All classes of antitumour drug have been studied: glucocorticoids, alkylating agents, topoisomerase poisons and antimetabolites. Bcl-2 expression influences the onset of drug-induced apoptosis in all cases. This represents the imposition of a true pleiotropic drug resistance, of the type observed in refractive human tumours. The resistance to drugs provided by Bcl-2 supports the idea, outlined above, that the cytotoxicity of these drugs is not due to the damage they impose *per se*, but to how that damage is coupled to apoptosis. Put simply, this is resistance to death. Those tumours arising from cells with a high threshold for the engagement of cell death, e.g. carcinomas arising from epithelia, will be inherently resistant to drugs, whereas the bone marrow will not – a picture reflecting clinical reality. This does not mean that the more classical mechanisms of drug resistance, such as DNA repair, are not important determinants of sensitivity: it is critical that sufficient damage occurs and is present for sufficient time to be 'sensed'. Similarly, efflux of a drug by the multidrug resistance protein, the inactivation of a reactive alkylating agent by elevated glutathione concentrations and modulation (e.g. overexpression) of a critical enzyme, such a thymidylate synthase, will inhibit the formation of a damage 'signal'.

There are a number of *bcl-2*-like genes whose products suppress apoptosis (e.g. *bcl-XL*, and *bag-1*), and partners which promote it (*bcl-XS, bad* and *bak*). It is not clear whether there is redundancy between these partners *in vivo* – does *bax* interact with *bcl-XL*, for example? Recent data on the production of 'knockout' mice does not support this idea: the *bcl-2*-null animals survive to birth, although they have some developmental problems and become immunodeprived, whereas the *bcl-XL* animals die *in utero*, with developmental abnormalities in the central nervous system. Preliminary data on the *bax* knockouts suggests it has no effect on development. Currently, data on the expression patterns of these genes in human tumours is being assembled and the place of each in a hierarchy of determinants of survival will be determined. Strategies to increase the expression of the proteins that promote apoptosis, so that their concentration is raised to a point where they are in excess of suppressor partners, are critical since these promoters of cell death appear to be close to the apex of the initiation of apoptosis. Strategies to remove the suppressors of cell death, such as Bcl-2, are equally important. Neither is an easy task, since once again (see p53 above) there is no clear idea of the biochemical activity of these proteins. Assays which measure heterodimerization may prove to be useful in the discovery of agents which disrupt binding (e.g. of Bax to Bcl-2, see Figure 21.4). In addition, antisense strategies against *bcl-2* have proved effective; these may be particularly useful in follicular lymphomas where *bcl-2* is translocated and the antisense is targeted to breakpoint fusion sequences.

21.7 SIGNAL CASCADES ARISING FROM GROWTH FACTORS AND THE EXTRACELLULAR MATRIX PROMOTE SURVIVAL: ABERRANT SIGNALLING MAY PREVENT DRUG-INDUCED CELL DEATH

It was outlined above that the death default position inherent to all cells is held in abeyance by provision of signals from neighbouring cells, by circulating growth/survival factors, by signals from components of the extracellular matrix and by the ratios of death/survival proteins like Bax and Bcl-2. Aberrations in these survival signalling pathways that occur during carcinogenesis may raise the survival capability of a cell: this will permit it to sustain further damage, promoting carcinogenesis and, it would be predicted, increase its resistance to cytotoxic therapy. One such aberration of signalling is provided by the translocation of the *abl* gene in chronic myeloid leukaemia (CML) to the *bcr* locus. The *bcr–abl* gene product acts as a membrane-associated tyrosine kinase. This protein tyrosine kinase acts in a similar manner to the viral oncogene v-*abl*. In studies of v-*abl*, it was shown that the signals it provided were purely those of survival. Thus, in a cell line which is entirely dependent for survival on the presence of interleukin-3, removal of the cytokine induces apoptosis. However, if v-*abl* is artificially expressed in these cells, interleukin-3 can be withdrawn without the cell death. Most pertinently, when v-*abl* is active in these cells (a temperature-sensitive form is used so that it can be switched on and off with a shift in temperature) and the cells were treated with either hydroxyurea or melphalan, all the DNA damage is produced but the cells do not die. The resistance to drugs provided by v-*abl* has more recently been observed with in experiments using *bcr–abl*, where it may act in CML to provide an escape from therapy and progression to blast crisis.

The survival signal from v-*abl* has been analysed in some depth: the activity of this tyrosine kinase elevates the production of diacylglycerols from the plasma membrane. These are activators of the ubiquitous protein kinase C enzyme (pkC). Careful analysis showed that a single isoform of pkC (pkC βII) is activated when v-*abl* ensures survival. If this activity is important for survival and the provision of drug-resistance, then inhibition of protein kinase C should abrogate the survival signal and permit the drug-induced DNA damage to engage apoptosis. Calphostin C is a specific inhibitor of protein kinase C: it completely restored the drug sensitivity of cells expressing v-*abl*. It is not feasible to use calphostin C in the clinic but the approach shows that inhibitors of survival signalling can be developed which might be specific for the menu of signals used by a particular cell.

The concept that signalling pathways exist that signal specifically for survival should not be surprising: nerve growth factor provides such signals to postmitotic nerves. Growth factors may provide both growth and survival signals. For example PDGF provides certain fibroblasts with a potent growth signal whereas insulin-like growth factor-1 (IGF-1) is a very poor mitogen but provides a strong survival signal. The pharmaceutical industry has a large effort in the field of cell signalling antagonists with potential use in cancer; it is possible that these programmes might produce specific antagonists for survival signalling, a strategy which is preferable to antagonizing mitogenic signalling, which will be of limited efficacy in low-growth fraction tumours.

21.8 TARGETING CELL DEATH PATHWAYS SHOULD BE AN EFFECTIVE STRATEGY, INCLUDING TUMOURS WITH LOW GROWTH FRACTIONS

The torrent of novel findings about the regulation of cellular survival and cell death is providing a new framework, not only for thinking about how tumours arise (cells which can avoid self-deletion after damage) but also for treatment. If aberrant survival, permissive for proliferation, is a basis for carcinogenesis then it is logical that we seek strategies which aim to modify cell survival. This permits the targeting of both high and low growth fraction tumours, an avenue not open to those seeking new classes of antiproliferative drugs, aimed at the cell cycle, for example.

21.9 FURTHER READING

Dive, C. and Hickman, J. A. (1991) Drug-target interactions: only the first step in the commitment to a programmed cell death? *Br. J. Cancer*, **64,** 192–196.

Fisher, D. E. (1994) Apoptosis in cancer chemotherapy: crossing the threshold. *Cell*, **78,** 539–542.

Hickman, J. A., Potten, C. S., Merritt, A. J. and Fisher, T. C. (1994) Apoptosis and cancer chemotherapy. *Proc. Roy. Soc. Lond. B*, , **345,** 319–325.

Reed, J. C. (1994) Bcl-2 and the regulation of cell death. *J. Cell Biol.*, **124,** 1–6.

22	# Tumour angiogenesis as a therapeutic target

Rhys T. Jaggar and Adrian L. Harris

All solid tumours must develop a blood supply in order to grow and recent evidence suggests that this may also be true for acute leukaemias. Hence, targeting the tumour vasculature is an attractive anticancer strategy. A knowledge of the physiological control of angiogenesis and how such processes are subverted during tumour growth will be required for rational design and usage of antiangiogenic drugs.

22.1 ANGIOGENESIS IS A TIGHTLY CONTROLLED PHYSIO-LOGICAL PROCESS INVOLVING BOTH ACTIVATORS AND REPRESSORS

Angiogenesis, the growth of new blood vessels, occurs during normal embryogenesis and during the wound healing process. It also occurs in certain pathological conditions such as rheumatoid arthritis, diabetic retinopathy, psoriasis, atherosclerosis and the growth of solid tumours.

The angiogenic process may be split into four parts (Figure 22.1):

- degradation of the basement membrane surrounding the pre-existing blood vessel;
- endothelial cell division;
- endothelial cell migration;
- anastomosis of the new vessels to pre-existing ones to form a new vascular bed.

Although anastomosis is still incompletely understood, during the last 10 years there has been a great increase in understanding the processes of basement mem-

Molecular Biology for Oncologists. Edited by J. R. Yarnold, M. R. Stratton and T. J. McMillan. Published in 1996 by Chapman & Hall, London. ISBN 0 412 71270 9

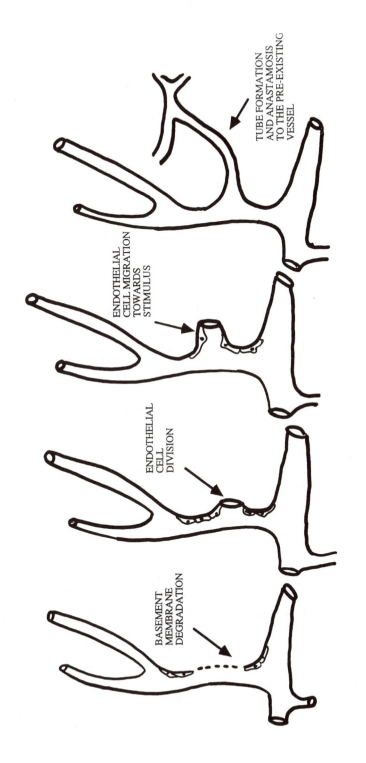

Figure 22.1 The process of angiogenesis.

brane degradation and endothelial cell migration and division. It is now clear that there are several naturally occurring molecules capable of stimulating the migration and/or growth of new blood vessels (**angiogenic factors**). However, in normal adults, the activity of these factors is masked by the over-riding presence of other molecules that inhibit angiogenesis (**antiangiogenic factors**). Under certain circumstances, these inhibitory signals are over-ridden in order to allow vascular development or remodelling. These include the vascularization of the brain or limb development during fetal growth; wound healing; or the proliferative phase of the endometrium during the female menstrual cycle. As such, this system is similar to the clotting cascade: normally quiescent but capable of rapid activation.

22.2 TUMOUR GROWTH IS DEPENDENT ON ANGIOGENESIS

It is now over 20 years since Judah Folkman at Harvard proposed that solid tumour growth is absolutely dependent on angiogenesis: growth of tumours beyond a diameter of $\approx$ 2–4 mm does not occur without the acquisition of a vascular bed. In clinical terms, angiogenesis has recently become an important prognostic indicator for cancer progression. Several studies have correlated the vascular density of tumours or quantitative blood vessel counts (which are indirect measures of the rate of vessel formation) with disease progression and/or metastasis and/or survival in cancers of the breast, stomach, prostate, colon, lung and bladder adenocarcinomas and thin malignant melanoma. The methodology of assessment is quite critical, since angiogenesis is not uniform in tumours. It occurs at the growing edge of tumours in vascular 'hot spots' and it is these that correlate with prognosis.

This requirement for angiogenesis implies that tumours must somehow elicit endothelial cell division and migration into the tumour. The isolation and characterization of angiogenic factors produced in tumour tissues and secreted from tumour cell lines was a landmark in cancer research (see Folkman and Shing's review).

22.3 MANY ANGIOGENIC MOLECULES ARE EXPRESSED IN TUMOURS

Several *in vitro* and *in vivo* assays for identifying angiogenic factors have been developed. Several of these are described in section 22.11. Using these and other assays, a large number of molecules have been implicated over the last 15 years in stimulation of angiogenesis. The physiological relevance of some of these is still unclear, as several may be indirectly angiogenic by stimulating production or activity of true angiogenic molecules.

Table 22.1 lists molecules currently accepted as important physiological mediators of tumour angiogenesis. The genes for these polypeptide angiogenic factors have now been cloned and antibodies have been raised, allowing the analysis of

RNA and protein expression within tumours and in surrounding normal tissue.

Currently, for most tumour types, no complete picture of the expression of all known angiogenic molecules is available. However, in all tumour types studied, transcriptional up-regulation of at least one angiogenic factor has been demonstrated in late-stage disease.

- **Platelet-derived endothelial cell growth factor** (PdECGF) has been shown to be up-regulated in many breast and ovarian tumours, squamous cell carcinomas and invasive bladder carcinomas.
- **Vascular endothelial growth factor** (VEGF), an endothelial cell-specific mitogen, is up-regulated in many gliomas, as well as superficial bladder tumours and breast tumours.
- **Acidic fibroblast growth factor** (aFGF) has been shown to be up-regulated in invasive bladder carcinoma.
- **Basic fibroblast growth factor** (bFGF) is up-regulated in Kaposi's sarcoma, renal cancers and some glioblastomas.
- **Interleukin 8** (IL-8) is up-regulated in non-small-cell lung cancer.

The method of activation of angiogenic factor gene expression is now being elucidated. Although such work is still in its early stages, a few pertinent observations are listed below.

- Hypoxia, which has been shown to occur within many tumours, is a potent inducer of VEGF expression in breast cancer cells and vascular smooth muscle cells.

Table 22.1 Angiogenic factors activated in human cancer

Angiogenic factor	Activity of molecule	Up-regulation in tumours
Basic fibroblast growth factor (bFGF)	Heparin-binding growth factor	Kaposi's sarcoma; glioblastoma
Acidic fibroblast growth factor (aFGF)	Heparin-binding growth factor	Invasive bladder cancer
Vascular endothelial growth factor (VEGF)	Heparin-binding growth factor	Glioblastoma; breast carcinoma; rhabdomyosarcoma; superficial bladder cancer
Platelet-derived endothelial cell growth factor (PdECGF/TP)	Thymidine phosphorylase enzyme	Breast carcinoma; ovarian carcinoma; invasive bladder carcinoma
Interleukin-8	Cytokine that activates monocytes; direct growth factor?	Non-small-cell lung carcinoma
Angiogenin	RNase?	Adenocarcinoma

- Interleukin 8, a cytokine molecule expressed in many non-small-cell lung car-cinomas, has been shown to activate monocytes to produce other angiogenic factors, and may therefore be an indirect activator of angiogenesis as well as a direct endothelial cell growth factor.
- bFGF protein is converted from a cell-associated (inactive?) form to an exported (active?) form during the development of aggressive fibrosarcomas in *BPV-1* transgenic mice.
- Mutation of the *p53* tumour suppressor gene, one of the commonest changes found in a wide spectrum of cancers, can induce up-regulation of VEGF expression in synergy with a tumour promoter phorbol ester, TPA, at least in a mouse fibroblast tissue culture system.

There is great scope at present for rigorous studies of angiogenic factor RNA expression and protein activity in tumours of a variety of types and the mechanism of activation where appropriate. This may aid the development of appropriate antagonists of angiogenic factors. It may also be of predictive value in designing future treatment schedules, as the antiangiogenic therapeutic tools reach the clinic.

22.4 NATURALLY OCCURRING INHIBITORS OF ANGIOGENESIS HAVE RECENTLY BEEN ISOLATED

For many years now, it has been recognized that certain tissues, notably joints and the eye, are almost completely avascular and that vascularization of these tissues causes pathological effects. These tissues have therefore been a starting point for the identification of angiogenesis inhibitors. A relatively large number of naturally occurring inhibitors of angiogenesis have now been purified and, where appropri-ate, their genes have been cloned.

Many of these inhibitors act by the inhibition of the proteinases necessary for the degradation of the basement membrane surrounding endothelial cells: exam-ples are TIMP1 (**t**issue **i**nhibitor of **m**etallo**p**roteinase), TIMPII, CDI (**c**artilage-**d**erived **i**nhibitor) and PAI-1 (**p**lasminogen **a**ctivator **i**nhibitor).

However, inhibition of the action of endothelial cell growth factors may also be a common mechanism. Four examples of this are as follows.

- Soluble bFGF receptors exist in serum which may well bind to free bFGF and curtail the action of the angiogenic factor.
- Soluble VEGF receptors, produced using recombinant DNA technology, inhibit tumour growth in mouse tumour models.
- α_2-macroglobulin, a potent inhibitor of angiogenesis, prevents the binding of VEGF to its receptor.
- Heparin, an angiogenic stimulus, is able to prevent complexing of VEGF to α_2-macroglobulin.

A third mechanism of action of angiogenesis inhibitors is the induction of endothelial cell apoptosis: the cytokine tumour necrosis factor α (TNF-α), which shows potent antitumour activity when injected adjacent to tumours in mice, works in this way. The molecule **angiostatin** (a fragment of plasminogen), is an inhibitor expressed in primary mouse Lewis lung carcinomas (a mouse tumour model system) which prevents the growth of lung metastases. This inhibition appears to be due to induction of apoptosis in the endothelium of developing micrometastases. This observation may be of relevance to the clinical observation that occasional patients show rapid growth of metastasis within a few months of surgery to their primary tumours.

A number of other cytokines such as IFN-α and IFN-γ are also antiangiogenic.

The key questions concerning these naturally occurring inhibitors at present are their tissue distribution under normal conditions and how their expression or activity changes, if at all, during tumour development.

22.5 SEVERAL THERAPEUTIC DRUGS INHIBIT ANGIOGENESIS

There are now several antiangiogenic molecules either in routine clinical use or in Phase I clinical trials.

- **Interferon-α** is currently used as a treatment for haemangiomas and Kaposi's sarcoma and several trials are under way in combination with chemotherapeutic agents in ovarian, stomach and hepatocellular cancer.
- **Suramin**, which prevents binding of bFGF to its receptors, has been used in trials for prostate cancer, and the recent development of analogues with lower toxicity promises further developments.
- The fumagillin analogue **AGM-1470/TNP 470**, which inhibits endothelial cell proliferation, is being used for trials in patients with Kaposi's sarcoma.
- **BB94**, which inhibits type IV collagenase and other stromal proteases, is now in Phase I trials for pleural effusion and ascites. A new oral analogue is also being studied.

Finally, **thalidomide** has recently been shown to be an inhibitor of angiogenesis and is starting clinical trials in the USA for brain tumours. A clinical trial for ovarian cancer is currently being designed by the Imperial Cancer Research Fund in Oxford, UK.

With the greater understanding of the angiogenic process, the development of designer antiangiogenic drugs is now possible. There is thus hope that a rapid improvement in the quality of these drugs may be possible in the next decade.

22.6 IT MAY BE POSSIBLE TO BLOCK THE BINDING OF TUMOUR ANGIOGENIC FACTORS TO THEIR SPECIFIC RECEPTORS

It is clear that most, although not all, of the tumour angiogenic factors are growth factors. These elicit their responses via binding to specific cell-surface transmembrane receptors. Hence, molecules that can interfere specifically with this interaction may be useful antiangiogenic molecules. In this regard, the extracellular ligand-binding domain of receptors (termed 'soluble receptors') has been shown in many cases to act as a competitive inhibitor. Similarly, small peptides from these domains are also active. Whether such molecules will be effective when delivered systemically remains to be seen.

22.7 ANTIBODIES TO TUMOUR VASCULAR ENDOTHELIUM MAY BE USED AS TARGETING VEHICLES FOR DRUGS OR AS ANTIANGIOGENIC MOLECULES IN THEIR OWN RIGHT

It is well known that tumour endothelium exhibits properties distinct from normal endothelium. Several research groups have therefore sought to identify tumour-endothelial-specific cell-surface proteins which could be used as targets for therapeutic drugs.

The first demonstration of antibody-mediated destruction of the tumour vasculature was described by Burrows and Thorpe. By expressing interferon-γ in tumour cells, they were able to induce *MHC* class I gene expression on the surface of tumour endothelial cells in xenografted tumours. By linking ricin to an antibody to *MHC* class I, they were able to demonstrate specific destruction of a solid tumour.

Two papers have recently appeared which identify $\alpha_v\beta_3$-integrin (a member of a family of proteins that mediate cell–cell adhesion) as a marker of proliferating as opposed to quiescent endothelial cells (Chapter 10). Importantly, antibodies or peptide antagonists to this integrin were able to inhibit human tumour xenograft growth and angiogenesis, albeit in an artificial assay (tumours grown on the chorioallantoic membrane of chick embryos). The mode of action of such inhibition was the stimulation of apoptosis of endothelial cells. The development of antibodies to tumour endothelium represents a key strategy in the delivery of cytotoxic therapies to tumours.

Several antibodies to angiogenic molecules have been developed and have been shown, at least in mouse tumour model systems, to inhibit tumour growth. Thus, antibodies to VEGF have been shown to inhibit glioblastoma growth; antibodies to bFGF decrease growth of colon carcinoma and glioblastomas; and antibodies to angiogenin decrease the growth of adenocarcinomas. It will be necessary to modify such antibodies for human use, before they can be used in clinical trials, since mouse monoclonal antibodies will elicit an immune response in humans and chronic administration of therapy is likely to be necessary (Chapter 23). However, it can be envisaged that such molecules may well enter into phase I clinical trials within the next 5 years. The pharmacological properties of such antibodies may be the key determinant in the success of such strategies.

22.8 IT MAY BE POSSIBLE TO TARGET GENES FOR TOXIC PROTEINS OR PRODRUGS TO THE TUMOUR VASCULATURE

Expression of toxic proteins or prodrug activating proteins in the tumour vasculature following delivery of their genes represents another therapeutic approach. As yet, there remain several technical hurdles to be overcome before such methods can be used in the clinic.

The chief problem for such a strategy is the delivery system (Chapter 24). The use of retroviruses allows selective delivery to proliferating cells, a useful characteristic given the quiescent nature of normal endothelium. However, three critical problems remain.

- the production of sufficiently high titre viruses (current limits are 10^9 particles/ml), which probably need to be of at least the order of 10^{12}–10^{13}/ml to allow efficient uptake by tumour vasculature following systemic delivery;
- the engineering of endothelial-specific protein expression within the context of a retroviral vector, or the modification of the viral envelope to target infection to endothelial cells;
- the development of therapeutic virus particles that are resistant to the neutralizing action of human complement.

Using other virus systems – adenovirus or adeno-associated virus (AAV) – viral titres of 10^{12}–10^{13}/ml are now routinely available. However, such viruses also infect quiescent cells and are thus currently more suited to gene replacement therapies (such as the expression of Factor IX in haemophiliacs) rather than destructive therapies. If incorporation of tumour-specific expression mechanisms becomes possible within these viral vectors, then they may be of use in antitumour therapies.

An alternative delivery system may be the encapsidation of the DNA expression vector within liposomes. Again, methods for targeting such liposomes to tumour endothelium must be developed before trials may be contemplated. Such methods may include the incorporation of suitable antibodies into the lipid formulation. Liposome delivery systems must also incorporate methods for the prevention of DNA destruction within the endosome of the cell following fusion of the lipid to the endothelial cell. The use of endosome disruption proteins found naturally in adenoviruses is currently being developed.

22.9 CLINICAL TRIALS WILL NEED TO INCORPORATE PHYSIO-LOGICAL MEASURES OF ANTIANGIOGENIC EFFECT AND TEST COMBINATION THERAPIES

Selection of suitable patients for clinical trials of antiangiogenic drugs is just as critical as good drug design. Since patients who respond poorly to conventional therapy often have high tumour angiogenesis, they may be suitable candidates for such trials. The use of diagnostic tests prior to treatment, such as measurement of angiogenic factor levels in tumour, blood or urine; or blood flow measurements before surgery may also be important in determining patient suitability for a particular treatment regimen.

Such trials should aim to monitor a particular physiological property such as reduction in blood flow, markers of endothelial damage or decreased production of angiogenic factors, since inhibition of angiogenesis alone may not produce tumour regression, but rather inhibit further growth.

Effective therapy will undoubtedly require combination treatments. There is experimental evidence in animal models that antiangiogenic drugs combined with several different types of anticancer drugs act synergistically without increasing toxicity. One possible trial design may be the treatment of patients with antiangiogenic drugs between courses of chemotherapy: this may increase periods of remission following treatment. Another attractive combination in this regard is the use of antiangiogenic agents combined with hypoxia-activated drugs, since regions of most solid tumours show marked hypoxia.

In conclusion, the increasing understanding of tumour angiogenesis offers great promise for novel treatment regimens. Whether tumours will prove as resistant to antiangiogenic therapies as they have to chemo- and radiotherapy is a key question in cancer treatment.

22.10 ACKNOWLEDGEMENTS

We wish to thank Drs Richard Isaacs, Tim O'Brien and Emmanuel Huguet for critical comments on the manuscript and Miss Rangana Choudhuri for artwork in Figure 22.1. The authors acknowledge the financial support of the Imperial Cancer Research Fund.

22.11 APPENDIX: ASSAYS FOR THE DETECTION OF ANGIOGENIC FACTOR ACTIVITY

Chick chorioallantoic membrane

The first commonly used method developed for the detection of angiogenic factor activity utilizes the chick chorioallantoic membrane (CAM). A window is made in the eggs 4 days after they are laid and some albumin is removed to lower the window level. The following day, the angiogenic compound is laid on the membrane and 3 days later the vasculature formed within the membrane is analysed.

Rat sponge assay

A more recent assay involves the implantation of a polyester sponge into the back of a rat. This is connected to a cannula, which allows the injection of radioactive xenon. The rate at which the radioactive xenon is cleared from the sponge has been shown to correlate with the development of a vascular supply into the sponge. By adding angiogenic compounds the rate of radioactive xenon clearance over a 10-day period can be compared to baseline levels. At the end of an experiment, sponges may also be examined histologically for evidence of blood vessel development (Plate 3). The figure shows the structure of a sponge into which the angiogenic factor thymidine phosphorylase has been introduced (a) and an untreated sponge (b).

A third, highly potent assay utilizing the rabbit cornea has recently been used successfully by many workers.

Transfection of angiogenic factor genes into MCF-7 cells

While the three methods described above involve adding purified factors to an assay system, transfection of angiogenic factor genes into the breast carcinoma cell line MCF-7 has recently provided an assay for angiogenic factors in the stimulation of tumour growth, when such cells are xenografted into the flank of athymic mice. The success of this system is based on the fact that the parental MCF-7 cells form slow-growing, poorly vascularized tumours in the absence of any transfected angiogenic factor.

Expression of the transfected angiogenic factor has been correlated with increased rates of tumour growth and, in the case of vascular endothelial growth factor, with an increased vascular density and vascular 'hot spots'. This method is of greatest use in identifying as angiogenic factors proteins whose conventional purification is difficult, or in the confirmation of activity of known angiogenic molecules in a tumour angiogenesis assay.

22.12 FURTHER READING

Auerbach, W. and Auerbach, R. (1994) Angiogenesis inhibition: a review. *Pharmacol. Ther.*, **63**, 265–311.

Brooks, P. C., Montgomery, A. M. P., Rosenfeld, M. *et al.* (1994) Integrin $\alpha_v\beta_3$ antagonists promote tumour regression by inducing apoptosis of angiogenic blood vessels. *Cell*, **79**, 1157–1164.

Burrows, F. J. and Thorpe, P. E. (1993) Eradication of large solid tumours in mice with an immunotoxin directed against tumour vasculature. *Proc. Nat. Acad. Sci. USA*, **90**, 8996–9000.

Folkman, J. and Shing, Y. (1992) Angiogenesis. *J. Bio. Chem.*, **267**, 10931–10934.

O'Reilly, M. S., Holmgren, L., and Shing, Y. (1994) Angiostatin: a novel angiogenesis inhibitor that mediates the suppression of metastases by a Lewis lung carcinoma. *Cell*, **79**, 315–328.

23 | Antibody technology has been transformed

Robert E. Hawkins and Kerry A. Chester

23.1 RECOMBINANT DNA TECHNOLOGY IS IMPROVING ANTIBODY-BASED THERAPIES IN CANCER

The development of hybridoma technology by Köhler and Milstein 20 years ago generated monoclonal antibodies with defined specificities which could be produced in unlimited quantities. Since then, antibodies with some degree of tumour specificity have been produced and tested in trials of antibody imaging and therapy. These have shown that antibodies localize well to tumours for imaging purposes, but useful therapeutic effects have been harder to demonstrate. However, recent success has been reported using unmodified antibodies as adjuvant therapy for colorectal cancer and using radiolabelled antibodies in radiosensitive tumours (lymphomas). More antibodies are in phase III clinical trials and results should be available before long. Alongside these clinical developments there has been a revolution in the way antibodies are isolated and manufactured. These methods, based on advances in molecular biology, allow the production of second-generation antibody-based molecules. This chapter reviews the clinical issues in antibody targeted cancer therapy and describes changes in antibody technology that are likely to find useful clinical applications.

23.2 ANTIBODIES CAN BE USED TO DELIVER PHARMACEUTICALS, OR TO KILL CELLS IN THEIR OWN RIGHT

Antibodies are versatile targeting molecules, which can be used to deliver radionuclides, enzymes, genes, drugs or toxins to target cells or to act as biologically

Molecular Biology for Oncologists. Edited by J. R. Yarnold, M. R. Stratton and T. J. McMillan. Published in 1996 by Chapman & Hall, London. ISBN 0 412 71270 9

active molecules in their own right. Natural antibodies cause cell death by antibody-dependent cellular cytotoxicity or complement-mediated cell lysis. In addition, antibodies can be used to block the interaction of ligands, such as growth factors and hormones, with their receptors. They can also be used to interfere with tumour growth by blocking cell–cell or cell–substrate interactions. Since specific antibodies can be generated to most target antigens the clinical potential for application of antibodies to cancer therapy is enormous. There are, however, difficulties with therapy using antibodies (or any macromolecules). Major issues are the relative scarcity of suitable target antigens, the poor tumour penetration by macromolecules, the immunogenicity of the reagents and the suboptimal affinity of the antibody for the target. These are discussed below, and understanding them should lead to the design and production of new and improved antibody-based targeting molecules.

23.3 PROMISING NEW TARGET ANTIGENS INCLUDE CELL SURFACE CARBOHYDRATES AND THE TUMOUR VASCULATURE

Selecting a suitable target antigen is an essential first step in design for any antibody-based therapy. For example, if the target antigen is expressed unevenly in a tumour, a therapeutic agent which causes the death of surrounding cells (the 'bystander effect') is preferable. The classes of tumour antigens that make potential targets for antibody therapy are shown in Table 23.1.

Specificity is important because even weakly cross-reactive normal tissues may cause problems, especially if they are more accessible to the antibody. There are some tumour-specific targets such as the idiotype in lymphomas, or mutant cell-surface proteins increasingly identified in other tumours. However, in the absence of a well-defined tumour-specific antigen, the relative abundance of tumour anti-

Table 23.1 Target antigens for anticancer monoclonal antibodies

Desirable characteristics	Examples
Unique to tumour	Immunoglobulins T-cell receptors Mutated cell surface proteins
Relative abundance in tumour	Oncofetal antigens Growth factor receptors Hormones Dead cell markers Altered carbohydrate groups
Confined to tumour and non-essential normal tissues	Differentiation antigens
Stromal targets	Endothelial activation markers Fibroblast activation markers

gen may provide adequate selectivity for some applications. This is especially true in cases where the forms expressed in normal tissue are relatively inaccessible to antibody, as with carcinoembryonic antigen (CEA), which is expressed mainly on the luminal surface of normal intestinal glands. In any case, the cross-reactivity should be well defined and the therapeutic molecule designed so that it is less toxic to the cross-reactive tissue types. Promising targets include plasma membrane carbohydrates which are relatively specific, abundant antigens on many tumours. Another attractive target is the tumour vasculature, which is readily accessible and essential for tumour growth.

23.4 MACROMOLECULES DIFFUSE SLOWLY IN THE EXTRAVASCULAR SPACE

Our understanding of tumour vasculature and the penetration of macromolecules into solid tumours is incomplete but some principles are clear. Studies with various high-molecular-weight dextrans demonstrate that the tumour neovasculature is more permeable than normal blood vessels and this allows the leakage of macromolecules from the vessels. However, once extravasated they penetrate the tumour parenchyma slowly and inefficiently. This also applies to antibodies and must clearly be considered when designing targeting molecules.

23.5 IMMUNE REACTIONS TO ANTIBODIES ARE REDUCED BY REPLACING RODENT COMPONENTS WITH HUMAN SEQUENCES

Rodent antibodies are immunogenic in humans, resulting in progressively shorter half-life of injected antibody with repeated dosage. Although rare in practice, this can also result in toxic side-effects such as serum sickness or anaphylaxis. Production of chimeric or fully reshaped antibodies (humanization) allows this problem to be reduced or avoided, but rapid methods of making human antibodies directly are needed (see below).

23.6 AFFINITY AND AVIDITY ARE IMPORTANT FOR ANTIBODY–ANTIGEN BINDING

The specificity of antibody binding is one thing, but the affinity of binding to antigen is also important. Antibody interaction with antigen is usually described in terms of **affinity**, a term expressing the probability of two molecules binding each other. The interaction is also influenced by the probability of separation, so the net interaction is a composite of the kinetic 'on-rate' and 'off-rate'. These considerations are clearly complex and there are few experimental data to help define the optimal characteristics. In addition to affinity and 'off-rate' there are other features of the binding process which can be exploited. Natural antibodies are at least bivalent (i.e. have at least two antigen-binding sites), and this feature strengthens binding. Affinity is a function of univalent interactions but the binding of polyvalent antibody results in a much more stable interaction – known as the avidity effect. This can result in large increases in functional affinity (up to 1000-fold for an IgG molecule compared to a single Fab fragment) but depends critically on the density of the target antigen in relation to the spacing of the antibody heads. Natural antibodies exploit this feature, but by careful design it may be possible to manufacture molecules that are small and penetrate well without loss of avidity (see below).

23.7 SMALL HUMANIZED ANTIBODIES ARE MORE DIFFUSIBLE AND LIKELY TO AVOID IMMUNE RESPONSES

Genetic engineering allows the creation of many antibody-derived molecules, shown in Figure 23.1 in relation to their natural counterparts.

The ideal antibody fragment for therapy would be small for optimal penetration, and have high affinity and specificity for its antigen. The smallest molecule which binds in the same way as the native antibody is the Fv comprising the VH and VL (Figure 23.1). With a 15-amino-acid linker between the VH and VL the molecule is known as a single chain Fv (scFv) and is more stable (Figure 23.1). The molecule should also be humanized enough to avoid immune responses during repeated therapy. To overcome tumour antigen heterogeneity and the difficulty of reaching all cells, the effector arm of the targeting molecule should also have a bystander effect. Finally, given the relatively low proportion of injected dose reaching the tumour, the preferred effector mechanism should incorporate some amplification and should be non-toxic in the delivered form. The extent to which the genetic manipulation of antibodies will allow us to approach this ideal is discussed below.

23.8 BACTERIOPHAGE LIBRARIES CAN BE USED TO ISOLATE HUMAN ANTIBODY FRAGMENTS

The expression of antibodies on the surface of bacteriophage (bacterial viruses) has allowed the rapid development of this approach to making antibodies. The polymerase chain reaction (PCR) is used to artificially replicate rearranged v-genes from the DNA of polyclonal B cells (each B cell providing a uniquely rearranged v-gene). Once amplified, these gene fragments can be inserted into the bacteriophage genome and expressed as protein on the viral surface. The process is manipulated to ensure that each bacteriophage particle incorporates a single antibody gene fragment, and the collection forms a gene library. The display of antibody on the viral surface means that bacteriophage can be used directly to

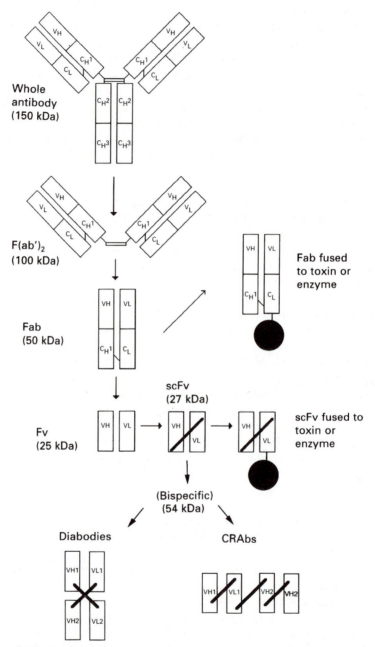

Figure 23.1 The various formats of antibody-derived fragments compared with native immunoglobulin. Molecular size is important for tumour penetration and in determining the circulating half-life. The Fab and scFv can also be produced as bispecific antibodies. Diabodies and CRAbs may be expressed as bispecific fragments without further manipulation. These fragments can be produced in eukaryotic cells or bacteria.

select for specificity and binding characteristics viz. high affinity and slow 'off-rate'. The antigen concentration in the system is varied to influence the initial selection of antibodies with high affinity, or to improve existing antibodies. After initial selection, various methods of mutagenesis can be used to increase the variability in antibody gene sequences and generate secondary libraries prior to the selection of variants with improved specificity and binding characteristics.

Antibodies generated by this system using the B cells of immunized mice have favourable characteristics (higher affinity and better specificity) compared to conventional monoclonal antibodies. They also perform well in imaging studies in mice and humans. For making human antibodies the method has further advantages. By using synthetic v-gene repertoires or v-gene sequences from non-immune humans, we can make and affinity-mature human antibodies to many different antigens just like the natural immune system (Figure 23.2).

As the technology advances and the v-gene libraries become larger, the quality and number of the antibodies produced improve. It thus seems possible that the process of isolating and testing useful human antibodies may eventually be done from a 'single-pot' library. Other methods of making human antibodies include generation of transgenic mice carrying the human immunoglobulin locus – again the phage system will enhance and simplify these methods.

A further advantage of the phage system is that it permits the selection of antibodies in the form of DNA sequences, which can be modified and expressed in a number of ways. For example, it allows the attachment of novel effector mechanisms, including direct expression of the gene *in vivo*, to deliver therapeutic antibody. The basic types of antibodies are shown in Figure 23.1 and some of the possible methods of using such fragments are discussed below. Overall the phage

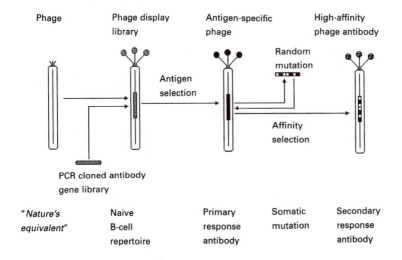

Figure 23.2 Antibodies *in vitro*. An overview comparing the phage system with that of the immune system.

Table 23.2 Advantages of using bacteriophage to make antibodies for therapy

- Human antibodies can be made directly
- Large numbers of antibodies can be made
- Selection for high affinity (and slow 'off-rate') is possible
- Antibody genes are obtained directly allowing expression in many different forms

system seems destined to become the method of choice for making antibodies (Table 23.2).

23.9 ANTIBODIES ARE BEING ENGINEERED TO IMPROVE THEIR BIODISTRIBUTION

Poor penetration of antibody into a solid tumour mass remains one of the greatest barriers to effective therapy. One approach to improving this is to use small antibody fragments. scFvs have been shown to permeate more rapidly and deeper into tumours than whole IgG. Although allowing better penetration, scFvs are rapidly cleared through the kidneys, resulting in a short serum half-life. In addition, retention of scFvs in the kidney is a common observation, and although the extent may differ with different scFvs the factors controlling this are not clear. Unmodified scFvs may therefore be convenient for tumour imaging where images can be obtained in patients within hours rather than the days required for whole antibodies, but for therapy there are other important considerations. For example, scFvs are univalent so they must rely on intrinsic affinity rather than avidity to remain bound in the tumour. Certainly there is an indication that high affinity antibodies are advantageous for targeting and that they can improve antitumour activity and survival in animal models. In addition, by careful design it may be possible to make use of avidity effects but still have a small molecule which penetrates well. For example, by shortening the 15 amino acid linker molecule between VH and VL sequences, scFvs can be encouraged to dimerize and form effectively bivalent molecules (called diabodies) with better avidity. Chelating recombinant antibodies (CRAbs) offer another approach, comprising two scFvs with specificity for different epitopes on the same molecule. They may be particularly interesting if they increase functional affinity and specificity for their target antigen. The structure of diabodies and CRAbs in relation to native immunoglobulin is shown in Figure 23.1.

23.10 DIFFERENT APPROACHES TO ANTIBODY THERAPY HAVE SHOWN PROMISING CLINICAL RESULTS

A number of effector molecules have been linked to antibodies to produce targeted cell killing (Figure 23.3). The relative merits of each are compared in Table 23.3.

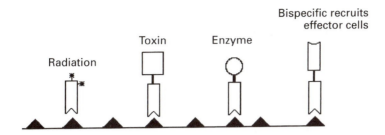

Figure 23.3 Current methods of cancer therapy that employ antibodies to guide cell destruction. A number of methods of antibody-mediated cell destruction are currently used. Radiation and toxins are directly cytotoxic to targeted cells and radiation can also have the benefit of a bystander effect. Targeted enzymes rely on a second phase of therapy with prodrugs and bispecifics are dependent on effective recruitment of the host immune defence. The relative merits of each approach are compared in Table 23.3.

Of these effectors, drug conjugates and immunotoxins kill only the targeted cell (single hit), while others such as natural antibodies and radioactivity exhibit a bystander effect. The latter is attractive since it is not possible to reach every tumour cell with current antibody targeting agents. For patients with radiosensitive tumours, such as lymphomas, antibody-guided ionizing radiation can result in prolonged remissions, although there may be considerable toxicity. Occasional responses have also been reported in patients with less radiosensitive tumours, such as colorectal cancer, and the approach is effective in nude mice bearing human tumour xenografts. The basic problem is how to target a sufficient radiation dose to the tumour without systemic toxicity – altering the size, affinity and avidity of antibody molecules should improve this.

Natural effector mechanisms may also provide a bystander effect and if enough cells are killed, the patient's natural immune mechanisms may be able to attack the remaining tumour. Nevertheless, problems of accessibility suggest that small volume disease is the most appropriate target for native antibodies. It is encouraging that a randomized trial using a native murine antibody as adjuvant therapy in patients with colorectal cancer has shown a survival advantage comparable to that obtained with chemotherapy, but with less toxicity. Such treatment could provide an entirely new form of therapy for common tumours. Bispecific antibodies may be more effective in recruiting effector cells. Newer methods of producing bispecific antibodies will simplify production and allow widespread testing, and may allow design of new fragments with better biodistribution. Bispecific antibodies may be combined with other ways of harnessing cell-mediated immunity, such as the use of superantigens linked to antibodies or the use of antibodies linked to cytokines such as IL-2. Furthermore, these natural mechanisms may be used in combination with systemic cytokines and perhaps with adoptive T-cell immunotherapy to further stimulate the antitumour response.

Table 23.3 Comparison of antibody effector mechanisms

Feature	Effector mechanism				
	Natural antibody	Radiation	Toxin conjugate	Drug conjugate	ADEPT
Bystander effects	Yes	Yes	No	No	Yes
Immunogenic	No[a]	No[a]	Yes[b]	No[a]	Yes[b]
Toxicity of unbound form	Low	High	Low[c]	Low[c]	Low[d]
Amplifies signal	Yes	No	No	No	Yes

All antibody components are human or humanized.

Toxicity will be dependent on the antigen target and the nature of antibody cross-reactivity.

[a] Antibody responses can occur to the idiotype of a natural antibody and to the coupling agent, if used, for radiation and drugs.

[b] Current constructs involve toxins/enzymes that are non-human and thus immunogenic but newer forms may be made from human proteins.

[c] Toxicity partly depends on the method of linkage used. In vivo instability has been a problem with some methods. For toxins this may be overcome by using fusion proteins. Vascular leak syndrome can occur with some conjugates.

[d] The toxicity depends on the prodrug. There may be some activation of drugs in normal tissues, resulting in toxicity.

Immunotoxins can be effective in animal models but have been problematic in patients. The efficacy of immunotoxins may be facilitated by protein-engineering, but the lack of a bystander effect will limit their use to tumours where cells are accessible and to targets expressed on most, if not all, tumour cells. These considerations also apply to drug conjugates but here there is a further problem of getting sufficient drug to the target. This may be improved by the use of more potent drugs. It is likely that the method of drug linkage to antibody, and the fate of the antibody after binding to its target, will be important considerations. For instance, an internalizing antibody conjugated to doxorubicin with a thioether linker has proved to be a very effective therapeutic agent against human carcinomas in nude mice.

Finally, the ADEPT (antibody-directed enzyme-activated prodrug therapy) system is attractive (Chapter 25). In this approach, an antibody is used to target an enzyme to the tumour and retain it there while the antibody–enzyme conjugate clears from normal tissues. A non-toxic prodrug is then given and this is activated by the enzyme to produce a cytotoxic drug at the tumour site. The ADEPT system has the attraction that it includes an element of amplification because the targeted enzyme can produce many molecules of the active drug within the tumour. There is also a strong bystander effect because the final cytotoxic molecule is small, allowing diffusion throughout the tumour. The obvious drawbacks are the need to develop appropriate and non-immunogenic drug/antibody–enzyme combinations. Results of the first ADEPT clinical trial have shown encouraging responses in patients with colorectal cancer receiving antibody to CEA conjugated to carboxypeptidase G2 to activate a benzoic acid mustard prodrug. Fusion proteins based on the scFv fragment and small monomeric enzymes can be made and these smaller reagents should have better ability to penetrate tumour deposits and avoid the difficulties inherent in large scale chemical conjugation.

23.11 ANTIBODIES MAY BE USEFUL FOR TARGETING GENE THERAPIES

Gene therapy offers entirely novel ways of using antibodies if high efficiency targeted gene delivery can be achieved (see also Chapter 24). For example, intracellular antibodies may be expressed and may be able to inhibit tumour growth in highly selective ways if directed at appropriate targets. For tumours known to be dependent on growth factors or hormones, genes encoding neutralizing antibodies delivered to cells at the tumour site may produce prolonged inhibition of growth. Antibodies directed against VEGF (vascular endothelial growth factor), which supports angiogenesis in the growing tumour, may be a generic example. VEGF antibodies may only have to be produced in the tumour vasculature, thus avoiding the significant barrier of tumour penetration. Antibodies may find further use to target the delivery of genes. This has otherwise been achieved using liposomes, viruses or DNA for gene delivery to specific tissues, but the size of such particles means that access to tumour tissues is severely limited.

23.12 ACKNOWLEDGEMENTS

Robert Hawkins is a Cancer Research Campaign Senior Clinical Research Fellow.
Kerry Chester is supported by the Ronald Raven Chair in Clinical Oncology Trust.

23.13 FURTHER READING

Chester, K. A., Begent, R. H. J., Robson, L. *et al.* (1994) Phage libraries for generation of
  clinically useful antibodies. *Lancet*, **343**, 455–456.
McCafferty, J., Griffiths, A. D., Winter, G. and Chiswell, D. J. (1990) Phage antibodies:
  filamentous phage displaying antibody variable domains. *Nature*, **348**, 552–554.
Press, O. W., Eary, J. F., Appelbaum, F. R. *et al.* (1993) Radiolabeled-antibody therapy of
  B-cell lymphoma with autologous bone marrow support. *N. Engl. J. Med.*, **329**,
  1219–1224.
Reitmuller, G., Schneider-Gadicke, E., Schlimok, G. *et al.* (1994) Randomised trial of mon-
  oclonal antibody for adjuvant therapy of resected Duke's C colorectal carcinoma.
  Lancet, **343**, 1177–1183.
Winter, G., Griffiths, A. D., Hawkins, R. E. and Hoogenboom, H. R. (1994) Making anti-
  bodies by phage display technology. *Annu. Rev. Immunol.*, **12**, 433–455.

Current and future aspects of gene therapy for cancer

24

Jonathan D. Harris and Karol Sikora

Over the past 5 years over 300 patients have received foreign DNA. Some have received genes that allow scientists to follow the location of the foreign DNA (and therefore cell populations) while other genes have been delivered to carry out certain function in cells. What was once a dream has now become a reality: the chance of curing cancer by genetic intervention seems more and more possible. The new problems posed by gene therapy have brought new challenges to molecular biology and have also crossed the conventional specialities into which modern high-technology medicine has become divided bringing together expertise from separate areas. There are currently over 100 ongoing or approved trials for gene therapy throughout the world with approximately two-thirds of these for cancer treatment. Although the types of strategy used in these trials are varied, one major problem remains; the delivery of genes to the right tissue or cell type. Targeted expression is therefore a key issue which must be faced by scientists and clinicians if gene therapy is to be fully integrated in the clinic over the next few years.

24.1 CANCER ARISES FROM ABNORMALITIES AT THE MOLECULAR LEVEL

Cancer is most often a somatic cell disorder where the genetic make-up of these cells has been altered, resulting in abnormal patterns of growth control. Oncogenes and tumour suppressor genes are part of the normal genome and these are vital in the transfer and processing of physiological signals from outside the cell to the

Molecular Biology for Oncologists. Edited by J. R. Yarnold, M. R. Stratton and T. J. McMillan. Published in 1996 by Chapman & Hall, London. ISBN 0 412 71270 9

nucleus. Any alteration to these sensitive mechanisms may result in the conversion of a cell to a cancerous phenotype. This is, however, an oversimplification of the cancer process, since for no type of cancer is the molecular pathogenesis precisely understood and it is likely that a series of cooperative events is required for tumour formation. During the coming decade and the next, it is hoped that much of the human genome will have been sequenced, allowing comparisons regarding the genetic make-up of normal cells and those cells with abnormal growth control to be made. Furthermore, the mechanisms of the expression of certain genes responsible for differentiation and proliferation will hopefully be elucidated.

24.2 GENETIC-BASED THERAPIES RELY ON THE FUNDAMENTAL GENETIC DIFFERENCE BETWEEN NORMAL AND CANCER CELLS

Cancer therapy aims to destroy as much of the tumour as possible without damaging normal tissue to any significant extent. Conventional therapies present a variety of problems such as drug resistance in chemotherapy, inaccessible areas for tumour surgery, destruction of non-tumour tissue, etc. Since malignancy is primarily a genetic process, involving lesions at the molecular level, novel gene-based therapies may be developed. These would have an advantage over current therapeutic regimens – high tumour selectivity. Prognostic applications of the genetic analyses of oncogene alterations have already begun, such as in breast and ovarian cancer (*erbB2* amplification) and lung adenocarcinoma (*ras* mutations). Further analysis of these and other oncogenes, as well as tumour suppressor gene alterations, will almost certainly reveal other targets for different forms of genetic intervention at the molecular level.

24.3 GENE THERAPY FOR CANCER – PAST AND PRESENT

Gene therapy, the transfer of exogenous genetic material into an organism in order to provide therapeutic benefit, is now over 5 years old. Clinical trials for cancer gene therapy may be divided into two main groups: gene marking (or tagging) and gene therapy. Gene marking involves the transfer of a foreign gene into the cells of a patient enabling those cells to be detected at a later stage by standard molecular biological techniques. Gene therapy, as the name suggests, involves the transfer of a therapeutic gene into a patient in order to augment a cellular protein's function, or to provide cells with new functions, such as attacking tumour cells or stimulating the immune system to destroy the tumour.

24.4 GENE THERAPY FOR CANCER – CLINICAL TRIALS USING GENE MARKING

Gene marking is surprisingly not a new idea. At the beginning of the century scientists used agar containing dyes to pursue the fate of cells during embryological development. However, such dye-staining studies suffered from a 'dilution' effect whereby, over time, cell division diluted out the dye. With the advent of recombinant DNA and gene transfer techniques, the genetic tagging of cells circumvented this dilution problem.

The first gene 'therapy' trial took place in 1989 and involved the tagging of tumour-infiltrating lymphocytes (TILs). These cells can be isolated from solid tumours, cultured *in vitro* in the presence of interleukin-2 and then reinfused into the patient in the hope that they will home in on and destroy the tumour. Unfortunately, TIL therapy is fairly unsuccessful and only 35% of patients show any measurable effect. Thus the tagging of the TILs could provide clues as to the localization and survival of the TILs in the body following reinfusion. Marker genes may be easily transferred into TILs during the *in vitro* part of the therapy. When these were returned to the patient, one could detect their fate using molecular biological methods. The gene used for the first gene-marking study for melanoma was the bacterial gene that codes for neomycin phosphotransferase (Neo^R), which renders transduced cells resistant to the neomycin analogue G418. The use of neomycin not only allows gene tagging to be carried out when the gene is included in a viral or plasmid gene transfer construct but also allows enrichment of successfully transduced cells containing the Neo^R gene. TILs were isolated from melanoma patients and cultured *in vitro* in the presence of IL-2. While growing in culture, they were transduced with a retroviral vector carrying the Neo^R gene. By subsequent selection in G418-containing medium, only cells that had taken up and were expressing the foreign DNA grew. After an appropriate time, the transduced TILs were infused back into the patients and at various times post-infusion, peripheral blood cells were isolated from the patients and isolated DNA was tested for the presence of the Neo^R gene by the polymerase chain reaction. It was seen that in all of the 10 patients studied, marked cells were detected during the first 3 weeks post-infusion. Marked cells could be detected even up to 60 days in one or two patients. Some of these patients were also being given IL-2 but this was stopped 3 weeks after infusion of the recombinant cells. This may suggest that IL-2 acts as a stimulator of TIL production *in vivo*. Since this first trial in 1989, other additional gene-marking clinical protocols using TILs have begun.

A number of other cells have been used. For example, it is known that a combination of radiotherapy and chemotherapy followed by (autologous) bone marrow transplantation is effective for a number of different cancers. However, the aggressive treatment has a high relapse rate, the cause of which is unknown. It has been suggested that this is possibly due to residual malignant disease remaining after chemotherapy and radiation treatment. It could also be that some tumour cells contaminate the purged bone marrow that is subsequently used. This latter possibility may sound surprising since the bone marrow is usually taken during remission. However it still remains possible that there might be some tumour cells present, since the purging process is not completely effective. The genetic tagging of bone marrow prior to autologous bone marrow transplantation may allow clinicians to detect any possible contaminating cells present during relapse. There are a number of such ongoing studies both for neuroblastoma and for acute lymphocytic, acute myeloblastic and chronic myelogenous leukaemia.

The first gene-marking protocol using bone marrow cells was initiated in 1991; this involved transduction of bone marrow removed from AML and neuroblastoma patients with a retroviral vector carrying the Neo^R gene. Using the polymerase chain reaction, the investigators showed the efficiency of transduction to be approximately 10%. The tagged marrow was mixed with normal marrow prior to reinfusion. In two of the patients who underwent the procedure, marked cells were observed following relapse. It was clear that the marrow purging was not completely efficient. The use of gene marking can therefore be used as an indicator of the purging efficiency. Gene marking for bone marrow cells has also begun in patients suffering from chronic myelogenous leukaemia and multiple myeloma and other trials are ongoing in which different purging techniques are being compared using gene marking.

24.5 CLINICAL CANCER GENE THERAPY TRIALS

This work and the development of gene marking has led to many gene therapy trials for cancer. One such adoptive immunotherapy trial has involved the transduction of TILs from melanoma patients with retroviral vectors expressing tumour necrosis factor (TNF). Previous studies using TNF therapy were unsuccessful, because toxicity from the TNF limited the concentration of the biological agent around the tumour. Preliminary gene therapy studies demonstrated that dose-limiting toxicities were not seen when TNF-transduced TILs were used. Upon biopsy, tumours had undergone coagulative necrosis, suggesting that the TNF had been responsible for killing the cancer cells. The use of cytokines for cancer treatment by gene therapy forms almost half of the ongoing trials. A number of promising experiments using animal models have shown that by transferring either one cytokine, most commonly IL-2, or two cytokines, such as IL-4, GM-CSF, IL-12 or γ-interferon, tumour regression and protection from subsequent tumour challenge can result.

Another trial is using engineered cells to stimulate the immune system against the tumour. A group at the University of Michigan is using a liposome-mediated gene delivery system to deliver foreign genes directly into tumour skin nodules. There is mounting evidence that suggests that tumours do not present potential antigens efficiently, since many have down-regulation of MHC class I molecules. It has been shown that upon stimulation of expression of MHC class I molecules in animals an immune response can be mounted against a previously unrecognized tumour. These animals can also resist subsequent challenge of the same tumour that has not been engineered to express MHC. The Michigan group is attempting to transfer the co-stimulatory molecule B7 directly into melanoma skin nodules. Recently published results from the trial demonstrate that the procedure is well tolerated and some tumour reduction has been noted in a limited number of patients.

24.6 SUICIDE GENE TRANSFER INTO CANCER CELLS

One of the most common types of gene therapy trial for cancer is the use of pro-drug-activating enzymes to cause destruction of the tumour cells (Chapter 25). In these strategies, a non-mammalian gene is delivered into cells. This gene encodes an enzyme capable of converting a harmless prodrug into a toxic metabolite, which then kills the cell. The most common enzyme used in clinical trials is the Herpes simplex virus thymidine kinase, which can sensitize cells to ganciclovir through the phosphorylation of the prodrug to agents that interfere with DNA synthesis.

A trial using this technique to treat adult glioblastoma began in December 1992. Fibroblast cells capable of producing retroviral particles that express Herpes simplex virus thymidine kinase enzyme gene (*HSVTK*) were injected stereotactically into the patients' brains. Since retroviral vectors require cell division to integrate their DNA, the viruses would only deliver their payload into the rapidly dividing glioblastoma cells. After a specified time, patients were given the ganciclovir prodrug, which hopefully would be converted to cytotoxic intermediates by the thymidine kinase enzyme. The investigators have recently been granted approval to carry out a similar trial in children suffering from brain cancer.

Interestingly, it has been reported that not all the tumour cells need to contain the *HSVTK* gene to be killed by this method. In fact, as few as 5–10% of transduced cells in a population of tumour cells may be enough to cause regression upon exposure to ganciclovir. The cause of this so-called 'bystander effect' or 'kiss of death' has puzzled scientists. The cause may be explained by a number of possibilities. It could be due to the escape of toxic intermediates from *HSVTK*-expressing cells, which could then enter non-transduced cells and kill them. It could also be that the thymidine kinase enzyme is somehow transported into non-transduced cells. However, both of these possibilities are probably unlikely given the size and charge of the agents concerned. A more likely explanation is that the phosphorylated intermediates in *HSVTK*-expressing cells are transported into non-transduced cells. It has been suggested that gap junctions, which are normally involved in cell to cell communication, might play a part in this process. They allow molecules through with a molecular weight of less than 1000, too big for the thymidine kinase but adequate to allow transport of the prodrug or phosphorylated metabolites.

24.7 GENE REPLACEMENT TO CORRECT GENETIC ABNORMALITIES IN CANCER CELLS

We know that tumours can be caused by errors in certain oncogenes or tumour suppressor genes which cause the uncontrolled proliferation associated with cancer. For example, in approximately 35% of lung cancers, mutations in the K-*ras* oncogene has been observed and in over 60% of these tumours there are errors in the coding sequence of the tumour suppressor gene *p53*. Jack Roth and his colleagues at the M. D. Anderson Centre in Texas have recently begun gene therapy trials against lung cancer in an attempt to correct these genetic abnormalities. The investigators are to use retroviruses to deliver foreign genes directly into the lungs of the patients. In those patients with K-*ras* mutations, the foreign gene is an antisense K-*ras* construct. The expression of the specific sequences should, in theory, block the transcription of the mutant sequence and the proliferation of the tumour may be ablated. In patients with *p53* mutations, functional copies of *p53* will be delivered by the retroviruses. The *in vitro* and preclinical work has been promising and it is hoped that the trial may be an exciting new area of clinical gene therapy.

24.8 TARGETED GENE DELIVERY

One of the biggest problems in gene therapy is targeted delivery. As with any type of cancer treatment, the aim is to destroy as much of the cancer as possible without destroying the normal tissue. How can one target genes to certain cell types? This may be achieved by either delivering the genes into certain types of cells (**transductional targeting**) or by having genes expressed only in the cancer cells (**transcriptional targeting**). The trial involving the delivery of thymidine kinase gene using retroviruses to treat glioblastoma is an example of transductional targeting. One can use the physiology of the rapidly growing tumour cells to our advantage by using a retrovirus that requires cell division for gene delivery. Therefore, the normal quiescent cells of the brain are unaffected by the prodrug activation. One can also target viruses to certain cell types. Retroviruses have molecules that interact with receptors on the target cell surface. These envelope molecules are responsible for determining which cells may be infected. If one could change the envelope genes to mimic ligands for tumour cell surface receptors, one could produce a virus that would only infect tumour cells, a tumour-specific retrovirus. Although this technology is still in its infancy, recent reports have shown that it is possible to alter the specificity of retroviral vectors and deliver their genes specifically into certain types of cell (Chapter 22).

Targeted expression of genes in certain cells (transcriptional targeting) is another potential way of specifically ablating tumour cells. Here we rely on the expression of genes in particular tissue or tumour types. The switches that control the expression of the genes in the tissue or tumour are often damaged and remain turned on, instead of being controlled. We can use these molecular switches to our advantage by creating an artificial switch, which could then be delivered into these cells; the specific expression of suicide genes, for example, would only occur in those cells where the switch is turned on all the time, i.e. the tumour cells. Such strategies have already been described for a number of different cancers.

24.9 SAFETY AND ETHICAL ASPECTS OF GENE THERAPY

Gene therapy is essentially much like any other kind of therapy, and has its inherent risks. Therefore, any risk to the patient must be balanced by corresponding benefit. If no alternative method of therapy is available and the likelihood of gain is small, then it will still be possible to carry out the therapy on that patient (since some benefit is better than none). Therefore gene therapy must be assessed by the same criteria used for other clinical protocols. In addition, since this type of therapy is new, only clinical trial data can provide information regarding the efficacy of the treatment and may reveal any unexpected complications.

There is no doubt that gene therapy will remain a controversial therapy for some time. There are the patients for whom benefit may be gained that might not be available using conventional treatments. On the other hand, there is the sinister possibility of some workers abusing gene therapy to genetically engineer humans of a desired form, the 'Brave New World' scenario. This is compounded with the controversy surrounding the potential for germline therapy, although no protocols have been proposed yet. It is possible that in the future one may alter the characteristics of the unborn child and this may lead to abuse of the method, for example changing desirable traits, such as hair colour, skin colour and possibly personality. There is much current debate surrounding this particular type of gene therapy and it is likely that this will continue for some time.

24.10 FUTURE PROSPECTS FOR GENE THERAPY

Taking all the information described in this chapter into account, it is clear that although many different types of gene therapy are under way, this novel treatment is still in its infancy and has a long way to progress before fully effective therapies are implemented. Gene therapy has come a long way since its first tentative steps in 1989 and the industry is growing at an alarming rate. New technologies and gene transfer methods are finding their way into the clinic for Phase I studies to treat cancer. Until efficient methods of gene transfer and targeting of the genes to tumours or certain cell types are developed, gene therapy is unlikely to become a new modality of cancer treatment. It is expected that, over the next few years, gene therapy may be offered in combination with the already established modalities of treatment in the clinic and from early clinical studies; however, it is clear that gene therapy is here to stay and future developments in this field may provide a cure for some types of cancer.

24.11 FURTHER READING

Dranoff, G. and Mulligan, R. C. (1995) Gene transfer as cancer therapy. *Adv. Immunol.*, **58**, 417–454.

Ettinghausen, S. E. and Rosenberg, S. A. (1995) Immunotherapy and gene therapy of cancer. *Adv. Surg.*, **28**, 223–254.

Gutierrez, A. A., Lemoine, N. R. and Sikora, K. (1992) Gene therapy for cancer. *Lancet*, **339**, 715–721.

Harris, J. and Sikora, K. (1994) Human genetic therapy. *Mol. Asp. Med.*, **14**, 451–546.

Roemer, K. and Friedmann, T. (1992) Concepts and strategies for human gene therapy. *Eur. J. Biochem.*, **208**, 211–225.

Van Beuschem, V. W. and Valerio, D. (1992) Prospects for human gene therapy, in *Transgenesis* (ed. J. A. H. Murray), John Wiley, Chichester, pp. 284–321.

Directed enzyme prodrug therapies: ADEPT and GDEPT

25

Caroline J. Springer

25.1 SELECTIVE ATTACK ON TUMOUR CELLS CAN BE ACHIEVED WITH ANTIBODIES

A major goal in cancer treatment has been to target toxic agents to tumour cells selectively, whilst sparing normal cells from damage. Antibody (Ab) conjugates have been utilized to provide such selectivity. Despite enormous strides forward in the clinical utility of chemoimmunoconjugates, immunotoxins and radioimmunoconjugates, their efficacy has been hampered by a variety of factors. For example, antigen is often expressed heterogeneously within tumours, so that antibody conjugates cannot bind uniformly. Also, antibody conjugates have high molecular mass so they penetrate inefficiently into tumours, leading to low levels of intratumoral accumulation. In addition, there is often poor internalization of chemoimmunoconjugates and immunotoxins (which is required for activity) or they may be destroyed in lysosomes and there may be inefficient release of the cytotoxic agent from them. Moreover, in order that the activities of both components of the conjugate will not be compromised, only a finite number of molecules of drug or toxin may be conjugated to antibodies. This limits the number of toxic molecules delivered to the same order of magnitude as that of the antibody reaching the tumour.

There are two new methods of overcoming the drawbacks of these antibody conjugates, while retaining the advantage of selectivity. These systems comprise

Molecular Biology for Oncologists. Edited by J. R. Yarnold, M. R. Stratton and T. J. McMillan. Published in 1996 by Chapman & Hall, London. ISBN 0 412 71270 9

a two-step approach called ADEPT (antibody-directed enzyme prodrug therapy), GDEPT (gene-directed enzyme prodrug therapy) or VDEPT (virally-directed enzyme prodrug therapy). Each approach separates the selective from the cytotoxic function, which has benefits over a one-phase chemoimmunoconjugate, immunotoxin or radioimmunoconjugate.

25.2 ADEPT EXPLOITS THE SELECTIVITY OF ANTIBODIES AND THE ACCESSIBILITY OF SMALL PRODRUGS

In ADEPT, the selective component is the antibody of an Ab–enzyme conjugate that binds antigen preferentially expressed on the surface or in the interstitial spaces of the tumour cells. In the first phase, the Ab–enzyme complex is administered and allowed to accumulate at the tumour site. Time is allowed for the clearance of the conjugate from blood and normal tissues. In the second phase a latent non-toxic prodrug is administered, which is converted selectively by the enzyme associated with the antibody at the tumour into a low molecular mass toxic drug. The diffusion within the tumour of these small cytotoxic agents thus generated is more favoured compared to those of the higher molecular mass immunoconjugates which leads to greater tumour access of the drug. There are a number of aspects to be examined when considering ADEPT in comparison with immunoconjugates

Table 25.1 Advantages and disadvantages of ADEPT

Advantages	Disadvantages
1. There is an amplification effect in that each enzyme molecule is able to convert many prodrug molecules	1. The Ab–enzyme conjugate is immunogenic
2. The specificity of the antibody leads to increased activity for malignant cells	2. The two-step system is complex
3. The conjugate is not required to internalize	
4. The released active drug has a low molecular mass; it is therefore able to diffuse readily within the tumour	
5. The concentration of the drug released in the tumour is higher than with direct injection of drug alone	
6. There is a bystander effect of drug released by conjugate on antigen-positive cells, which may diffuse to kill antigen-negative cells	
7. The concept has proved to be feasible at the clinical level	

and conventional therapies. The advantages and disadvantages of ADEPT are out-lined in Table 25.1.

Specific requirements are made of the enzymes used in ADEPT. The most important are summarized in Table 25.2. For utility in ADEPT, prodrugs must be designed to be effective substrates for the enzyme of choice. Only selected link-ages can be cleaved by each specific enzyme. There are a number of requirements to be made of the prodrug which are described in Table 25.3.

Table 25.2 Requirements for the enzymes used in ADEPT

Enzymes should be:

- able to catalyse a cleavage reaction
- distinct from any circulating endogenous enzymes
- able to effect high catalytic activity
- active in physiological conditions

Table 25.3 Properties required for the prodrug in ADEPT

The prodrug should:

- be less cytotoxic than its corresponding active drug
- be a suitable substrate for the activating enzyme under physiological conditions
- be chemically stable under physiological conditions
- be highly diffusible through the tumour
- have a large differential cytotoxicity between itself and its active drug
- have good pharmacological and pharmacokinetic properties

25.3 ADEPT CAN BE EFFECTIVE IN THE LABORATORY AND CLINIC

Most ADEPT prodrugs are derived from well known anticancer agents or their close counterparts as model molecules. Some enzyme prodrug combinations have demonstrated efficacy *in vivo* in ADEPT systems and they are now described.

One ADEPT approach uses alkylating agent prodrugs in combination with the activating bacterial enzyme carboxypeptidase G2 (CPG2). *In vivo* studies of this system with the mustard prodrug CMDA demonstrated its ability to achieve both the ablation of chemoresistant choriocarcinoma human tumour xenografts in nude mice and growth delays of ovarian human tumour xenografts. An antibody con-jugate of CPG2 has also been used in ADEPT studies with a breast carcinoma xenograft model in conjunction with the CMDA prodrug. This ADEPT system induced sustained dose-dependent regressions and stasis in nude mice bearing the breast carcinoma tumours. Control chemotherapy with conventional drugs at maximum tolerated doses in the same tumour models proved to be ineffective.

A pilot scale clinical trial in patients with colon carcinoma was initiated by Professor K. D. Bagshawe at Charing Cross Hospital, London. It used an anticarcinoembryonic antigen (CEA) Ab with the CMDA prodrug and served to demonstrate the clinical feasibility of ADEPT. In order to accelerate clearance of the Ab–CPG2 conjugate from the blood, a clearing agent of a galactosylated antiCPG2 Ab was used as an additional intermediate step in ADEPT. This works by forming immune complexes which are removed by the galactose receptors on the liver. Eight of the 17 patients receiving ADEPT protocols with adequate dosage were assessable and five of these showed partial remissions or mixed responses.

Further studies in the design of prodrug systems for activation by CPG2 were aimed at preventing leakage of active drugs back from the tumour. For this approach active drugs with short half-lives were desirable. Recently a new compound 4-N,N-bis(2-iodoethyl)amino-phenyloxycarbonyl-L-glutamic acid was developed which fulfils this criterion. The half-life of the active drug released by cleavage with CPG2 is of the order of few seconds, thus minimizing any potential leakage back from the tumour. In ADEPT *in vivo* this prodrug produced long lasting regressions in a colorectal carcinoma tumour xenograft. This compound, called ZD2767, is a new prodrug candidate for full clinical evaluation of ADEPT.

Another approach uses Ab conjugates of alkaline phosphatase (AP) to convert the prodrug etoposide phosphate to etoposide. Experimental therapies have been performed on nude mice bearing human lung tumour xenografts. Administration of the Ab–AP was followed approximately 2 days later by the prodrug and this has resulted in a significant antitumour response. Prodrugs for use with the AP enzyme have also been used successfully and other prodrug/enzyme systems have been developed.

25.4 GDEPT AND VDEPT TARGET GENE EXPRESSION TO TUMOUR CELLS

There are many obvious similarities between ADEPT and GDEPT. The basic difference between the design of the two systems is that, whereas in ADEPT the foreign enzyme should be directed to the target cells by an Ab, in GDEPT and VDEPT the gene for the foreign enzyme should be expressed in the target cells. Thus GDEPT (and VDEPT) consists of a two-phase therapy, which requires firstly that a gene producing a non-endogenous enzyme is delivered to the target. In the second step, a prodrug is injected, which is designed so that is will be converted selectively by the enzyme expressed at the target. Ideally the gene will be placed under the control of a tumour specific promoter, such that it will be expressed selectively in the target cells and not in normal cells.

GDEPT has many of the advantages that have been described for ADEPT. For example, in Table 25.1, points 1,4,5,6 apply also to GDEPT although it shares the disadvantage of being a complex two-stage system.

In order to achieve success in GDEPT a number of factors must be considered. Firstly, the gene should be expressed only in the tumour cells. Secondly, expression in the cancer cells should be as high as possible. Unfortunately when injected systemically it is unlikely that more than 10–20% of tumour cells will express the gene. Therefore, a bystander effect is required whereby the prodrug is cleaved to an active drug that kills not only cells expressing the foreign enzyme but also cells that are not expressing enzyme. This means that less than 100% infection of cells can still give total cell kill. It has been demonstrated, *in vivo,* that when as few as 2% of the tumour cells are infected with foreign enzyme and subsequently treated with suitable prodrug long-term survivors have been obtained. Therefore an expression efficiency of 10–20% should be enough to achieve 100% cell kill in tumours and efficiencies of 1–5% are considered sufficient for a therapeutic response.

The enzymes generated in GDEPT should have the same characteristics as those described in Table 25.2 for ADEPT. The prodrug requirements for GDEPT are also the same as those used in ADEPT (Table 25.3). In addition, the prodrug must be able to cross the tumour cell membrane for activation by intracellularly expressed enzyme. Thus lipophilic prodrugs are required in order to penetrate across cell membranes or prodrugs could be synthesized that are taken up by cells by active transport. It is believed that the prodrugs so far in use (nucleoside analogues, 5-fluorocytosine, cyclophosphamide) are taken up by cells by passive diffusion.

25.5 A NUMBER OF GDEPT APPROACHES HAVE SHOWN EFFICACY

Purine prodrugs are active only when they are cleaved in dividing cells. They are cleaved by the herpes simplex virus thymidine kinase (HSVTK). HSVTK catalyses the phosphorylation of the purine nucleoside analogues acyclovir, ganciclovir and 1(2-deoxy-2-fluoro-B-D-arabinofuranosyl-5-iodo uracil) (FIAU), which are poor substrates for the mammalian nucleoside monophosphate kinase.

Brain tumours were the first choice for GDEPT, since neurones and other non-tumour cellular elements in the brain are non-dividing. Also, vascular endothelial cells that respond to angiogenesis signals are present in the vicinity of the tumour, so their destruction is desirable. The blood–brain barrier is considered another advantage in this model. It was demonstrated that established glioma tumours in Fisher rats could be cured when transfected with the *TK* gene and subsequently treated with GCV.

Cytosine deaminase (CD) is a bacterial enzyme (from *E. coli*) that is not present in human cells. This enzyme converts the non toxic 5-FC to 5-fluorouracil (5-

FU), which is then transformed by cellular enzymes to a potent pyrimidine antimetabolite. Studies were carried out to develop a therapeutic model for colorectal carcinoma using the CD/5-FC system. In the first step, it was demonstrated that significant antitumour effects could be obtained *in vivo* with 5-FC treatment when tumour colorectal cells were transfected with the *CD* gene. It was shown that only a small percentage of the cells within the tumour were required to express CD in order to generate a significant antitumour effect with 5-FC treatment.

The use of cyclophosphamide (CP) and isophosphamide (IP) were considered for GDEPT since they are activatable by the enzyme cytochrome-P450 (CYP). Rats were injected with CYP-transfected tumour cells. Treatment with CP, which was started 7 days after transplantation, led to complete tumour growth inhibition.

25.6 ADEPT AND GDEPT ARE READY FOR CLINICAL EVALUATION

The results for the preclinical data for ADEPT and GDEPT suggest an encouraging future for these new therapies. Promising clinical results from the pilot scale clinical ADEPT trial have already been reported and another trial is currently being performed. Approval for GDEPT trials has been granted to treat brain and ovarian tumours using either retroviral or DNA transfer of the *HSVTK* gene. The results are eagerly awaited.

25.7 ACKNOWLEDGEMENT

Thanks are due to the Cancer Research Campaign for their financial support.

25.8 FURTHER READING

Bagshawe, K. D. (1994) ADEPT and related concepts. *Cell Biophys.*, **24/25**, 83–91.
Bagshawe, K. D., Sharma, S. K., Springer, C. J. and Antoniw, P. (1995) Antibody-directed enzyme prodrug therapy (ADEPT): pilot scale clinical trial. *Tumour Targeting*, **1**, 17–29.
Huber, B. A., Richards, C. A. and Krenitsky, T. A. (1991) Retroviral-mediated gene therapy for the treatment of hepatocellular carcinoma: An innovative approach for cancer therapy. *Proc. Nat. Acad. Sci. USA*, **88**, 8039–8043.
Niculescu-Duvaz, I. and Springer, C. J. (in press) Gene-directed enzyme prodrug therapy (GDEPT): choice of prodrugs, in *Advanced Drug Delivery Reviews: Enzyme Prodrug Therapy*, (ed. K. D. Bagshawe), Elsevier Science, Amsterdam.
Springer, C. J. and Niculescu-Duvaz, I. (1995) Antibody-directed enzyme prodrug therapy (ADEPT): a targeting strategy in cancer chemotherapy. *Curr. Med. Chem.*, **2**, 687–706.
Springer, C. J., Dowell, R., Burke, P. J. *et al.* (1995) Optimization of alkylating agent prodrugs derived from phenol and aniline mustards: A new clinical candidate prodrug (ZD2767) for antibody-directed enzyme prodrug therapy (ADEPT). *J. Med. Chem.*, **38**, 5051–5065.

Oligonucleotide therapies | 26

Val Macaulay

26.1 GENE EXPRESSION CAN BE ARTIFICIALLY BLOCKED

Better understanding of gene regulation and expression are important in fundamental terms and also in suggesting new avenues for treatment. This chapter will
outline the principal ways in which oligonucleotides can be used to block gene
expression, the problems associated with each approach and some of the ways in
which these obstacles may be overcome.

Genes are organized into upstream regulatory regions, which include recognition sequences for the binding of transcription factors and other DNA-binding proteins, and downstream regions, which include exons (coding sequences) and
introns (non-coding). The processes involved in gene expression are shown in
Figure 26.1.

The human genome contains about 70 000 genes, and since all cells contain the
same genetic information, it might be thought that a genetic treatment would be
non-specific. However only a subset of genes is expressed in any given cell type,
and the pattern of expression varies widely between cells. Certain genes are overexpressed in malignant cells, so treatments aimed at blocking expression of those
genes should be manifest, if not specifically, then at least selectively, by the malignant cells. In addition, specific genes may be mutated in the cancer cells. This may
be a point mutation (e.g. activating *ras* mutations) or a gene rearrangement, with
production of a unique fusion protein (e.g. $p210^{bcr-abl}$ in chronic myeloid
leukaemia, CML). These provide target sequences specific to the cancer cell. In
selecting suitable targets it is important to remember that just because gene activation may have contributed to transformation, it does not necessarily mean that
sustained gene expression continues to be required for maintenance of the malignant phenotype. Therefore it is important to show that any given manoeuvre not

Molecular Biology for Oncologists. Edited by J. R. Yarnold, M. R. Stratton and T. J.
McMillan. Published in 1996 by Chapman & Hall, London. ISBN 0 412 71270 9

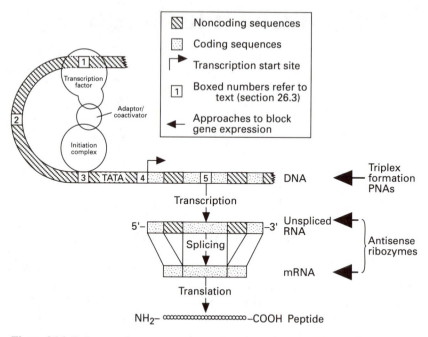

Figure 26.1 Pathways of gene expression and potential sites of inhibition. The gene is shown here as a block, but it is of course double-stranded DNA in a double helix (duplex). Each strand is made up of a sugar-phosphate backbone with bases attached, and these bases are purines (guanine, adenine) or pyrimidines (cytosine, thymine). The two strands are complementary, that is, there is specific base-pairing between C and G, and between A and T. The strand which specifies the sequence of amino acids is called the 'sense' strand and is conventionally written as the top line, with the information running left to right, i.e. 5' (upstream) to 3' (downstream). The processes involved in gene expression are: **transcription initiation**, a complex and highly regulated process, sometimes involving DNA bending, in which transcription factors and other DNA binding proteins interact with upstream regulatory sequences leading to activation of the initiation complex; **transcription** – making an RNA copy from DNA (RNA is single-stranded, and has the same sequence as the sense strand of DNA except that uracil replaces thymine); **splicing** out intronic sequences to form mature messenger RNA (mRNA); **translation** – the information ('genetic code') is read as groups of three bases (codons), each representing an amino acid. The mRNA attaches to a ribosome and provides a template with which transfer RNAs hybridize to bring in amino acids in the correct sequence to make the protein. Boxed numbers refer to mechanisms by which gene expression can be modulated by triplex formation. (Source: modified with permission from Thuong and Hélène, 1993.)

only reduces expression of the protein in question but also has significant phenotypic effects on the target cell population. There are several ways in which gene expression can be blocked experimentally:

- **targeted disruption of the gene itself**, by replacing it with a homologous DNA sequence containing an inactivating mutation (homologous recombination); this causes inheritable gene 'knockout', and is used experimentally to assess the role of individual genes in development and tumorigenesis;
- **inhibition of one of the steps involved in gene expression**, without introducing permanent changes in DNA sequence.

The latter approach can be achieved in various ways with oligonucleotides, i.e. short synthetic pieces of nucleic acid. In theory, these could be either DNA or RNA, but RNA is extremely unstable, so in practice an oligonucleotide almost always means a short piece of DNA. To block the expression of a specific gene it is necessary to know at least part of its sequence. An oligonucleotide need be only 15–19 bases long to be specific for a single sequence in the human genome. Oligonucleotides can be used to block gene expression in one of two ways:

- **indirectly**, using an oligonucleotide sequence that adsorbs transcription factor by mimicking a transcription factor binding site, thereby reducing the amount of free factor (protein) available to bind DNA and stimulate gene transcription; so far this 'decoy strategy' is in its infancy, but it should theoretically be possible to switch a gene on or off by targeting proteins that activate or repress gene transcription;
- **directly**, by interaction with DNA or RNA to block gene expression at the level of transcription or translation (Figure 26.1).

This chapter focuses on approaches to inhibit gene expression by direct interaction with DNA or mRNA. Processes that can be targeted include:

- growth
- differentiation
- apoptosis
- metastasis
- sensitivity to drugs/radiation
- angiogenesis
- immune or endocrine pathways.

Ideally, agents should be stable (i.e. resist intracellular and serum nucleases), get taken up uniformly into cells to achieve adequate intracellular levels, gain access to the target (DNA or RNA) and be capable of forming stable complexes that are able to disrupt the target molecule/process under physiological conditions. Antisense agents and ribozymes are designed to target mRNA and prevent its translation into protein. Triplex formation ('the anti-gene approach') is designed to target double-stranded DNA, blocking transcription. Peptide nucleic acids (PNAs) also target DNA; they were designed as triplex-forming agents, but in fact exert their effects in a slightly more complex way, by strand displacement.

26.2 ANTISENSE OLIGONUCLEOTIDES TARGET mRNA FOR DEGRADATION BY RNase ENZYMES

The antisense approach involves targeting specific RNA sequences with the aim of blocking translation of the message into protein. Most studies have chosen to target mature cytoplasmic mRNA, but it is also possible to target nuclear RNA, before the intron sequences have been spliced out, in which case the intron–exon junctions can be targeted to increase specificity.

Antisense agents can be either oligonucleotides or RNA transcripts generated in the cell following transfection of an antisense expression construct. In either case the antisense agent is designed so that its sequence is complementary to the target mRNA. Initially, it was thought that when hybridization occurred between the mRNA (sense strand) and complementary sequence (antisense oligonucleotide), translation was blocked at that site. In fact, it seems that the major mechanism responsible for the antisense effect of oligonucleotides involves RNase H, an enzyme involved in DNA replication. The mRNA-oligonucleotide heteroduplex provides an excellent substrate for the selective cleavage of the native, sense mRNA strand by RNaseH (Figure 26.2).

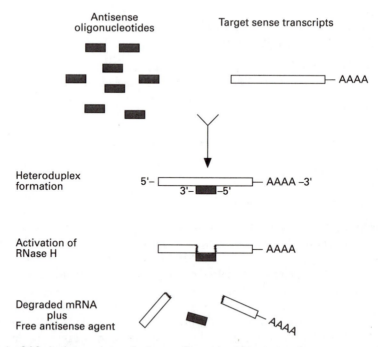

Figure 26.2 Antisense agents. Antisense oligonucleotides are complementary to, and hybridize with, target transcripts to form a duplex. This triggers activation of RNase H, which digests the native, sense mRNA strand. Thus the mRNA is inactivated, and the antisense agent is free to bind to another target molecule. This is probably the major mechanism by which antisense effects are mediated, though there may be others.

When the RNA has been degraded, the oligonucleotides are released and can bind to additional copies of the target mRNA. It is less certain what mediates the effects of antisense RNA generated *in situ* following stable transfection of antisense constructs (see below). Here, formation of sense–antisense hybrids in the nucleus may lead to degradation of both components of the duplex. There is also experimental evidence for reduction in translation of surviving sense mRNA.

Antisense effects can be easily assessed in cultured cells

Antisense effects can be assessed in cultured cells simply by adding oligonucleotide to the tissue culture medium. The short pieces of DNA are taken up readily by endocytosis. In general, depending on the type of oligonucleotide used, *in vitro* effects are seen at nanomolar to micromolar levels. The effects can be demonstrated by showing a reduction in level of specific protein translated from target transcripts using Western blotting or immunoprecipitation. It should also be shown that this has the predicted effects on the phenotype of the cell (e.g. morphology, growth, susceptibility to apoptosis, etc.). Where the protein is required for enhanced growth rate, antisense treatment may be shown to slow the growth of treated cells *in vitro*; some studies have also shown effects *in vivo* following inoculation of oligonucleotide-treated cells into mice.

The antisense approach can target any transcribed gene

Experiments using *in vitro* translation systems indicate that for many genes the best target is at the translation initiation codon (AUG, which encodes methionine). This is near to the site of assembly of the ribosome complex, a process which involves recognition of the 5'-methyl-G cap of mature mRNA. Antisense effects appear to be more potent if the oligonucleotide includes a short dC region at its 3'-end that can hybridize with the methyl-G cap of the target. Apart from this there are no intrinsic sequence limitations to the antisense approach, unlike ribozymes and triplex formation, so in principle any expressed gene can be targeted in this way. Ideally the target mRNA and protein should have a short half-life so that interruption of translation results quickly in a biological effect. For example, c-*myc* mRNA and protein both have half-lives of less than 1 hour. It is not necessary to inhibit protein production totally to achieve a biological effect.

Experimentally, the antisense approach has been used to demonstrate a critical role for a variety of genes in a range of cell lines derived from haemopoietic and solid tumours. These include genes encoding the following:

- **Secreted or cell surface proteins**, including growth factors CSF1, interleukins 2, 4, 6, basic fibroblast growth factor (bFGF), insulin-like growth factors 1 and 2 (IGF-1 and −2), transforming growth factor alpha (TGF-α), growth factor receptors including EGFR, c-ErbB2, Type I IGFR and c-Fms, as well as MDR1, ICAM-1 and Bcr–Abl. The use of antisense oligonucleotides against *bcr–abl* illustrates the targeting of unique sequences resulting from chromosomal rearrangements. The t(9:22) rearrangement involves the translocation of the proto-oncogene *abl* to the breakpoint cluster region (BCR) on chromosome 22. This results in overexpression of *abl* tyrosine kinase, and high levels of these transcripts are found in most patients with CML and Philadelphia-chromosome-positive acute lymphoblastic leukaemia (ALL). Oligonucleotides complementary to the breakpoint junction of *bcr–abl* transcripts reduce levels of the gene product (p210) and selectively inhibit the growth of Philadelphia-positive leukaemic cells. Infusion of these oligonucleotides has been shown to prolong the survival of SCID mice with leukaemia. This approach is now moving into clinical trials aimed at *ex vivo* purging of bone marrow of patients with CML.

- **Cytoplasmic proteins**, including N-, K- and H-*ras*, Hsp27, aromatase and the growth stimulatory subunit (RIα) of cAMP-dependent protein kinase. Antisense effects against *ras* genes have been shown despite the slow turnover rate of *ras* transcripts and protein. Several studies have used cells carrying activating point mutations and expressing high levels of mutant $p21^{ras}$. Antisense oligonucleotides designed to target the mutant sequence have been shown to reduce levels of $p21^{ras}$ and to inhibit growth *in vitro* and tumorigenicity *in vivo*.
- **Nuclear proteins**, including c-Myc, N-Myc, c-Myb, retinoblastoma, c-Jun, Bcl-2, NFkB, retinoic acid receptor α, p120 nucleolar antigen and DNA polymerase-α. *bcl-2* is involved in the t(14:18) translocation found in many follicular lymphomas. The normally dormant *bcl-2* gene is rearranged to a position next to the immunoglobulin heavy chain coding sequence. This leads to overexpression of *bcl-2*, associated with increased resistance to apoptotic death following treatment with drugs or radiation. In experimental studies *in vitro*, antisense oligonucleotides to *bcl-2* have been shown to reduce levels of protein encoded by target transcripts, and to enhance susceptibility to killing by conventional cytotoxic drugs.

Problems with the antisense approach remain to be resolved

The stability of unmodified oligonucleotides is low and they are rapidly degraded by serum nucleases. Much energy has gone into modifying their structure to make them more stable and resistant to nuclease attack while retaining their efficiency of hybridization and ability to induce RNase H. Uptake into the cytoplasm is often heterogeneous, and there may be problems gaining access to the desired intracellular compartment. Some of the approaches used to improve stability and cell uptake are discussed below and summarized in Table 26.3. Quantitative studies using stable transfection of cells with antisense expression vectors indicate that sense:antisense RNA hybrid duplex formation is inefficient, requiring 100- to 1000-fold excess of antisense for 50% of target transcripts to be present in duplexes.

When assessing the significance of antisense effects, it is important to bear in mind that exogenously added DNA at high concentration can be toxic to cells, and some modifications (e.g. phosphorothioate) can be more toxic than unmodified (phosphodiester) oligonucleotides. Some non-specific effects may be attributable to intracellular accumulation of nucleotides, the breakdown products of oligonucleotides. In some systems, oligonucleotides have been shown to bind to small molecules and proteins, either in a non-specific or sequence-specific way. Toxicity can also result from the hybridization of the antisense agent to mRNA sequences other than the predicted target.

26.3 THE DNA DOUBLE HELIX CAN INTERACT WITH A THIRD NUCLEOTIDE STRAND; THE TRIPLE HELIX (TRIPLEX) CAN INTERFERE WITH GENE EXPRESSION

Formation of a triple helix, or triplex, can occur when double-stranded DNA interacts with a third strand (Figure 26.3).

Triplexes are thought to occur naturally during recombination and transcriptional control events. Recent interest has focused on induction of local triplex formation between double-stranded DNA and synthetic oligonucleotides, aiming to block gene transcription. Triplex formation offers significant advantages over the antisense approach (Table 26.1), and offers the prospect that low dose administration could result in durable suppression of gene expression.

Triplex formation can inhibit gene expression in different ways depending on the nature of the target site (Figure 26.1). Inhibition could be a direct effect, if the third strand competes with transcription factors for duplex binding in upstream sequences (Figure 26.1, site 1). Alternatively, the affinity of DNA-binding proteins can be affected indirectly if triplex formation alters DNA conformation (site 2 in Figure 26.1). In coding sequences, triplex formation can block binding of the initiation complex (site 3), transcription initiation (4) or can cause premature stalling of RNA polymerase (5).

Oligonucleotides interact by hydrogen bonding in the major groove of the alpha helix in purine-rich regions

Watson–Crick base pairs in DNA possess several hydrogen bonding sites which are accessible in both the major and the minor grooves. In the major groove,

Table 26.1 The antisense and triplex approaches

Characteristics of target	Antisense	Triplex
Identity	mRNA	DNA
Location	Cytoplasm	Nucleus
No. molecules/cell	1000s	2
Regeneration rate	Rapid	Slow

thymine can form two hydrogen bonds with an adenine involved in a Watson–Crick AT base pair. These are known as 'Hoogsteen' hydrogen bonds, by analogy with the arrangement of hydrogen bonds observed by Hoogsteen (1959) in cocrystals of adenine and thymine. Protonated cytosine (C^+) can also form two Hoogsteen hydrogen bonds with guanine in a GC pair. Thus triple helices can form with two polypyrimidine strands hydrogen bonded to a polypurine strand (Figure 26.3). The triplex-forming pyrimidine strand is bound in the major groove, parallel to the purine-rich strand.

Triple helices can also form at homopurine–homopyrimidine (one strand purines, the other strand pyrimidines) sequences in double-stranded DNA by oligonucleotides containing G and T or G and A. Here, G binds to GC pairs and T or A to AT pairs. Third strands containing Gs and As bind antiparallel to the purine-rich coding strand of the target. The orientation of G/T third strands depends on the target sequence. Table 26.2 outlines the characteristics of the Pyr.Pur.Pyr and Pur.Pur.Pyr motifs.

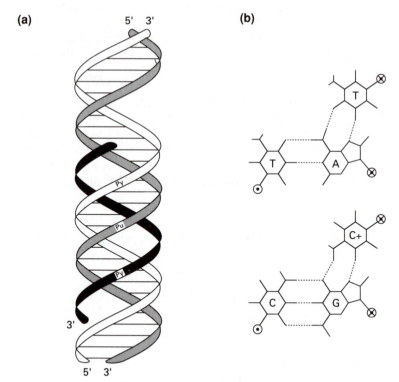

Figure 26.3 Triplex formation. (**a**) A schematic representation of a triple helix with the third strand (black) bound to a homopurine–homopyrimidine sequence of double helical DNA. (**b**) Base triplets formed by Watson–Crick AT and GC base pairs with T and protonated C (C+) respectively. (Source: redrawn with permission from Hélène and Toulmé, 1989).

Triplex formation is limited by low affinity and by the scarcity of suitable target purine-pyrimidine sequences

Currently only homopurine–homopyrimidine sequences represent suitable targets. These occur infrequently in clinically relevant genes, and are more frequent in upstream regulatory regions than in coding sequences. Potential target sequences can be extended by the use of non-natural bases (e.g. analogues of adenosine or guanosine), which may recognize thymine and cytosine in TA and CG base pairs. An alternative is to use a 'switchback' linker to allow the third strand oligonucleotide to interact with adjacent purine-rich tracts on both strands of the duplex. The required strand polarities are retained by combining the two major classes of base triplets, Pyr.Pur.Pyr on one strand and Pur.Pur.Pyr on the other. This approach has been used to target sequences in the human *p53* gene.

The rate of triplex formation is about 1000 times slower than the recombination rate between complementary strands to form a duplex. Electrophoretic mobility shift assays indicate that triplex formation, like the antisense effects described above, is an inefficient process. In theory a single third strand is required per target molecule but in practice about a 10^3–10^4 fold excess of third strands is required to convert 50% of the duplexes to triplexes. Once formed, however, the complexes are durable, lasting for approximately 12 hours. Formation of Pyr.Pur.Pyr triplets is pH-dependent, since cytosine is protonated only at low pH (pH 4–5). This has restricted use of the Pyr third strand model to acidic conditions or targets which have relatively few GC base pairs. Other factors that affect binding affinity include the length of the third strand oligonucleotide, the presence of base mismatches and the concentration of cations and spermine. Even under optimal conditions the melting temperatures for short (10–15 oligomer) unmodified pyrimidine oligonucleotides are in the range 10–30°C, so at 37°C the majority of triplexes dissociate into duplexes and free third strands.

Progress has been made in several areas to circumvent some of these obstacles. For example, the requirement for low pH is partially abrogated by substituting 5-methylcytosine (which is more readily protonated at higher pH), and can be completely overcome by using analogues such as 6-amino-2'-Q-methylcytosine, which

Table 26.2 Characteristics of two triplex motifs

	Pyr.Pur.Pyr	*Pur.Pur.Pyr*
Target sequence	Purine-rich	Purine (esp. G)-rich
Triplets	**T**.AT and **C+**.GC	**G**.GC and **A**.AT
Orientation of third strand	Parallel	Antiparallel
Dependent on	Low pH (~ 5)	Mg^{2+}
	Spermine	Spermine

have the correct hydrogen bond donors at physiological pH. Binding affinity can be increased by covalently linking intercalating groups such as acridine to the 5' end of third strand oligonucleotides. This allows triplex formation to direct site-specific intercalation, which in turn stabilizes the triplex, to the extent that the complexes are stable at physiological temperatures.

Psoralens are photoactive intercalating groups which on irradiation at 320–400 nm (UV) form stable covalent adducts with pyrimidine bases. A psoralen molecule linked to an oligonucleotide can be UV-cross-linked to one or both DNA strands, forming stable mono- or bisadducts. These can however be recognized and repaired by DNA repair enzymes.

Triplex formation is a useful experimental tool and a potential cancer therapy

Triplex-forming oligonucleotides have been used experimentally to introduce mutations into selected genes in mammalian cells. Psoralen-linked oligonucleotides can be cross-linked to target DNA sequences and form stable DNA adducts which are repaired by the cell. There is a high incidence of T.A to A.T transversions at the psoralen intercalation site, possibly because the adducts make it difficult for DNA polymerase to repair the lesion faithfully. Triplex-forming oligonucleotides can also be linked to strand-cleaving agents such as EDTA.Fe(II) chelate, which induces cleavage of double-stranded DNA in the presence of molecular oxygen and a reducing agent. This makes it possible to cut DNA in a site-specific manner without having to rely on naturally-occurring restriction sites for cutting by enzymes. This could be very useful for example in generating large DNA fragments for mapping and sequencing the human genome.

Triplex formation can be used to block the expression of genes important to the development of disease, including cancer. The triplex-forming agent itself can be used to interfere with gene function, or the third strand can be used as a vehicle for gene-targeted drugs. Many studies have focused on genes whose products confer aspects of the malignant phenotype. Several groups have targeted homopurine–homopyrimidine sequences in upstream regulatory regions of genes, including c-*myc* and the α-subunit of the IL2 receptor. In addition to *in vitro* effects, suppression of transcription has been shown in intact cells following brief culture in the presence of third strand oligonucleotides. These studies used the Pur.Pur.Pyr motif to favour triplex formation under physiological conditions. Other upstream sequences which have been targeted *in vitro* include promoter regions of the c-*erbB2*, rat *neu*, H-*ras* and human dihydrofolate reductase genes, as well as the interferon and progesterone response elements. A few studies have targeted transcription start sites or coding regions, including those of the human *MDR1* and aromatase genes.

26.4 PEPTIDE NUCLEIC ACID IS A DNA ANALOGUE WITH A PEPTIDE CHAIN IN PLACE OF THE SUGAR-PHOSPHATE BACKBONE

Peptide nucleic acid (PNA) is a DNA analogue in which the sugar-phosphate backbone is replaced by a peptide backbone consisting of (2-aminoethyl)glycine units (Table 26.3).

PNAs were originally designed in the hope of generating stable triplex-forming agents that would be resistant to nuclease attack. They are indeed stable in serum, and can bind extremely tightly with oligonucleotides or with double-stranded DNA. In fact the affinity for DNA is higher than that between the two strands of the duplex, and the PNA exerts its effects by displacing one of the duplex strands. The interaction between the PNA and the other duplex strand is stabilized by binding of a second strand of PNA to form a (PNA)2/DNA triplex (Figure 26.4).

The interaction is rapid, and forms complexes which are stable for several hours. It seems extraordinary that a chemical bond designed by nature for one

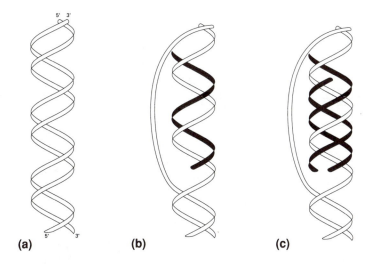

(a)   **(b)**   **(c)**

Figure 26.4   Peptide nucleic acids. (a) DNA duplex. (b) One PNA strand (shown in black) invades the duplex, binding to one of the DNA strands by Watson–Crick hydrogen bonding. The other DNA strand is displaced, forming a D-loop. (c) The PNA/DNA duplex is stabilized by binding of a second PNA strand to form a (PNA)2/double-stranded DNA complex. This is able to interrupt the passage of RNA polymerase, and so transcription is blocked at the site of PNA binding.

Table 26.3 Oligonucleotide modifications (Source: adapted with permission from Carter and Lemoine (1993) *Br. J. Cancer*, **67**, 869–876)

	Natural (phosphodiester)	Methyl phosphonate	Phosphorothioate	α-anomer	PNA
Solubility	Good	Less than PE[a]	Good	Good	Poor
Stability[b]	Very poor	Better than PE	Better than PE	Good	Very good[c]
Hybridization	Efficient	Less efficient	Efficient	Efficient	Very efficient[d]
Induction of RNase H	Good	Poor	Good	No good	Strand displacement
Triplex formation	Yes	Yes	No good[e]	Good	
Cell access	Good	Good	Poor	Good	Poor
Toxicity	Low	Low	Non-specific toxicity	Low	?Most
Structure					data *in vitro*

[a] PE = phosphodiester
[b] i.e. resistance to nucleases
[c] PNAs also resistant to peptidases
[d] Good only at salt concentrations below 50 mM
[e] No good for pyrimidine third strands but purine phosphorothioate oligos can form triplexes

purpose (i.e. to provide the links between amino acids in proteins) should turn out to be so effective at interacting with a completely different class of molecule.

PNAs can cause sequence-specific transcription arrest *in vitro*. Despite potent experimental effects, there are several problems to be overcome before they can be used successfully in biological systems. They are relatively insoluble in aqueous solution, intracellular access is poor, and binding to DNA is relatively poor at physiological salt concentrations.

26.5 RIBOZYMES ARE RNA SEQUENCES WITH NATURAL RIBONU-CLEASE ACTIVITY THAT RECOGNIZE AND CLEAVE mRNA

Ribozymes are short pieces of RNA (oligoribonucleotides) that have natural ribonuclease activity. They recognize and cleave mRNA at GUC sequences. The cleavage reactions presumably result because the specific RNA conformation brings reactive groups into close proximity. As can be seen from Figure 26.5, ribozymes have a defined secondary structure which is important for catalytic activity.

The correct folding is achieved by complementary base-pairing by parts of the RNA to other parts of the same strand. The catalytic activity of ribozymes can be directed at specific target transcripts by including an antisense region. When the target mRNA has been cleaved, the ribozyme is free to interact with another target molecule. As with the approaches discussed above, the efficacy of this approach depends on the ratio of effector to target molecules. The rate of catalysis is very low, and optimal effects require a great excess of ribozymes.

The assays used to show ribozyme effects include *in vitro* transcription assays, to demonstrate reduction in levels of full-length transcripts and appearance of truncated mRNAs. Immunoprecipitation and/or Western blotting can be used to show a reduction in levels of protein encoded by the target mRNA, and it is important to assess any phenotypic effects (morphology, growth) consequent upon this. Ideally, to confirm the specificity of these effects, it should also be shown that these same effects are not associated with use of a control ribozyme inactivated by point mutation.

Ribozymes have been used to target mutant H-*ras*, which, unlike its wild-type counterpart, contains the sequence GUC. This approach has also been used to target transcripts encoding PDGF-ß, pleiotropin, urokinase plasminogen activator receptor, *MDR1*, *bcr–abl*, GUC sequences within the HIV genome, and fusion transcripts associated with the t(15:17) translocation in acute promyelocytic leukaemia.

There are several problems associated with the use of ribozymes. The requirement for GUC restricts the sites which can be targeted in this way. Another disadvantage is that RNA is much less stable than DNA, being very sensitive to degradation by nucleases. Therefore it is very unlikely that RNA molecules will be suitable for administration as therapeutic agents to intact cells. To circumvent this problem, it may be possible to design stabilizing modifications to RNA.

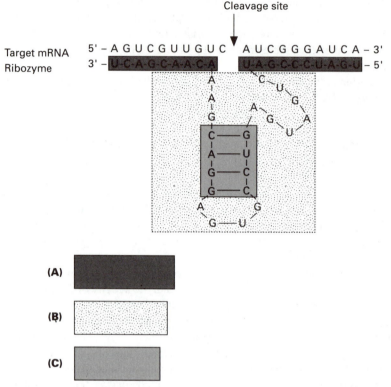

Figure 26.5 Ribozyme design. There are two main structural domains: the antisense region (**A**), i.e. complementary to target mRNA; conserved sequences in the ribozyme required for catalytic activity (**B**). Within this domain, (**C**) is the region of internal base-pairing that imposes the correct secondary structure.

Another way around the problem of instability, at least to test the effect within a defined cell population, is to construct an expression vector for transfection into cells, from which multiple copies of the ribozyme will be generated.

26.6 OLIGONUCLEOTIDES CAN BE STRUCTURALLY MODIFIED TO INCREASE THEIR STABILITY

The natural structure of DNA and unmodified oligonucleotides is a sugar-phosphate backbone with bases attached to the sugar ring in the so-called beta orientation. Oligonucleotides are taken up by cells by endocytosis, hybridize with high affinity with mRNA and efficiently induce RNase H. In experimental studies conducted in serum-free medium they are very effective antisense agents. However their half-life in serum or serum-containing media can be as short as 20 minutes.

This renders them unsuitable for use in experiments under physiological conditions and for consideration as therapeutic agents for clinical use.

There are various ways of modifying the backbone to enhance stability and resistance to nucleases (Table 26.3). Ideally this should be achieved without non-specific toxicity to the cell, and without prejudicing the affinity of interaction with target molecules and, for antisense agents, ability to induce RNase H. Several modifications are appropriate for antisense, especially phosphorothioate and methylphosphonate oligonucleotides. However these are less useful for triplex formation, and other stabilizing modifications are needed here. One simple solution is to block the 3'-end, for example with an amine group. Another more complex solution is to make oligonucleotides from α-nucleotides, in which the bases have the opposite orientation with respect to the sugar ring compared to naturally occurring ß-nucleotides. Synthesis of such α-anomers involves difficult chemistry and so far can be accomplished only for pyrimidine oligonucleotides (i.e. Cs and Ts only).

26.7 CELLULAR UPTAKE OF OLIGONUCLEOTIDES MUST BE IMPROVED IF THEY ARE TO BE CLINICALLY EFFECTIVE

The simplest and most direct method of administration *in vivo* is intravenous infusion. Oligonucleotides are rapidly cleared from the circulation and distributed to most tissues. Most of the dose is excreted in 2–3 days, predominantly in the urine. In the tissues *in vivo*, and in cell culture *in vitro*, oligonucleotides are able to get into cells. This occurs partly by passive diffusion across the cell membrane, but mostly by endocytosis. Oligonucleotides within endosomal vesicles are inside the cell but not free in the cytoplasm. Sequestration in endosomes renders the oligonucleotide subject to lysosomal fusion and degradation. Diffusion out of the vesicle must occur before antisense agents can interact with mRNA in the cytoplasm. In the case of triplex formation, or antisense agents directed at unspliced RNA, the target molecules are in the nucleus. Oligonucleotides are transported into the nucleus through nuclear pores, which are specialized structures in the nuclear membrane.

Intracellular uptake can be increased by linking the oligonucleotide to a hydrophobic group such as cholesterol, polylysine or porphyrin, or by liposomal encapsulation. However addition of cholesterol or other bulky groups can markedly reduce the efficiency of hybridization. In experimental systems, intracellular levels can be increased by microinjection, or using transfection techniques such as lipofection or electroporation.

26.8 ANTISENSE mRNA CAN BE SYNTHESIZED INSIDE THE CELL USING AN EXPRESSION VECTOR

When contemplating any of these approaches as therapy, the main problems are extracellular degradation by serum nucleases and inefficient and heterogeneous uptake into cells. As described above, various avenues are being explored to overcome these problems, but there is an alternative method which bypasses several problems at least in the experimental setting. This is to make an **expression vector**, a virus or plasmid that carries a foreign DNA sequence. After transfection into the cell population under study the vector drives expression of the desired transcripts. By inserting the cDNA 'back to front' (i.e. 3' to 5') in the vector, the promoter will drive expression of the non-coding 'antisense' strand. This approach has been limited to the production of relatively large transcripts which, according to their design, can function as antisense agents or ribozymes. Recently a strategy has been described that enables intracellular expression, not only of kilobase fragments of RNA, but also of short oligoribonucleotides which can function as triplex forming agents.

Depending on the type of promoter used, the construct can be designed to generate RNA continuously, or intermittently, under the control of some exogenous agent, usually either a temperature change or chemical agent, e.g. dexamethasone or heavy metal. Tissue-specific expression can be achieved by inserting a tissue-specific promoter into the vector, e.g. the tyrosinase promoter to drive expression in melanoma cells or the prostate-specific antigen promoter to drive expression in prostate cancer cells. This approach shares features with 'gene therapy', a term which encompasses any stratagem designed to introduce new genetic information into the cell (Chapter 24).

26.9 WILL OLIGONUCLEOTIDE THERAPY BE CLINICALLY USEFUL?

The approaches described here are undoubtedly valuable in the experimental setting to explore the role of specific genes which contribute to the malignant phenotype. With regard to their current clinical value, the antisense approach has proceeded to clinical trials as an *ex vivo* therapy to purge the bone marrow of patients with haematological malignancies. In Canada patients with AML/myelodysplastic syndrome have been treated with an phosphorothioate antisense oligonucleotide targeting *p53*. The therapy was given by continuous infusion, and was well tolerated.

There are many other applications which have great potential, but which have not yet been tested clinically. One such is the use of the antisense approach in glioblastoma cell lines, which express IGF-1 in an autocrine manner. The introduction of an antisense IGF-1 expression vector slowed the growth of transfected cells in culture and abolished their tumorigenicity in nude mice. When a small number of these modified cells were inoculated into animals, unmodified tumours at distant sites regressed. Clearly this model is relevant to the clinical problems posed by metastatic cancer, although it is not clear at present whether it will be applicable to other tumours in addition to glioblastoma. Another clinically relevant observation is that the effects of conventional chemotherapy can be enhanced by concurrent oligonucleotide treatments. Examples include targeting *MDR1* mRNA to enhance sensitivity to anthracyclines, antisense-mediated reduction of *bcl-2* expression to reverse the chemoresistance of non-Hodgkin's lymphoma cells to cytosine arabinoside and methotrexate, and the use of anti-*fos* ribozymes to enhance the sensitivity of ovarian cancer cells to cisplatin.

Triplex formation has some theoretical advantages over the antisense approach, but so far none of the experimental results reported have been followed up clinically. Because the affinity of triplex interactions is relatively low, applications of potential clinical relevance will require the use of oligonucleotides linked to intercalating groups to enhance stability. Photoactive groups such as psoralens, which can be cross-linked to DNA, will increase the chance of obtaining durable effects on gene expression. However the need for photoactivation will limit clinical use to superficial skin malignancies or *ex-vivo* marrow purging.

Much has been made of the potential toxicities and cost of these and other novel therapies. Side-effects are likely to result from the toxicity of the agents themselves, from the effects of inhibiting target processes/molecules in normal cells and from cross-reacting effects on other targets. At present the cost of preparing oligonucleotides in the amount and purity required for clinical use is formidable, but this problem will be overcome by scaling up production if justified by experimental results. However these considerations are irrelevant at the moment; the critical factor dictating clinical usefulness is whether molecular therapeutic approaches will have sufficiently potent antitumour effects in animal models and clinical trials to merit regular clinical use. In particular, for any given tumour type, how will the effects of these treatments compare with what can be achieved with conventional therapy? The answer to this question is unclear at present, but it is because of this consideration that many groups are pursuing new treatments in tumours such as renal cancer and melanoma, where current therapy offers little.

The present position of oligonucleotide-based therapies is roughly that occupied by monoclonal antibodies 10–15 years ago: valuable experimental tools, with great potential as therapy. Unfortunately this potential has not been realized for antibodies, which generally have contributed little to conventional treatment. The next 5–10 years should clarify whether molecular therapies will have greater success.

26.10 ACKNOWLEDGEMENTS

I am very grateful to Alan Ashworth, Paula Bates, Denis Talbot and my father, Jerry Kirk, for helpful suggestions.

26.11 FURTHER READING

Czubayko, F., Riegel, A.T. and Wellstein, A, (1994) Ribozyme-targeting elucidates a direct role of pleiotrophin in tumor growth. *J. Biol. Chem.*, **269**, 21358–21363.

Hélène, C. and Toulmé, J.-J. (1989) Control of gene expression by oligodeoxynucleotides covalently linked to intercalating agents and nucleic acid-cleaving reagents, in *Oligonucleotides–Antisense Inhibitors of Gene Expression*, (ed. J. Cohen), MacMillan Press, UK.

Murray, J. A. H. (ed.) (1992) *Antisense RNA and DNA*, Wiley-Liss, New York.

Nielsen, P. E., Egholm, M., Berg, R. H. and Buchardt, O. (1993) Peptide nucleic acids (PNAs): potential antisense and anti-gene agents. *Anticancer Drug Design*, **8**, 53–63.

Thuong, N. T. and Hélène, C. (1993) Sequence-specific recognition and modification of double-helical DNA by oligonucleotides. *Angew. Chem. Int. Ed. Engl.* **32**, 666–690.

Trojan, J., Johnson, T. R., Rudin, S. D. *et al.* (1993) Treatment and prevention of rat glioblastoma by immunogenic C6 cells expressing antisense insulin-like growth factor-I RNA. *Science*, **259**, 94–96.

Wagner, R. W. (1994) Gene inhibition using antisense oligodeoxynucleotides. *Nature*, **372**, 333–335.

PART THREE

Gene Structure and Function

Introduction to gene structure and expression | 27

Richard Wooster

27.1 INTRODUCTION

The aim of this chapter is to introduce aspects of gene structure and regulation. While it is not possible to give an exhaustive account, a number of topics are discussed that are pertinent to the current field of molecular oncology. The first of these is the structure of the DNA molecule and how it is organized into genes. The second concerns transcription and the factors responsible for its regulation. The third topic concerns the processing of transcribed RNA, and the final section discusses the topic of DNA polymorphism.

27.2 THE STRUCTURE OF DNA

Nucleic acids consist of a sugar, a phosphate and a nitrogenous base

Deoxyribonucleic acid (DNA) is a chain of chemically linked nucleotides. The nucleotides consist of a sugar, a phosphate and a nitrogenous base. The sugar and phosphate groups form the backbone of the DNA molecule. In DNA, the sugar is deoxyribose, while in RNA the sugar is ribose. In DNA, the nitrogenous bases are adenine (A), guanine (G), cytosine (C) and thymine (T). In RNA, uracil (U) replaces thymine (T). From their chemical structure, adenine and guanine are defined as purines, while cytosine, thymine and uracil are pyrimidines.

Molecular Biology for Oncologists. Edited by J. R. Yarnold, M. R. Stratton and T. J. McMillan. Published in 1996 by Chapman & Hall, London. ISBN 0 412 71270 9

The DNA molecule is a double-stranded helix

The DNA in a cell is double-stranded, forming a double helix where one nitrogenous base forms weak bonds with another on the opposite strand to form a **base pair** (bp). In general, adenine pairs with thymine and cytosine pairs with guanine and therefore one strand is complementary to the other. By convention, one end of the molecule is defined as 5' while the other end is 3', and the sequence is written in the 5' to 3' direction (5' and 3' derive from the numbering of the carbon atoms in the sugar molecule). When a gene is present within the DNA sequence, the sequence is written such that the DNA coding for the amino terminus of the protein is to the left of the page (the 5' end of the gene), while the carboxy terminus is to the right of the page (the 3' end of the gene).

DNA codes for proteins

The four bases that store the genetic information in DNA must code for the 20 amino acids that are used in proteins. A sequence of three bases, called a **codon**, is specific for a particular amino acid. There are a total of 64 different codons. Of these, three (TGA, TAG and TAA) encode signals that stop translation of mRNA into protein (**stop codons**). The remaining triplet codons encode the various amino acids. Some amino acids are coded by only one codon while others have up to six different codons. The codon ATG is unique in that it is present at the start of every coding region. This codon also codes for the amino acid methionine and is therefore also present within the coding regions.

The human genome contains 6×10^9 base pairs

The diploid human genome contains approximately 6 000 000 000 bp, which are divided into 46 chromosomes. Each genome contains two copies of each autosome (numbered from 1–22) and two sex chromosomes (called X and Y). The haploid human genome of 23 chromosomes contains about 100 000 genes. The average size of a gene at the DNA level is about 10 000 bp, i.e. 10 kb (Table 27.1).

Table 27.1 A comparison of the size of genes, the number of exons in the gene and the size of the processed messenger RNA

Gene	Number of exons	Gene size (kb)	mRNA size (kb)
Dihydrofolate reductase	6	31	2
DMD	> 60	2000	14
RB1	27	250	4.6
WT1	10	50	3
β-globin (mouse)	3	0.85	0.59

Therefore, only 30% of the genome is covered by genes. Furthermore, only 2–3 kb of the average gene codes for proteins. Thus, only 5–10% of the human genome codes for proteins. The known roles of the remaining 90–95% of the genome include the maintenance of the structure of the genome itself and the regulation of gene transcription (see below).

Genes contain exons and introns

Thousands of genes have been cloned and sequenced and it is possible to describe an idealized gene (Figure 27.1).

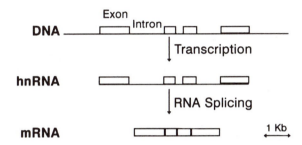

Figure 27.1 The intron/exon structure of a idealized gene. The alternating exon/intron structure of an idealized gene is shown at the top of the diagram. This gene is transcribed to produce the hnRNA shown and the introns are removed by RNA splicing.

The DNA 5' to the coding region usually contains sequences that control the transcription of the gene (see below). The gene itself is divided into two distinct elements, **introns** and **exons**. The exons contain the coding region of the gene. The introns are removed from the primary transcript (see RNA splicing, below) to leave the exons. The coding region has a start codon towards the 5' end of the gene with one stop codon at the 3' end of the coding region. The sequence between the start and stop codon constitutes an **open reading frame** (ORF), and this is flanked by 5' and 3' non-coding sequences. The number of exons and introns within a gene varies between 1 and over 60 (Table 27.1).

27.3 TRANSCRIPTION

The transcription of DNA produces RNA

DNA is transcribed into RNA by DNA-dependent RNA polymerases (RNA pol). These polymerases are part of a larger aggregate of proteins that move in a 5'-3' direction along the DNA from a point of initiation to a point of termination. There are three RNA polymerases called RNA pol I, II and III. RNA polymerase II transcribes genes that code for proteins (pol I transcribes ribosomal genes and pol III transcribes transfer RNA genes). The activity of RNA polymerase II, and hence the rate of transcription and expression, is tightly controlled. The control involves two major components: the presence of specific DNA sequences upstream or downstream of the gene, and proteins that recognize and bind to these specific sequences. The DNA sequences are called **promoter** and **enhancer sequences**. The proteins that bind to them are called **transcription factors**.

Promoter and enhancer regions are segments of DNA that are involved in the regulation of transcription

A **promoter** consists of a number of short conserved nucleotide sequences (motifs) located 5' (upstream) of a gene that functions by binding transcription factors (see below). The same DNA motif is often found in many different promoter sequences. For example, one of the motifs contains the consensus sequence TATA, called the **TATA box**. This four-nucleotide sequence is usually located 25 bp upstream (5') of the first base to be transcribed (the transcription start site). Other common nucleotide motifs include **GC** and **CAAT boxes**. While some of these sequences are required for transcription initiation, they are not all essential and the number and types of motifs adjacent to individual genes are variable.

An **enhancer** is a conserved sequence that may be positioned from a few hundred bases to tens of kilobases from the start point of transcription of a gene. In contrast to promoters, enhancer regions can be positioned either upstream or downstream of a gene (i.e. 5' or 3') and can function with the nucleotide sequence in either orientation.

Transcription factors are proteins that bind to specific DNA sequences within promoters and enhancers

Many proteins are capable of binding to DNA but not all these are transcription factors. For example, histone proteins bind to DNA throughout the genome and help to organize the architecture and packaging of the chromatin. By contrast, transcription factors bind specifically to DNA sequences within promoters and enhancers. Some transcription factors bind to most promoter regions and are essential for transcription. Others only bind to promoters and enhancers associated with particular genes and ensure tissue-specific expression of that gene.

Transcription factors often share amino acid sequences that are important for binding to DNA

There are a number of DNA-binding motifs in the protein sequences of transcription factors. The Wilms tumour suppressor gene product contains four repeating protein sequences that bind specifically to DNA. Each protein sequence incorporates a zinc ion that is hydrogen-bonded to two cysteine and two histidine residues. This sequence is referred to as the **zinc finger**. Other transcription factors containing zinc fingers include the steroid hormone receptors, e.g. the oestrogen receptor protein.

Another motif is the **leucine zipper**. This consensus contains a leucine amino acid every seven residues along the polypeptide chain of the transcription factor. It is thought that the leucines allow the transcription factors to form heterodimers and homodimers. Dimerization is thought to occur before the factors can bind to the DNA. One example of the leucine zipper structure is the formation of Jun–Jun homodimers and Jun–Fos heterodimers. Jun and Fos proteins are oncogene products that are components of an important complex of transcription factors known as AP1. Finally, the helix–turn–helix motif of the homeobox transcription factors is another common and important protein sequence which interacts specifically with regulatory sequences in DNA.

27.4 THE PROCESSING OF RNA

RNA splicing leads to the removal of introns from the transcribed RNA

The initial transcript of a gene is called **heterogeneous nuclear RNA (hnRNA)** and is composed of both introns and exons. During the process known as RNA splicing, the hnRNA is cut and subsequently rejoined, without the introns, to create messenger RNA (mRNA). The junctions between introns and exons contain short consensus signal sequence (Figure 27.2).

Figure 27.2 The consensus splice sequences. The intron/exon boundaries have been amplified to reveal the DNA sequence surrounding these junctions. The numbers below the DNA sequences show the percentage of known sequences that have this consensus.

These sequences act as a recognition signal for proteins that bind to these sites and perform the DNA cutting and splicing reactions. Naturally occurring or artificially introduced mutations within these conserved splice signal sequences reduce or remove the ability of the hnRNA to undergo splicing. Furthermore, the introduction of new splice signal sequences can lead to a change in the natural pattern of splicing.

Alternative splicing produces several mRNAs from a single precursor hnRNA

Most hnRNAs are processed to produce a single-sized mRNA. However, some hnRNAs are processed to produce multiple mRNAs, where the difference between them is the presence or absence of certain exons. For example, there is only one *WT1* (Wilms tumour) suppressor gene. However, two alternative splice sites lead to the possibility of four distinct *WT1* mRNAs coding for proteins with distinct functions. At one of these splice sites, the splicing reaction can lead to a 17 amino acid insert towards the N-terminus of the protein. At the other site, the splicing reaction can lead to the insertion of three amino acids between two of the zinc fingers in the DNA-binding region. All four variants are found in the tissues in which *WT1* is expressed, where they are thought to contribute to the normal function of the *WT1* gene product.

The degradation of mRNA is prevented by sequences at the ends of the RNA molecule

mRNA molecules terminate in a GT region, which is followed by the nucleotide sequence AAUAAA at the 3' end of the gene. The AAUAAA sequence is called a **polyadenylation signal**, and it stimulates a protein complex to add a chain of adenosines to the 3' end of the mRNA. This **polyA tail** is thought to prevent degradation of the molecule from the 3' terminus. The 5' terminus is protected by a 'cap' that constitutes a 5'–5' bond structure (the usual DNA and RNA bonding structure is a 5'–3' bond). The centre of the RNA message is still susceptible to degradation and processing.

27.5 DNA POLYMORPHISMS

Although the human genome is almost identical from one person to another, there are small variations in DNA sequences between individuals. These are usually in non-coding DNA, either in the extensive regions of DNA between genes or within introns. Several types of sequence variation are recognized, including restriction site polymorphism, minisatellites and microsatellites (see below). They have an important role in genetic linkage analysis and in the study of DNA deletions in tumours (e.g. loss of heterozygosity).

Restriction endonucleases can be used to detect restriction fragment length polymorphisms

Over 100 enzymes called **restriction endonucleases** have been isolated from bacteria and are used in the laboratory to cleave DNA at defined nucleotide sequences. If a single base sequence variation changes a restriction enzyme cleavage site, the enzyme will not cut the DNA there (Figure 27.3((a)).

It is therefore possible to detect differences in homologous DNA sequences between two alleles if a single base change in one allele falls within a restriction site. This type of polymorphism is known as a restriction fragment length polymorphism (RFLP). A single base polymorphism occurs approximately every 200 base pairs, although not all fall within restriction enzyme recognition sites.

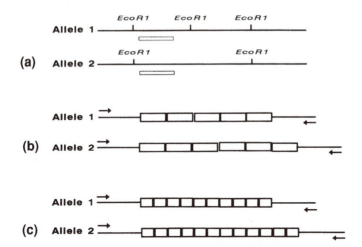

Figure 27.3 The structure of DNA polymorphisms. **(a) RFLPs.** The solid line repre-sents DNA sequence. *Eco*R1 is the restriction endonuclease that is used in this example. The vertical lines indicate the positions at which this enzyme will cleave the DNA. The differential cleavage pattern can be visualized using the probe indicated by the open box. **(b) Minisatellites.** The solid lines represent 'random' DNA sequences. The open box shows a repeat unit (6–50 bp) and this unit is present five times in allele 1 and six times in allele 2. These repeats can be amplified using the primers shown by the arrows and the PCR. **(c) Microsatellites.** As for **(b)** except that the repeat unit is smaller (2–5 bp) and the number of repeat units are 10 and 12 for alleles 1 and 2 respectively.

Minsatellites contain a variable number of tandem repeats

Minisatellites consist of short sequences of DNA repeated a number of times in tandem array (Figure 27.3(b)). The basic sequence can be from 6 bp to over 50 bp in length. The number of times the basic sequence is repeated varies from allele to allele (from 10 to more than 60 times) and gives rise to a high level of polymor-phism in the population. These repeats are called **minisatellites**, or variable num-ber of tandem repeats (**VNTRs**). They may be analysed by Southern blotting and DNA hybridization or by the polymerase chain reaction (Figure 27.3(c)). These repeats occur throughout the genome, although they tend to cluster towards the telomeres of chromosomes. In general, VNTRs are more polymorphic than RFLPs because the latter generally have only two forms.

Microsatellites are minisatellites that contain small repeat units

Microsatellites are VNTRs in which the tandem repeat sequence is from 1–5 bp long. One of the most widely exploited is the dinucleotide GT repeat (Figure 27.3(c)). These are analysed using the polymerase chain reaction. It has been esti-mated that there are 50 000–100 000 microsatellites in the human genome, of which about 2000 have been characterized, i.e. located. As with VNTRs, microsatellites are generally more polymorphic than RFLPs.

27.6 FURTHER READING

General

Lewin, B. (1990) *Genes IV*, Oxford University Press, Oxford.
Watson, J. D., Hopkins, N. H., Roberts, J. W. *et al.* (1987) *Molecular Biology of the Gene*, Benjamin Cummings, Menlo Park, CA.

Specific

Aoyama, N., Nagase, T., Sabazaki, T. *et al.* (1992) Overlap of the p53-responsive element and cAMP-responsive element in the enhancer of human T-cell leukaemia virus type I. *Proc. Nat. Acad. Sci. USA*, **89**, 5403–5407.
Gyapay, G., Morissette, J., Vignal, A. *et al.* (1994) The 1993–94 Genethon human linkage map. *Nature Genet.*, **7**, 246–339 (for microsatellites).
Haber, D. A. and Housman, D. E. (1992) Role of the WT1 gene in Wilms tumour. *Cancer Surv.*, **12**, 105–117 (for alternative splicing).
Little, M. H., Prosser, J., Condie, A. *et al.* (1992) Zinc finger point mutations within the WT1 gene in Wilms tumour patients. *Proc. Nat. Acad. Sci. USA*, **89**, 4791–4795.
Sharp, P. A. (1987) Splicing of messenger RNA precursors. *Science*, **235**, 766–771 (for RNA splicing).

Chromatin structure and nuclear function

28

Dean A. Jackson

28.1 IN HUMAN DIPLOID CELLS, TWO METRES OF DNA IS FOLDED IN A HIGHLY ORDERED WAY

The cells of higher organisms contain two major compartments, the **cytoplasm** and the **nucleus**, separated by a double nuclear membrane. The outer face of this membrane is continuous with a network of cytoplasmic membranes, the **endoplasmic reticulum**, and is often seen to be coated with ribosomes, while its inner face lies in close contact with a filamentous network of intermediate filament-related lamin proteins – the **nuclear lamina**. Communication between these two compartments is controlled by a specialized organelle, the **nuclear pore complex**. Typically, a cell might have between 500 and 2500 nuclear pore complexes, depending on its metabolic activity.

Nuclei, occupying some 10% of a cell by volume, are clearly seen by electron microscopy as rather dense, amorphously staining structures. They contain the cell's DNA, a linear double-stranded helical polymer composed of two purines, adenine and guanosine and two pyrimidines, cytosine and thymine. In diploid mammalian cells, 6×10^9 bp (base pairs) of DNA, measuring 2 m if extended, folds so as to occupy a nucleus measuring 5–10 μm across. To achieve this, the DNA/protein complex called chromatin must fold into a series of higher-order arrays. In its most compact state chromatin forms a characteristic number of discrete units, chromosomes; structures essential for cell division but not evident throughout interphase when chromatin is relatively decondensed.

In contrast to the organelle-rich cytoplasm, nuclei appear relatively ill-structured. The only evident compartments are two to five highly structured centres of

Molecular Biology for Oncologists. Edited by J. R. Yarnold, M. R. Stratton and T. J. McMillan. Published in 1996 by Chapman & Hall, London. ISBN 0 412 71270 9

ribosomal RNA synthesis and ribosome biogenesis (the nucleoli) and regional variations in chromatin density. The variation in chromatin density corresponds to inactive (heterochromatic) and actively transcribing (euchromatic) areas of the genome. It is now clear, however, that the apparently homogeneous nuclear interior is in fact highly structured and it appears that this organization is likely to be critical to different aspects of nuclear function.

Understanding how the genome is organized *in vivo* and how precisely this organization influences nuclear function will be of fundamental importance and will undoubtedly influence our understanding of gene expression, differentiation and malignancy. In addition, the ability to predict how genes might behave when introduced into a foreign nuclear environment will be valuable if we are to establish clinically viable gene therapy techniques.

28.2 THE DNA HELIX WRAPS AROUND A HISTONE PROTEIN CORE TO FORM NUCLEOSOMES, THE BASIC SUBUNITS OF CHROMATIN STRUCTURE

DNA is a linear polymer, rich in acidic groups that must be neutralized during folding in order to overcome charge repulsion. Basic, lysine- and arginine-rich, proteins called **histones** bind DNA to form the highly conserved fundamental subunit of chromatin structure, the **nucleosome**. The nucleosome comprises 146 bp of DNA wrapped in 1.75 turns around the histone core particle. Each core particle is an octamer of four 'core' histone proteins – H2A, H2B, H3, H4 – and is stabilized through interactions in the N-terminal domains of histones H3 and H4 (Figure 28.1).

Histones do not bind to specific DNA sequences though some binding preferences accommodate bending of DNA around the protein core. Nucleosomes are positioned at regular intervals along DNA and are separated by intervening DNA linking sequences which vary from 0–60 bp in length depending on the cell. Transcriptionally active or competent (capable of transcription given the appropriate signals) regions of the genome form open nucleosomal arrays of euchromatin (the 10 nm chromatin fibre) that can be visualized as 'beads on a string' by electron microscopy, where the beads are the nucleosomes and the string is the intervening linker DNA.

28.3 INACTIVE GENES ARE FURTHER CONDENSED INTO A 30 nm CHROMATIN FIBRE IN THE PRESENCE OF HISTONE H1

Transcriptionally active chromatin represents only a fraction, 10% or less, of the total chromatin in most cells. Most chromatin is not expressed and is packaged into a relatively condensed, transcriptionally inert 30 nm chromatin fibre, which is stabilized by a fifth histone, histone H1. When extended chromatin fibres are

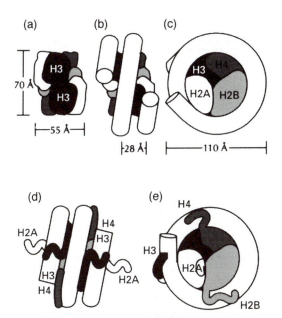

Figure 28.1 The nucleosome. Schematic illustration showing the locations of core histones in the histone octamer (**a**) and nucleosome core – (**b**), end view; (**c**), side view – with its 146 bp of DNA (1.75 turns). One possible distribution of the flexible amino-terminal domains is shown – (**d**) and (**e**): as (**b**) and (**c**) but now with two full turns of DNA. (Source: reproduced with permission from Smith, 1991.)

refolded *in vitro* structured arrays referred to as a 'solenoid' are often seen (Figure 28.2). In the nucleus, stretches of chromatin fibre within an average diameter of 30 nm can be identified, but these are generally irregular and appear to be defined by features of chromatin that are less precise than would be required to form solenoid-like repeats.

Differences in chromatin structure are reflected by the sensitivity of different parts of the genome to digestion with nucleases, typically pancreatic deoxyribonuclease (DNase I). Transcriptionally active or competent chromatin is relatively unfolded with exposed DNA, making it 'sensitive' to digestion with DNase I. In addition, upstream regulatory elements acquire a more dramatic DNase I 'hypersensitivity' that correlates with the precise location of sequence-specific DNA-binding proteins at functionally important sites.

The protein content of active chromatin differs from inactive chromatin in many ways. For example, complex combinations of different histone H1 subtypes (six are found in mammalian cells) found in different heterochromatic areas are partially replaced by the high mobility group (HMG) proteins in active chromatin.

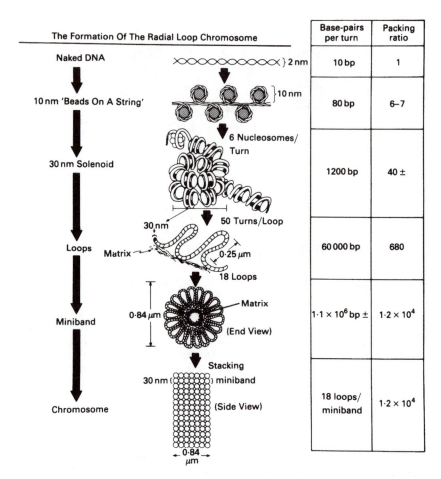

The Formation Of The Radial Loop Chromosome

	Base-pairs per turn	Packing ratio
Naked DNA / 2 nm	10 bp	1
10 nm 'Beads On A String' / 10 nm	80 bp	6–7
30 nm Solenoid / 6 Nucleosomes/Turn	1200 bp	40 ±
Loops / Matrix / 50 Turns/Loop / 18 Loops / 0·25 μm / 30 nm	60 000 bp	680
Miniband / Matrix (End View) / 0·84 μm	1.1×10^6 bp ±	1.2×10^4
Chromosome / Stacking miniband / 30 nm (Side View) / 0·84 μm	18 loops/miniband	1.2×10^4

Figure 28.2 Levels of chromatin condensation. A schematic illustration showing how the DNA duplex must fold in order to achieve the level of condensation seen in chromosomes. During interphase, chromatin loops of 60 kb (in this example) form through their interaction with components of the nuclear matrix. Within these loops chromatin might be folded into a condensed, 30 nm, fibre or the more open, 10 nm, structure characteristic of active genes. Dramatic architectural changes accompany chromosome formation, prior to mitosis. (Source: reproduced from Pienta, K. J. *et al.*, *J. Cell Sci.* **Suppl. 1**, 123–135 with permission of the Company of Biologists.)

28.4 THE CHROMATIN FIBRE INTERACTS WITH A NUCLEAR SUBSTRUCTURE TO FORM CHROMATIN LOOPS

The 10 nm fibre or 30 nm 'solenoid', depending on transcriptional status, is folded into chromatin loops, averaging typically 60–100 kb in length, which are attached at their bases to nuclear matrix proteins (Figure 28.2). With no means of estimating loop size *in vivo*, a variety of experimentally derived 'nuclear derivatives' have been used to determined this range. For example, processed nuclei called 'nucleoids' were prepared by extracting cells with a non-ionic detergent and 2 M NaCl and first demonstrated the existence of supercoiled DNA loops within interphase nuclei of eukaryotic cells. Later, insoluble 'nuclear matrices' and 'nuclear scaffolds' were identified to which the DNA loops are believed to attach at particular binding sites. DNA sequences 0.3–1 kb in length and rich in adenine and thymine bind the nuclear scaffold or matrix at 5–100 kb intervals, thereby creating the chromatin loops (Figure 28.2). Finally, HeLa cells analysed under isotonic conditions were shown to contain chromatin loops with an average length of 85 kb and broad size distribution covering 5–250 kb.

28.5 A HIGHER LEVEL OF CHROMATIN CONDENSATION IS REQUIRED FOR CHROMOSOME FORMATION PRIOR TO MITOSIS

When cells divide, the chromatin duplicated during S-phase is condensed further into chromosomes, with one chromosome set passing to each daughter cell. The further condensation that chromosome formation demands is achieved through additional levels of higher-order packaging, with loops folding into 100 nm coils which are further coiled or stacked into compacted chromosome structures (Figure 28.2). This occurs in response to a complex series of biochemical events that are coupled to the cell cycle. Prior to mitosis, solubilization of the nuclear membrane and chromosome condensation occur in response to phosphorylation events that correlate with the appearance of specific cell-cycle-dependent kinase activities.

Though less complex in terms of their protein composition, chromosomes maintain a distinct structure evident from the clearly recognizable banding patterns that can be seen after staining. Defined 'telomeres' (ends) and 'centromeres' (sites of microtubule attachment) each play important roles and have well characterized components.

28.6 NUCLEAR STRUCTURE AND FUNCTION ARE CLOSELY LINKED

The major function performed by the nuclei of eukaryotic cells is transcription of DNA into RNA. Other major functions are DNA replication and DNA repair, the latter required to ensure that potentially mutagenic lesions in DNA do not persist in the genome.

28.7 ONLY 2% OF THE GENOME ENCODES PROTEINS

The human genome of 3000 million base pairs encodes an estimated 100 000 genes. Remarkably, less than 2% of the genome is required to fulfil this coding potential; the estimated 10% of the genome that is transcribed includes non-coding introns which must be removed before the mature message (mRNA) can be transported to the cytoplasm. The organization of genes is quite variable: for example, the dystrophin gene is about 2 million base pairs long with more than 50 introns and a 17 kb mRNA while the globin genes are 1.5 kb long with two introns and a 0.6 kb mRNA. A total of 90% of the genome is non-coding. Whether this fraction performs any function is unclear – it may be simply 'junk DNA', but it could influence the folding and nuclear location of functionally important sequences.

Approximately 20% of a cell's genes are sufficient to provide the basic components of growing cells. The rest are required for the specialized functions performed by differentiated cells. In humans, about 250 morphologically distinct cells types correlate with the specialized cellular functions resulting from specific patterns of gene expression. In these different cell types, differences in the level of expression of a particular gene can be remarkable, reaching as much as 10^8-fold in extreme cases. This dramatic difference between active and inactive transcriptional states is achieved through the complex interplay of chromatin structure, transcription factor specificity and nuclear organization.

28.8 MULTIPROTEIN COMPLEXES CONTAINING TRANSCRIPTION FACTORS BIND DNA UPSTREAM OF GENES TO REGULATE THE ACTIVATION OF RNA POLYMERASES

Once an active chromatin state is established, the relatively decondensed euchromatin is accessible to transcription factors that together determine where RNA synthesis begins. The arrays of factors that contribute to this pre-initiation complex include proteins bound to many promoters, together with gene- or tissue-specific factors that bind characteristic DNA motifs depending on cell type. These transcription factors associate with a variety of adaptor proteins to form a multiprotein complex – containing perhaps 50 polypeptides in all – that provides a surface that binds the transcription apparatus, the RNA polymerase, so that transcription can begin.

Three categories of genes are transcribed by distinct, but related, RNA polymerases.

- **RNA polymerase I** is restricted to nucleoli and transcribes the multiple ribosomal genes encoding ribosomal RNA (structural molecules, not translated into protein); the primary transcripts are processed to give the mature 28S, 18S, and 5.8S ribosome-associated RNAs (S stands for Svedberg and is a measure of RNA size). Each human cell has about 500 copies of the ribosomal locus, of which 30–50% may be active in cycling cells. The number of active ribosomal genes is a useful indicator of metabolic activity and can be used to assess a cell's proliferative potential.

- **RNA polymerase II** is found throughout the nucleoplasm and transcribes all protein-encoding genes and some small structural RNAs.
- **RNA polymerase III** transcribes small, 5S and transfer RNAs (tRNA) and some small structural RNA.

All RNA polymerases are multiprotein enzymes with 12–16 subunits. Catalytic roles are performed by related, major subunits of about 200 kDa. Smaller subunits perform a variety of ancillary roles. Metabolically active cells have a maximum of 10^5 active RNA polymerase complexes at any time. Of these, 40%, 55% and 5% are polymerase I, II and III, respectively.

28.9 NEWLY TRANSCRIBED RNA IS PROCESSED TO mRNA BEFORE PASSING TO THE CYTOPLASM VIA PREFERRED ROUTES

The nascent transcript, called heterogeneous nuclear RNA (hnRNA), is modified in a number of ways before the mature mRNA passes to the cytoplasm. Shortly after transcription begins, a methyl-G nucleotide (referred to as the **5' cap**) is added to the 5' terminus of the transcript. This moiety functions in transport, translation and influences RNA stability. As transcription proceeds, intervening, non-coding intron sequences are spliced out so that the exons are contiguous in the mature mRNA. Finally, polyA polymerase adds a stretch of ≈ 200 adenosines to the 3' tail of the transcript, which reduces exonuclease damage at this end.

Once splicing and polyA synthesis are complete the mature mRNA is ready to be transferred to the cytoplasm. In a limited number of cases, *in situ* hybridization to nuclear RNA demonstrates RNA tracks, supporting the notion that transport occurs along preferred routes, probably determined by the nucleoskeleton. Inefficient splicing enhances the appearance of tracks in the most convincing case; when processing is efficient nascent transcripts are usually seen as small blobs adjacent to the corresponding genes, rather than as extended tracks. In these circumstances, transport may be so fast that molecules *en route* to the nuclear pores are too dispersed to be detected by *in situ* hybridization.

28.10 TRANSCRIPTION AND RNA PROCESSING OCCUR IN CLOSE PROXIMITY WITHIN A RESTRICTED NUMBER OF ACTIVE COMPARTMENTS

Recent experiments support the view that proteins performing DNA transcription and RNA processing are organized through an interaction with a nuclear structure.

Most significantly, it is clear that the active RNA polymerases are clustered at a limited number of transcriptional locations in the nucleus. In HeLa cells, about 50 000 active RNA polymerase II complexes are concentrated at 2500 active sites ranging in size from 40–200 nm (Figure 28.3(a)).

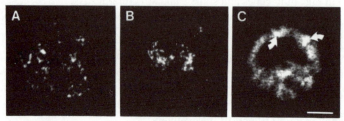

Figure 28.3   Visualizing sites of transcription and RNA processing. **(a), (b)** Encapsulated HeLa cells were permeabilized, incubated in a transcription mixture supplemented with Br-UTP and sites of transcription visualized using a mouse anti-Br antibody and second, anti-mouse antibody conjugated with FITC. Discrete sites of RNA polymerase II/III activity are scattered throughout the nucleoplasm **(a)**. When these are inhibited by pre-incubating with 250 μM/ml α-amanitin fewer nucleolar sites are seen **(b)**. **(c)** Regions rich in splicing components were stained using an auto-immune anti-sm antibody followed by a second, anti-human antibody conjugated with FITC, the nucleolus is unstained. **(a)–(c)** are 700 nm optical sections. The bar is 2.5 μm. (For further details see Jackson, D. A. *et al.*, *EMBO J.*, **12**, 1059–1065.)

In nucleoli, about 25 000 active RNA polymerase I complexes transcribe the ribosomal RNA genes at about 100 sites (Figure 28.3(b)); transcription takes place in a zone at the interface between morphologically distinct fibrillar centres (fc), likely RNA polymerase stores, and adjacent dense fibrillar components (dfc), which contain the active rRNA genes.

The 5'-capping of RNA transcript, splicing and polyA addition are all believed to take place within functional compartments at, or adjacent to, the nucleoplasmic transcription sites (Figure 28.3(c)).

28.11 FUNCTIONAL NUCLEAR COMPARTMENTS ARE ASSOCIATED WITH A NUCLEOSKELETON

Major structural elements within interphase nuclei are responsible for maintaining the spatial organization of these active sites. Different electron microscopy techniques have identified a nuclear skeleton (nucleoskeleton) composed of a complex array of 10 nm core filaments, with a 23 nm repeat, characteristic of intermediate filaments (Figure 28.4).

Candidate proteins such as the nuclear lamins and NuMA have been shown to form close associations with this nucleoskeleton, though a detailed understanding of the composition of these filaments awaits further characterization. This nucleo-

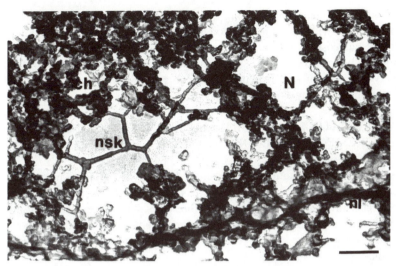

Figure 28.4 The nucleoskeleton. Electron micrograph of a thick resin-less section of an extracted HeLa cell. Cells were encapsulated, lysed, chromatin-cut with *Hae*III and 90% removed by electroelution (all procedures up to fixation took place in a physiological buffer). Samples were embedded in diethylene glycol distearate (DGD) and 250 nm resin-less sections were prepared. Residual chromatin clumps (ch) appear to be arranged along an underlying nucleoskeleton (nsk) of 10 nm thick intermediate filament-like filaments. A filamentous nuclear lamina (nl) marks the nuclear periphery. The bar is 100 nm. (See Jackson, D. A. and Cook, P. R. *EMBO J.*, **7**, 3667–3677 for further details.)

skeleton forms a network upon which active sites can assemble, forming centres for chromatin attachment that are a major feature of chromatin organization throughout the cell cycle. The structural influence of the nucleoskeleton ensures that the disposition of the active sites can persist even when most chromatin is removed and allows the preparation of a variety of operationally defined nuclear structures – such as the nuclear matrix – when chromatin-depleted nuclei are extracted in different ways.

28.12 WHEN GENES ARE TRANSCRIBED, THEY ARE CLOSELY ASSOCIATED WITH THE NUCLEOSKELETON

Nuclear compartments that perform transcription maintain their spatial organization when most chromatin is removed; the active centres must be organized through their interaction with structural nuclear elements. It is no surprise, therefore, that active genes are closely associated with the nucleoskeleton when most bulk chromatin is removed. Interaction between the nucleoskeleton-bound transcription sites and enhancer/promoter sequences is probably crucial to activating gene expression – in many instances transcription factors are found in nucleo-skeleton preparations. During transcription, active RNA polymerases remain associated with the nucleoskeleton, so that loops of active chromatin must move to the nucleoskeleton as transcription proceeds. The idea that sites of transcription can be influenced by nuclear architecture gives an explanation of the variability seen when genes are expressed from unnatural chromosomal sites.

28.13 CHROMATIN LOOPS FORM FUNCTIONAL DOMAINS THAT REPRESENT INDEPENDENT UNITS OF GENETIC ACTIVITY

At least a subset of chromatin loops correlates with functional domains defined classically by their sensitivity to pancreatic DNase I which preferentially digests 'exposed' DNA in unfolded, actively transcribed chromatin. For example, during erythrocyte maturation, decreasing sensitivity throughout the ß-globin domain correlates with a progressive condensation of chromatin and reduction in ß-globin expression.

An example of correlations between chromatin structure and function is illustrated by studies on the chicken lysozyme gene. Here, a region of DNase-I sensitivity extends over 21 kb with a lysozyme coding region of 4 kb located in the centre. Adenine- and thymine-rich sequences (called A-elements), which act as attachment sites to the nucleoskeleton form the 5' (upstream) and 3' (downstream) boundaries to this domain. These sequences thus appear to insulate internal genes from the inhibitory influences of adjacent (inactive) heterochromatin. Other classes of element, such as 'specialized chromatin structures' (SCS) and 'locus control regions' (LCR), first identified in the *Drosophila* heat shock and human β-globin loci, respectively, play roles in establishing appropriate expression from genes introduced into unnatural chromosomal sites.

28.14 EFFICACY OF EXPRESSION FROM GENES ARTIFICIALLY INTRODUCED INTO UNNATURAL CHROMOSOMAL SITES INFLUENCES PROSPECTS FOR GENE THERAPY

Our desire to treat genetic defects using gene therapy will inevitably demand an intimate knowledge of factors that influence expression *in situ*. It is often assumed that transcription factors and other proteins present in cells will direct appropriate gene expression. In many cases, this is not enough. The presence of a chromatin loop structure and its relationship to the nucleoskeleton introduce positional effects that have a dramatic effect on both the timing and extent of expression. As the positional effect is poorly understood, it would be valuable to develop a means of achieving stable and predictable expression for episomal (extrachromosomal) genes. A detailed knowledge of how gene expression is influenced by nuclear organization is likely to be crucial to the success of this approach.

28.15 DNA REPLICATION IS INITIATED AT SPECIFIC SITES IN THE GENOME

In an adult human, the majority of cells are terminally differentiated and no longer required to divide. In the minority that are still capable of cell division, it is absolutely essential that before mitosis is attempted the cell's entire diploid DNA content is duplicated; even small unreplicated regions can cause lethal or potentially mutagenic damage to DNA. In cycling cells, systems that control the initiation and progression/completion of S-phase have been described. Cell cycle machinery, notably cycle-dependent kinases (CDK) 4 and 6, together with cyclins D and E, direct a complex series of events that prepare the cell for S-phase.

At the start of S-phase, DNA replication is initiated at multiple sites in the genome in a highly regulated manner, both in terms of timing and location. The region of DNA replicated at one location, the replication unit or 'replicon' is approximately 75 kb in length in mammalian cells. Early in S-phase, about 5000 replicons are active. Synthesis is not uniformly dispersed throughout the genome; local replicons are activated as small clusters (with perhaps 5–10 replicons/cluster) perhaps reflecting chromosome structure (see below). How sites of initiation are selected in higher eukaryotes remains controversial. In yeast, specialized DNA sequences called 'autonomous replicating sequences' (ARS) are recognized which bind the multiprotein complexes required for replication. In mammalian cells, equivalent DNA sequences have not been characterized and the basis of initiation remains ill-defined. The lack of obvious initiation sequences in mammalian cells implies that other features of chromatin organization are critical to the initiation of replication in more complex organisms.

28.16 REPLICATION COMPLEXES ARE CLUSTERED IN REPLICATION 'FACTORIES' EACH OF WHICH CONTAINS 10–50 REPLICATION UNITS

The appearance of replication sites in mammalian cells (Figure 28.5 (a–d)) also supports the idea that nuclear or chromatin structure influences the efficiency and control of DNA synthesis.

Initially, diploid fibroblasts radiolabelled for short periods early in S-phase show fewer than 200 active replication centres, at a time when some 5000 replicons are active. As each replicon requires the synthetic activity of four polymerases (one lagging strand and one leading strand complex at each fork) these active sites contain an estimated 100 DNA polymerase complexes, operating simultaneously. Recently this observation has been supported by the characterization of morphologically discrete replication megacomplexes – 'replication factories' – that contain all the components required to perform synthesis of local clusters of chromatin domains. As all synthesis occurs within these factories the chromatin must move into the active site as elongation proceeds. Once initiated,

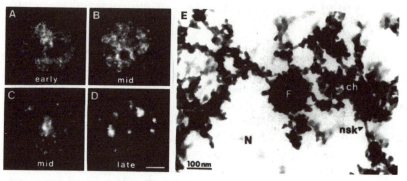

Figure 28.5 Visualizing sites of replication. **(A)**–**(D)** Encapsulated HeLa cells were permeabilized and incubated in a replication mix supplemented with biotinyl-16-dUTP, and sites of DNA synthesis were visualized using a goat anti-biotin antibody followed by a second, anti-goat antibody conjugate with FITC. Conventional fluorescence microscopy reveals characteristic patterns of incorporation **(A)**–**(D)** at specific times during S-phase. The bar is 2.5 μm. **(E)** After labelling as above, 90% of the DNA was removed and sites of biotin incorporation were stained with 5 nm colloidal gold. Samples were embedded in DGD and 250 nm resin-less sections prepared. Replication takes place in morphologically discrete replication factories (F) associated with the nucleoskeleton (nsk) and surrounded by residual chromatin clumps (ch). The bar is 100 nm. (See Hozak *et al.*, *J. Cell Sci.*, **107**, 2191–2202. **(A)**–**(D)** published with permission of Academic Press and **(E)** with permission of the Company of Biologists.)

replication proceeds according to a precisely defined programme with a series of characteristic labelling patterns and corresponding replication factory morphology. The appearance of these patterns is temporally linked, so that once S-phase is activated nuclear structure dictates that replication follows a predetermined programme, or 'replicon cascade'.

28.17 GENOME ORGANIZATION IS OF FUNDAMENTAL IMPORTANCE TO GENE REGULATION IN NORMAL AND MALIGNANT CELLS

Despite their apparently ill-structured appearance, it is now clear that eukaryotic nuclei are highly organized and that this organization can have a dramatic influence on the efficiency of nuclear function. At the two organizational extremes chromatin structure is well established. At one extreme, the specificity of histone–DNA interaction in nucleosomes is resolved to high resolution using X-ray crystallography, while at the other extreme we know that chromosomes condense with a specificity that allows each to be recognized, after staining, from one cell generation to the next. During interphase these chromosomal features are lost even though important features of chromosome structure remain; throughout interphase chromosomes maintain spatially isolated territories with the same linear arrangement of genes found during mitosis.

The major nuclear functions, transcription and replication, take place at a restricted number of sites within specialized nuclear compartments. This organization can influence function in a number of important ways. For example, the restricted spatial organization of components required for these processes ensures high local concentration of components with maximum efficiency. Subtle or dramatic variations in gene expression could result from the ability to set up active centres with different concentrations and combinations of individual components, with dedicated sites capable of performing specialized roles. Once established these active sites engage local chromatin domains at functionally important sites. This arrangement accounts for a major fraction of the chromatin loops (or domains) present in higher eukaryotic cells and provides one explanation of how nuclear context can influence the timing and extent of expression when genes are introduced into unnatural chromosomal sites. Understanding the basis of genome organization inside eukaryotic cells is clearly of fundamental importance to the control of appropriate (or inappropriate) gene expression during differentiation, development and malignancy.

28.18 FURTHER READING

Coverley, D. and Laskey, R. A. (1994) Regulation of eukaryotic DNA replication. *Annu. Rev. Biochem.*, **63**, 745–776.

Freeman, L. A. and Garrard, W. T. (1992) DNA supercoiling in chromatin structure and gene expression. *Crit. Rev. Eukaryot. Gene Express.*, **2**, 165–209.

Jackson, D. A. (1991) Structure-function relationships in eukaryotic nuclei. *BioEssays*, **13**, 1–10.

Smith, M. M. (1991) Histone structure and function. *Curr. Op. Cell Biol.*, **3**, 429–437.

Spector, D. L. (1993) Macromolecular domains within the cell nucleus. *Annu. Rev. Cell Biol.*, **9**, 265–315.

29 | The PCR revolution

Rosalind A. Eeles, William Warren and Alasdair Stamps

29.1 INTRODUCTION

The polymerase chain reaction (PCR) is an *in-vitro* method that uses enzymatic synthesis to amplify, exponentially, specific DNA sequences (Figure 29.1).

Devised by Mullis and refined by Saiki *et al.* in 1985, it originally used a DNA polymerase that was not heat-stable, so fresh enzyme had to be added before each cycle. The PCR revolution followed the development of computerized thermal cyclers, which automatically heat and cool the samples, and the introduction of a thermostable *Taq* polymerase isolated from algae (*Thermus aquaticus*) living in the hot springs of Yellowstone National Park.

The technique is powerful enough to amplify one copy of a specific DNA sequence millions of times. Prior to its introduction, amplification of a particular segment of DNA could only be achieved by labour-intensive and time-consuming cloning using bacteria. This could take several weeks, whereas PCR takes hours. The importance of the technique is reflected by the exponential increase in the number of publications relating to PCR, from three in 1986 to 1700 by 1990.

29.2 PRINCIPLES OF THE POLYMERASE CHAIN REACTION

The polymerase chain reaction is a cyclical process of heating and cooling to denature, anneal and enzymatically amplify DNA

The standard reaction uses two oligonucleotide primers that are complementary to and hybridize with opposite DNA strands flanking the region of interest in the target DNA. The primers are generally around 20 nucleotides in length, sufficiently long to be unique within the genome (Figure 29.2).

Molecular Biology for Oncologists. Edited by J. R. Yarnold, M. R. Stratton and T. J. McMillan. Published in 1996 by Chapman & Hall, London. ISBN 0 412 71270 9

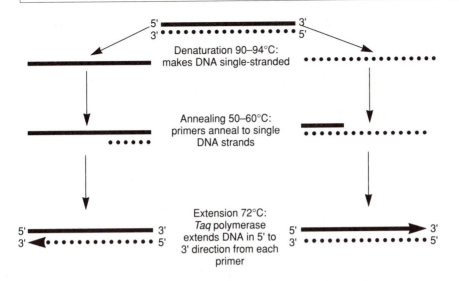

Figure 29.1   Schema of a single cycle in the polymerase chain reaction.

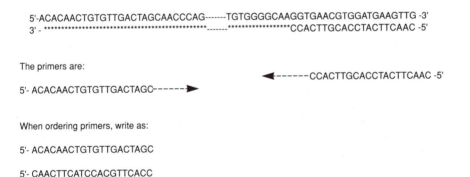

Figure 29.2   Selection of primers for a polymerase chain reaction. The top line represents a single-stranded DNA sequence for amplification (complementary DNA strand not shown). The break indicated in the sequence represents anything from 100 to several thousand base pairs. One primer position is shown, with asterisks indicating the nucleotides to be added. Both primer sequences are displayed below.

The reaction (shown diagrammatically in Figure 29.1), consists of the following steps:

- **template denaturation** at 90–94°C for 0.5–2 min; this separates the two DNA strands;

- **primer annealing** at 50–60°C for 0.5–1 min; the primers anneal to the template (the annealing temperature depends on the nucleotide composition of the primers);
- **extension** at 72°C; the new DNA strands are synthesized from the primers, complementary to the single-stranded template DNA to which the primer has been hybridized.

These steps are then repeated. Thus, the newly synthesized DNA strands also become available as templates for a further round of DNA synthesis in the next cycle of the reaction. The DNA is therefore amplified exponentially, theoretically 2^n times where n is the number of cycles, although in practice the efficiency is not 100%. Typically, 30–40 cycles are performed and, although it is easier to amplify small fragments of a few hundred base pairs, up to 10 kb can be amplified. However, the larger the amplification product, the greater the number of shorter non-specific sequences that may also be amplified.

Problems: the major drawback is contamination

Because it is so sensitive, the greatest problem with PCR is contamination of new reactions by products from previous reactions. The potential scale of the problem is illustrated by the following: if the products of a 0.1 ml polymerase chain reaction contaminated an Olympic-sized swimming pool, a 0.1 ml aliquot from the pool would contain 40 amplifiable molecules. Contamination can be minimised by aliquoting solutions and using separate pipettes and solutions for pre-and post-PCR experiments. Some experimenters use positive displacement pipettes or plugged pipette tips, which are now available to eliminate contamination due to the production of aerosols. Many laboratories also have designated areas for handling PCR products. Negative controls should always be included in each PCR experiment to monitor potential PCR contamination, and the control reactions should be set up last (Figure 29.3).

1 2 3 4 5 6 7 8 9 10 11 12 13 14 15 16 17

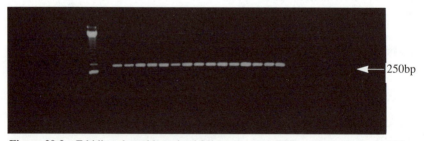

Figure 29.3 Ethidium-bromide-stained 2% agarose gel. PCR products, each of 250 bp in lanes 3–17. Lane 2 is a negative control (no DNA to demonstrate that there is no PCR contamination). Lane 1 is a DNA marker of digested bacterial DNA, digested into pieces of known size to act as a size marker.

29.3 APPLICATIONS OF THE POLYMERASE CHAIN REACTION

Amplification of DNA segments for further analysis: a substitute for cloning

PCR provides a rapid method of acquiring large quantities of a DNA segment located between two known primer sequences. It can be applied to genomic DNA, cDNA (complementary to mRNA) or to various cloned sequences. Subsequently, the amplified segment can be subjected to a wide variety of procedures with many different endpoints. In the context of cancer research, it is frequently sequenced or subjected to a number of procedures, e.g. single-stranded conformational polymorphism analysis (SSCP, see Chapter 3) that allows detection of mutations that may be implicated in oncogenesis. Alternatively, the product can be blotted onto a filter and hybridized to probes. If the amplified segment contains a polymorphic region, the procedure can be adapted for linkage analysis of heritable predisposition to cancer or for studies of loss of heterozygosity (Chapter 3).

Isolation of new genes using low stringency PCR for finding gene families

Related genes, both between species and within one species (e.g. the tyrosine kinase family of genes), have shared (conserved) sequences. To discover new, but related genes, degenerate primers can be designed which correspond to all possible nucleotide sequences that code for the protein in question, and further degeneracy can be introduced in order to encompass related protein sequences. Using a lower annealing temperature with conventional primers is another way of amplifying 'related' sequences.

PCR can amplify specific DNA sequences from small amounts of material

Because of its huge amplification potential, the PCR may be used to isolate specific segments of DNA from tiny amounts of tissue. Indeed, at its most sensitive, sequences can be amplified from single cells, from single human hairs or blood spots. More commonly in medical practice, it is applied to samples taken at amniocentesis or chorionic villus sampling, allowing antenatal diagnosis of a variety of diseases.

PCR can selectively detect DNA sequences not normally present in the tissue being tested

Viruses: viral DNA can be detected by PCR for diagnosis, such as in cases of the acquired immunodeficiency syndrome where the patient is still antibody negative, or in babies born to HIV-positive mothers who will have acquired antibodies across the placenta, but may not necessarily be infected with the virus. The presence of viral DNA in precancerous cervical lesions has also been detected using PCR.

Translocations: PCR can be used for the diagnosis or detection of minimal residual disease. Certain chromosomal translocations commonly occur in follicular non-Hodgkin's lymphomas (t(14;18)) and chronic myeloid leukaemia (Philadelphia chromosome positive: t(9;22)). In follicular non-Hodgkin's lymphomas, the translocation involves the immunoglobulin heavy chain region on chromosome 14 and the *bcl-2* region (a putative oncogene) on chromosome 18. The primers used hybridize to the regions flanking the translocation and will therefore only amplify the intervening DNA when the translocation is present. This can be used both for diagnosis and the detection of minimal residual disease, even in the presence of histologically negative bone marrow. Dilution experiments show that PCR assays can nonetheless detect a single lymphoma cell amongst one million normal cells. This is more than a 10^4-fold improvement in sensitivity over Southern blotting methods. In the case of chronic myeloid leukaemia, the breakpoint on chromosome 9 can occur over a large area and the potentially large fragments cannot be amplified. This problem can be overcome by basing the amplification on mRNA, rather than genomic DNA. This is performed on cDNA synthesized from mRNA by reverse transcriptase.

PCR allows analysis of highly degraded DNA samples

Many DNA samples of forensic or anthropological interest are highly degraded, either because they are very old or have been maintained under suboptimal conditions. The PCR will, however, selectively amplify the minority of intact DNA fragments in such specimens that carry both primer sequences. Thus, one can sometimes salvage information from extremely degraded material that may be many thousands of years old. In the context of forensic medicine, it has become possible to identify the individual from whom a degraded specimen originated by PCR across highly polymorphic regions known as microsatellites. Using several of these segments, one can build up a profile that is almost unique for any individual person. In the context of cancer research, DNA from histopathology archives has become available for analysis. The process of fixation in formalin and embedding in paraffin blocks for pathological examination degrades DNA considerably. However, the PCR often works effectively upon DNA extracted from this type of specimen and thus widens dramatically the range of cancers and other tissues available for study.

Gene expression

PCR can be used to investigate gene expression particularly when levels of mRNA are very low. Initially, mRNA must be converted into cDNA by reverse transcription. Subsequently, the cDNA becomes the substrate for amplification. Careful use of the appropriate controls can make this procedure quantitative or semiquantitative. One problem that arises is distinguishing PCR product amplified from the cDNA from product that may derive from residual traces of genomic DNA in the RNA preparation. This can be overcome by placing the two primer sequences in adjacent exons, which are separated by a large intron. In this situation, the PCR is unlikely to be effective on the genomic DNA because the segment to be amplified is too large. Because the intron is removed during the formation of mRNA, however, amplification from the cDNA which is synthesized directly from the mRNA will proceed unhindered.

29.4 FUTURE DEVELOPMENTS

The PCR revolution has occurred because of the application of computer technology and a thermostable DNA polymerase to the cyclical amplification of DNA. Its power of amplification from minute samples has opened up enormous possibilities for rapid diagnosis, forensic science and research. Further refinements of existing diagnostic methods to incorporate the benefits of PCR will lead to even wider use of this technique in the next few years.

29.5 ACKNOWLEDGEMENT

This work was supported by the Cancer Research Campaign and the Royal Marsden NHS Trust.

29.6 FURTHER READING

Eeles, R. A. and Stamps, A. (1993) *PCR: The Technique and its Applications*, R. G. Landes, Texas.

Erlich, H. A. (ed.) (1989) *PCR Technology. Principles and Applications for DNA Amplification*, Academic Press, New York.

Innis, M. A., Gelfand, D. H., Sninsky, J. J. and White, T. J. (eds) (1990) *PCR Protocols. A Guide to Methods and Applications*, Academic Press, New York.

Rodu, B. (1990) The polymerase chain reaction: the revolution within. *Am. J. Med. Sci.*, **299**, 210–216.

Wright, P. A. and Wynword-Thomas, D. (1990) The polymerase chain reaction: miracle or mirage? A critical review of its uses and limitations in diagnosis and research. *J. Pathol.*, **162**, 99–117.

30 | Techniques of molecular biology

Philippe J. Rocques

30.1 INTRODUCTION

The term 'molecular biology' is used to describe a set of techniques concerned with the study of genes and their functions, principally by manipulation of nucleic acid molecules (DNA and RNA). Only a broad overview of selected methods is presented here; some more exhaustive accounts are suggested in the further reading section.

30.2 NUCLEIC ACIDS CAN BE MODIFIED *IN VITRO* USING PURIFIED ENZYMES

Restriction endonucleases cut double-stranded DNA molecules at sites defined by specific base sequences

It is very useful to be able to cut DNA at sites of known base sequence and this is accomplished using **restriction endonucleases**. These are proteins purified or cloned from prokaryotic species in which they protect the host cell from invasion by foreign DNA. The restriction endonucleases most commonly used in molecular biology are enzymes that recognize a palindromic sequence, called a **restriction site**, of four, six or eight bases and cleave the DNA by breaking phosphodiester bonds on both strands somewhere within the site.

Restriction sites four bases long statistically occur more frequently in a randomly chosen DNA than do longer sites, and it follows that four-cutter restriction

Molecular Biology for Oncologists. Edited by J. R. Yarnold, M. R. Stratton and T. J. McMillan. Published in 1996 by Chapman & Hall, London. ISBN 0 412 71270 9

fragments are much smaller than those which would be derived from the same initial DNA by digestion with an enzyme having a longer recognition sequence.

In general, six cutters tend to be most useful for Southern blotting and four cutters are used for making genomic libraries.

DNA polymerase synthesises new DNA strands

DNA polymerase uses deoxynucleoside triphosphate precursors to elongate a short nucleic acid primer bound to a complementary template strand. All polymerases synthesize their products in the 5' to 3' direction. Polymerases frequently occur as components of multienzyme complexes which also possess exo- and endonucleolytic activity. In the case of reverse transcriptase, the *in vitro* template can be RNA or DNA.

Polymerases are used in DNA sequencing, the polymerase chain reaction (PCR) and labelling of DNA by incorporation of radioactive nucleotides. The associated nuclease activities also have applications. Reverse transcriptase is used in the synthesis of cDNA from RNA.

DNA ligase joins double-stranded DNA molecules end to end

DNA ligase joins DNA molecules by catalysing the formation of phosphodiester bonds between double-stranded DNAs with 3' hydroxyl and 5' phosphate termini in the presence of ATP. DNA ligase is used in library construction and subcloning among other applications.

DNA kinase and phosphatase respectively add phosphate groups to and remove them from 5' termini of DNA molecules

DNA kinase and phosphatase are enzymes which respectively phosphorylate and dephosphorylate the 5' terminal position of a DNA, and can therefore be used to manipulate the competence of populations of DNAs to be ligated *in vitro*. DNA kinase is also used for end-labelling of short DNAs (oligonucleotides) by transfer of a radiolabelled phosphate group from ATP.

30.3 ELECTROPHORESIS, BLOTTING AND HYBRIDIZATION ARE TECHNIQUES FOR THE ANALYSIS OF POPULATIONS OF NUCLEIC ACIDS

Electrophoresis allows separation of a population of DNA and RNA molecules according to size and conformation

Nucleic acids carry net negative charge, so they migrate through agarose or acrylamide gels in the direction of a positive electrode. The gel mobility of a double-stranded molecule decreases with increasing molecular weight (size). The situation is more complex for single-stranded molecules, since these can form secondary structure as a result of intrastrand base pairing. The gel mobility of a single-stranded molecule is therefore a function of both size and secondary structure conformation, the latter being base-sequence-dependent. A gel based on polyacrylamide provides greater resolution than is achieved with agarose. Such gels are used for sequencing (when it is necessary to separate large DNAs differing in size by a single nucleotide) and single-strand conformational polymorphism (SSCP) analysis. Sequencing gels and RNA gels are run under denaturing conditions (that is at high temperature and/or with formamide) to minimize secondary structure formation.

RNA or DNA in a gel can be visualized by staining with ethidium bromide and UV transillumination. Running nucleic acids of known sizes alongside the sample allows approximate determination of the sizes of bands, and it is possible to physically excise a portion of agarose containing a band of interest and extract DNA from it.

Single-stranded conformational polymorphism (SSCP) analysis is a method for detecting mutations in defined regions of DNA as electrophoretic mobility variants

SSCP analysis is carried out in two steps. In the first, a region of DNA in which mutations are sought is amplified by PCR. The PCR product DNA is then denatured (made single-stranded) by heating in formamide and electrophoresed on an acrylamide gel under non-denaturing conditions. This allows the formation of intrastrand hydrogen bonds producing secondary structure. PCR products amplified from genomic DNA containing a point mutation do not adopt the same conformation as wild-type products. The two products therefore have different gel mobilities, and the mutants can be identified by comparison with known normal controls following gel autoradiography.

SSCP detection works for PCR products up to about 500 bp, so that analysing a whole gene for mutations can require the use of multiple pairs of PCR primers in order to span the entire locus. It is not currently known exactly what proportion of all point mutations actually result in altered conformation, but it has been estimated that approximately 80% of mutations are detected by SSCP. This compares quite favourably with other far more labour-intensive methods of mutation detection.

Hybridization is the formation of a double-stranded DNA between two single strands

Double-stranded DNA *in vivo* usually consists of two exactly complementary strands. If the base pairs in a few positions are not complementary then a duplex can still be formed, but one of reduced stability. Annealing together of two strands which may be exactly or nearly complementary is called **hybridization**.

Southern and Northern blotting allow identification of particular components of a nucleic acid population following electrophoresis

Blotting preserves the distribution of nucleic acids resulting from electrophoresis in a permanent form accessible to further analysis. Following agarose electrophoresis, the separated DNAs or RNAs are transferred by capillary action onto a nylon membrane, to which they are covalently linked by UV irradiation, to make a Southern or Northern blot, respectively. In the case of Southern blotting, the DNAs to be separated are frequently fragments obtained by restriction enzyme digestion, whereas the aim in Northern blotting is to work with undegraded RNA. Blot hybridization followed by autoradiography allows determination of the size of nucleic acid species complementary to a radiolabelled specific nucleotide sequence or probe.

A Southern blot of digested total genomic DNA can be used to detect gene amplification or deletion and chromosomal rearrangements, since these result in altered band sizes and/or intensities following probing. Northern blotting is used to analyse gene expression at the level of transcription, revealing for example the size and tissue distribution of mRNA hybridizing to a probe.

Chromosomal *in situ* hybridization is a powerful tool for physical mapping of DNA probes along a chromosome

A labelled probe can be hybridized to a spread of fixed metaphase or interphase chromosomes, allowing determination of the chromosomal localization of sequences defined by the probe. This is technically easier with metaphase chromosomes, but interphase spreads give better resolution because the chromosomes are more elongated. This method can be used, for example, to map the positions of marker probes relative to the breakpoint of a chromosomal translocation.

30.4 MOLECULAR CLONING IS THE PRODUCTION OF MANY IDENTICAL COPIES OF A CHOSEN DNA MOLECULE

Cloning vectors facilitate the insertion of foreign DNA into a cell and ensure its subsequent replication and sometimes expression

Most vectors have a number of features. They contain sequences ensuring their stable propagation in a population of dividing host cells. They provide some means of selecting for host cells containing vector, e.g. by conferring antibiotic resistance. They may contain elements called **polylinkers** with multiple restriction sites into which foreign DNA is ligated following restriction enzyme digestion. Polylinkers are frequently flanked by promoters, allowing expression of cloned DNA.

Introduction of foreign DNA into bacterial cells requires temporary disruption of the cell membrane, achieved either by heat shock or by electric shock

The most usual way of transforming *E. coli* cells is by calcium chloride treatment followed by heat shock. This temporarily makes the cells competent to take up foreign DNA. There is an alternative to this called **electroporation**, in which DNA enters cells following transient disruption of the cell membrane by electric shock. Electroporation can be far more efficient, but requires optimization of conditions.

Plasmids are extrachromosomal DNA molecules which can be modified for use as cloning vectors

Plasmids are extrachromosomal DNA elements that occur naturally in a variety of organisms. Bacterial plasmids are closed circular DNAs far smaller than the bacterial chromosome, capable of independent replication within the bacterial cell and frequently conferring antibiotic resistance on the host. The plasmids used in molecular biology have been extensively altered by the addition of some or all of the features described above to make them more efficient for cloning purposes.

Bacterial viruses known as phage can also be exploited for cloning

A **bacteriophage**, or simply a **phage**, is a virus that infects and replicates in bacterial cells. Like plasmids, naturally occurring phage has been extensively modified for cloning purposes. The main difference between cloning in phage and plasmid lies in the method of introduction of the vector into the host cell. Transformation and electroporation are relatively inefficient processes, and do not work satisfactorily for DNA of size greater than about 20 kb. Far larger pieces of DNA can be inserted into bacteria using phage because use is made of the natural phage infection system.

Cosmid vectors are hybrids between phage and plasmids and have larger cloning capacities

Cosmids are vectors having properties of both phage and plasmids. They make use of the phage infection system, allowing insertion of very large pieces of DNA, and are able to propagate as plasmids once within the cell. The phage lambda packaging system will accept recombinant DNA ranging in size from about 20–50 kb. Because of this constraint, cosmids have had most of the phage genome removed to make room for larger inserts. Cosmids allow cloning of very large DNAs, ranging in size up to about 50 kb.

Yeast artificial chromosome vectors allow cloning of very large pieces of DNA

Yeast artificial chromosomes (**YACs**) allow cloning in yeast host cells of pieces of DNA comparable in size to normal yeast chromosomes. In practice, the process of preparation of intact high molecular weight DNA sets an upper bound of about 1 Mb, and most YAC clones range from 200–500 kb. YACs consist of cloned DNA flanked by two vector arms.

These must contain a centromere and yeast origin of replication (ARS) to ensure replication and segregation of the YAC, and each must have a telomere (a special sequence found at the ends of all linear chromosomes ensuring the integrity of terminal regions during replication). In addition, there is often some means for isolating short stretches of DNA from each end of the insert. These end clones allow determination of the orientation of a YAC relative to a chromosome by *in-situ* hybridization, and are also used for chromosome walking in YAC libraries. YACs are proving extremely useful in long range mapping of large regions of the human genome.

Libraries are random collections of fragments of genomic or cDNA from which selected DNAs can be cloned

A library is a collection of DNA fragments from which those of interest can be isolated or cloned. There are two main types of library, genomic and cDNA libraries.

Genomic libraries contain fragments of genomic DNA, generally obtained by restriction enzyme digestion. They contain many inserts that are not derived from protein coding sequence and which may be from regions of the genome very distant from the nearest gene. Complementary DNA, or cDNA, libraries contain DNA inserts obtained by reverse transcription of mRNA extracted from the cells or tissue of interest.

Identification of clones in libraries is carried out by hybridization of labelled nucleic acid probes

This section describes the basic method of plaque purification of clones from phage libraries. The library is used to infect a bacterial culture, which is then plated on agar. Infected bacteria lyse, releasing phage particles, which similarly kill surrounding cells. This results in a plaque containing many identical phage but no living bacteria. There are many such plaques on each agar plate, one corresponding to each bacterial cell initially infected. A replica of the plaque distribution is obtained by direct transfer of phage from the plates onto nylon membranes. The phage DNA is denatured and fixed to the membranes. The filters are then hybridized with a labelled DNA probe and autoradiographed, and the position of spots on the photographic film allows identification of the position of phage clones of interest. These are physically excised from the plate as plugs of agar containing phage, and subjected to one or two more rounds of plating and screening at lower density until a single clonal phage plaque can be picked.

Genomic DNA libraries are used principally for mapping work

Genomic libraries are mainly used for mapping projects, including attempts to clone genes from regions of the genome linked to known phenotypes. A probe might initially be shown by linkage analysis to map near a suspected disease gene. In order to move closer to the locus of interest, overlapping but distinct genomic clones are isolated from the library. Some of these new clones extend further towards the locus of interest. Repeated rounds of screening result in a contig (contiguous region) of clones spanning a portion of genome from the initial probe to the target locus. This approach is called **chromosome walking**.

cDNA libraries are used for studying the structure and function of protein coding regions

cDNA libraries are used for isolating coding sequence from a gene of interest. Reverse transcription of mRNA may be primed either using an assortment of random DNA hexamers or using oligo dT primers. Randomly primed libraries contain a collection of inserts not starting at any defined position within a given transcript. Oligo dT primed libraries have inserts which start at the 3' end of every message, since the primer anneals to the polyA tail.

Cloned cDNAs are used in conjunction with a variety of other techniques, including sequencing, mutagenesis and expression studies, all of which are described below.

DNA cloned in a plasmid can be subjected to a variety of experimental procedures such as mutagenesis and sequencing

Many cDNA and most genomic clones are too large for convenient manipulation. This problem is solved by inserting a fragment of the clone into a plasmid vector. The subcloned fragment can be analysed with relative ease.

Once a plasmid has been inserted into bacteria, a large culture can be grown starting from a single colony on a plate. This ensures that the culture is clonal. Large amounts of plasmid can be extracted from the culture.

A common method for investigating the function of a cloned cDNA or regulatory region is to introduce controlled mutations into the DNA. The functional effects of mutagenesis (deletions, insertions or base substitutions) can be assayed using, for example, gel retardation or expression systems.

30.5 DNA SEQUENCING IS CARRIED OUT BY THE DIDEOXY SEQUENCING REACTION

This is a method for determining the sequence of bases in a DNA molecule. It is based on the fact that DNA polymerase can use 2'-3' dideoxynucleotides (ddNTPs) as substrates for incorporation into a nascent DNA strand, but cannot then elongate the strand further because of the lack of a 3'-OH terminus. DNA synthesis is carried out from a specific primer in the presence of a template (the DNA to be sequenced), all four deoxynucleotides (dNTPs), and a single ddNTP at far lower concentration. One or more of the dNTPs is radiolabelled and four such reactions are performed, one for each ddNTP. Strand elongation proceeds until stochastic incorporation of a ddNTP. Each reaction therefore eventually contains strand elongation products of different lengths, all starting at the primer and finishing at a dideoxynucleotide. The four sequencing reactions are run on a denaturing acrylamide gel in separate adjacent lanes. After autoradiography, the four lanes taken together contain a band for each base in the template. Because the ddNTPs incorporated are complementary bases in the template strand, the lane in which a band occurs indicates which base is present in a given position in the DNA being sequenced. The desired sequence can therefore be read directly from the autoradiograph. This is often achieved with a digitizer so that the sequence is immediately available in a computer file.

30.6 EXPRESSION OF CLONED GENES IS A DIRECT METHOD FOR STUDYING GENE FUNCTION

Once a cDNA has been cloned and sequenced, it is often necessary to analyse its function further. Possible investigations range from *in vitro* work on the biochemistry of normal or mutant protein to determining the effects of the gene when expressed in various spatiotemporal patterns in transgenic mice.

Expression of a cloned cDNA in cultured cells can give valuable insights into the role the gene is playing *in vivo*, and allows determination of its interactions with other cellular components. Retroviral vectors are commonly used for such work.

Transgenic organisms allow the study of gene function or dysfunction in the context of the whole organism

The term 'transgenic' is used to describe an organism which has had its genome artificially altered, either by insertion of a gene (a transgene) or by disruption of an endogenous gene. Transgenic animals provide opportunities for investigating how genes, including oncogenes, function in the context of whole organisms.

Embryonic stem cells can be genetically altered in culture and will colonize embryos into which they are injected

Embryonic stem (ES) cells are cultured from mouse inner cell mass (ICM) or delayed blastocyst. They proliferate when grown on layers of irradiated feeder cells or in the presence of a defined differentiation inhibitory activity (DIA). In the absence of these, they differentiate, forming embryoid bodies. ES cells form tumours containing several differentiated tissue types when implanted into syngeneic mice. Most importantly, ES cells injected into blastocysts colonize the resulting embryo, giving a high frequency of chimerism, which can extend to the germ cells. These chimeric mice can develop into fertile adults. The recently developed embryonic germ cells (EG) may provide a means of obtaining germline chimerism in the majority of animals from which ES cells cannot be isolated.

Transgenic mice are constructed by either using embryonic stem cells or directly injecting DNA into a pronucleus

There are two principal methods for the construction of transgenic mice. One involves the manipulation in culture of the genome of an ES cell, followed by its injection into a blastocyst as above. Transgenes show simple Mendelian transmission, so breeding from germ cell chimeric mice ultimately results in progeny homozygous for the novel genetic trait.

Another method of producing transgenics is to inject DNA into one of the pronuclei of a fertilized egg. The injected DNA typically integrates in many tandem copies at a random location.

Transgenic mice will be increasingly important in studying oncogene function

Two recent examples of the use of transgenic technology in cancer research are briefly mentioned. The Philadelphia chromosome associated with human CML and ALL results from a t(9;22)(q34;q11) chromosomal translocation. Sequences at the translocation breakpoint direct production of a Bcr–Abl fusion protein having abnormal tyrosine kinase activity. Injection into fertilized mouse eggs of a construct encoding this Bcr–Abl fusion under metallothionine promoter control results in mice that die of AML or ALL 10–58 days after birth. This clearly supports a causal relationship between the Philadelphia chromosome and human leukaemia.

Transgenic mice have been made that entirely lack any functional *p53* gene, that is they are homozygous-null mutants for *p53*. This was accomplished by first making a *p53*-heterozygous-null ES cell. A targeting construct containing a disrupted segment of *p53* was inserted into ES cells by electroporation. Various methods were used to select for homologous recombination between the construct and an endogenous *p53* locus, and a suitable ES clone was used to generate chimeric mice. These transmitted mutant *p53* to their offspring at 50% frequency, and intercrossing the resultant heterozygotes produced 25% homozygous-null mutants as expected. The $p53^-/p53^-$ mice were apparently normal except for a strong predisposition to neoplastic disease. These findings suggest that while there is genetic redundancy in the normal function of $p53$, $p53$ homozygous mutation is a first step towards malignancy.

30.7 FURTHER READING

Old, R. W. and Primrose, S. B. (1985) *Principles of Gene Manipulation*, 3rd edn, Blackwell Scientific, Oxford.

Sambrook, J., Fritsch, E. F. and Maniatis, T. (1989) *Molecular Cloning*, 2nd edn. Cold Spring Harbour Laboratory Press, Cold Spring Harbor, NY.

Watson, J. G. *et al.* (1987) *Molecular Biology of the Gene*, 4th edn, Benjamin Cummings, Menlo Park, CA.

Glossary

Collated by Philip Mitchell, Philippe Rocques and John Yarnold

Terms marked with an asterisk are themselves entries in the glossary.

Actin   Abundant protein that forms actin filaments in all eukaryotic cells. The monomeric form is sometimes called globular or G-actin; the polymeric form is filamentous or F-actin.

Active genes   Genes transcribed in the cell type in question. Some genes are active in all cells (housekeeping genes), while others are specific to one cell type (e.g. haemoglobin gene in erythrocytes). Associated conformational features include looser duplex winding around nucleosomes*, less histone* H1 binding and situation in euchromatin.ß

Active site   Region of an enzyme surface to which a substrate molecule binds in order to undergo a catalysed reaction.

Adduct *see* **Drug adducts**

Alkylation   Process of covalent binding of reactive alkyl group of drug (e.g. CH_2Cl) to biological molecules (DNA bases or proteins) that have an excess of electrons.

Allele   Alternative sequences at a locus (can be coding or non-coding).

Alpha helix (α helix)   Common structural motif of proteins in which a linear sequence of amino acids folds into a right-handed helix stabilized by internal hydrogen bonds between backbone atoms.

Amplification   Increase in copy number of a chromosomal region, typically by tandem* duplication.

Anchorage dependence   The dependence of normal cells on an appropriate surface/substrate on which to grow in culture.

Annealing   Complementary base pairing of homologous single strands of nucleic acids, e.g. attachment of primers* to denatured target DNA in PCR.

Antisense   A nucleotide sequence (RNA or DNA) complementary* to the coding sequence (sense strand*).

Apoptosis   An active mechanism of cell death in which DNA degradation and nuclear destruction precede loss of plasma membrane integrity and cell necrosis.

Autocrine   A mechanism of growth stimulation involving the binding of a growth factor* secreted by a cell to its own plasma membrane receptors, *cf.* **paracrine**.

Bacterial transformation   The uptake of foreign DNA by a bacteria, which may result in, for example antibiotic resistance or some other phenotype.

BSO   Buthionine sulphoximine, an inhibitor of γ-glutamyl cysteine synthetase; leads to depletion of GSH*.

Cadherin   Member of a family of proteins that mediate Ca^{2+}-dependent cell–cell adhesion in animal tissues.

Carcinogen   A physical or chemical agent which has a causal role in the development of cancer (these are often, but not always mutagens*).

CDC genes   Cell division cycle proteins involved directly in the control of the cell cycle; for example CDC 2 (p34) is a kinase* enzyme.

cDNA   DNA complementary* to RNA and synthesized from it by reverse transcription*.

cDNA library   A collection of DNAs produced by reverse transcription* from the mRNA* of a cell population of interest and inserted into a suitable vector*.

Cellular oncogene   Proto-oncogene* altered by mutation*, which leads it to acquire an altered cellular function that contributes to carcinogenesis.

Centimorgan   Unit representing a recombination* frequency of 1%, i.e. approximately 1 million base pairs.

Centromere   Specific DNA sequences that attach the chromosome to the mitotic spindle during M phase.

Chaperone (molecular chaperone)   Protein that helps other proteins avoid misfolding pathways that produce inactive or aggregated states.

Chromatin   Chromosomal DNA together with a variety of associated proteins, the most abundant of which are histones*. *See* **DNA chromatin**.

Cleavable complex   DNA–drug adduct* or intercalation which stabilizes topoisomerase* II binding to DNA and which, on deproteination, reveals strand breakage.

Cloning   The generation of multiple identical copies of a DNA sequence by replication* in a suitable vector*, e.g. phage* or plasmid*.

Codon   A triplet of nucleotides coding for one amino acid.

Complementary bases   Bases that are hydrogen-bonded specifically in DNA duplex or in DNA/RNA heteroduplex.

Complementation   Said to occur when a cell deficient in a particular function, e.g. repair of UV DNA damage, is restored to normal function by addition of foreign DNA, by a technique such as cell fusion. This process has been used to show that a number of different gene defects can be involved in single human DNA repair disorders, e.g. xeroderma pigmentosum.

Constitutional deletion   Deletion inherited in the germline* from one or other parent and present in every cell of the body (usually examined in lymphocytes), *cf.* **Somatic deletion**.

Contact inhibition Inhibition of cell division in cell culture by cell–cell contact.

Copy number The number of copies of a gene present in a cell.

Cosmid Cloning vector used to carry large segments of DNA into and out of cells; derived from bacteriophage lambda.

CpG islands Sequences 'rich' in the dinucleotide CpG; often in the 5' region of genes and possibly involved in transcriptional regulation.

Cyclic AMP A ubiquitous intracellular messenger* synthesized from ATP by plasma-membrane-bound adenylate cyclase (activated by a G protein*).

Cyclin Protein that periodically rises and falls in concentration in step with the eukaryotic cell cycle. Cyclins activate crucial protein kinases (called cyclin-dependent protein kinases) and thereby help control progression from one stage of the cell cycle to the next.

Cytogenetics Analysis of chromosome structure.

Cytokines Proteins that act as intercellular signals to coordinate the immune response.

Deletion Loss of a segment of a chromosome *(see* **Constitutional deletion** and **Somatic deletion**).

Denaturation of DNA Melting (separation) of the complementary* strands, caused by high temperature or chemical conditions, usually reversible.

Denaturation of protein Loss of higher-order structures caused by high temperature or chemical conditions, usually irreversible.

Determination A commitment to follow a given developmental lineage (pathway).

Differentiation An increase in specialization towards a specific function.

DNA chromatin Structural packing of DNA double helix into nucleus. Hierarchy of helix coiled around histones (to form a nucleosome*); twisting of these into a 30 nm wide fibre; supercoiling of this fibre into loops attached to the nuclear matrix and coiling of loops/matrix into chromosome bands.

DNA cloning A technique involving the integration* of specific DNA sequences into a self-replicating element (plasmid* or virus) that reproduces itself in bacteria to generate large numbers of identical copies.

DNA library Collection of different cDNAs* or fragments of genomic* DNA propagated in a cloning vector* (phage* or plasmid*) from which specific sequences can be isolated (cloned*).

Domain Portion of a protein that has a tertiary structure of its own. In larger proteins each domain is connected to other domains by flexible regions of polypeptide.

Dominant negative mutation Mutation that dominantly affects the phenotype by means of a defective protein or RNA molecule that interferes with the function of the normal gene product in the same cell.

Dosage effect The effect on a cell's morphology/behaviour from having more or less than the usual diploid number of normal genes.

Double minutes Chromosomal fragments visualized on metaphase spreads, associated with amplification*, e.g. multidrug resistance (MDR), *myc* oncogene.

Downstream   Beyond the 3' end of a gene sequence, *cf.* **upstream***.

Drug adducts   DNA modified by covalently bound drug or reactive sidegroup of drug.

***Eco*RI**   A restriction endonuclease originally isolated from a strain of *E. coli.*

Endocrine   A mechanism of growth stimulation involving the secretion of a growth factor* which binds to a specific cell receptor* after diffusing through the circulation.

Epitope   Antigenic determinant (there may be several per molecule).

Eucaryotic cell   Cells which have a nucleus (yeast, mammalian cells), *cf.* **prokaryotic cells** (bacteria).

Euchromatin   The decondensed form of chromatin* typical of the interphase nucleus.

Exons   Transcribed sequence not spliced out of mature RNAs (*cf.* **introns**).

G protein   A GTP binding protein involved in transmitting a signal from a cell surface receptor to an intracellular effector.

Gel electrophoresis   Technique for separating molecules of DNA, RNA or protein according to relative size and charge by passing them through a porous gel matrix under the influence of an electric field.

Gel shift or 'retardation' analysis   Technique used to detect specific binding of a protein to a particular sequence of DNA. The resulting DNA–protein complex migrates more slowly in gel electrophoresis than 'unbound' DNA, i.e. it is retarded.

Gene   A region of genomic DNA specifying the coding and controlling sequences for the expression of a protein or RNA product.

Genetic engineering   The processes by which genes can be isolated from cells and manipulated (e.g. mutated*) *in vitro* before reintroduction into the same or different species.

Genome   The genetic complement of a cell organelle, species, etc., e.g. nuclear genome, mitochondrial genome, *Drosophila* genome.

Genotype   The hereditary information encoded in nucleic acid; *cf.* **phenotype**.

Germ line   The lineage of germ cells (which contribute to the formation of a new generation of organisms), as distinct from somatic cells (which form the body and leave no descendants).

Germ-line deletion *see* **Constitutional deletion**

Germ-line mutation   Mutation inherited from one or other parent and present in every cell of the body.

Glutathione-*S*-transferases   Family of enzymes responsible for conjugation reactions involving GSH*.

Growth factor   Proteins that bind to specific cell surface growth factor receptors* and modify cell growth, e.g. EGF, PDGF.

Growth factor receptor   Proteins that span the plasma membrane with an extracellular growth factor* binding domain and an intracellular signalling domain that is activated by growth factor binding.

GSH   Reduced glutathione; conjugation to GSH* may be an important step in detoxification for a number of drugs.

Heterochromatin   Regions of highly condensed chromatin* in the interphase nucleus visible with a light microscope, *cf.* **euchromatin**.

Heterozygote   Diploid cell or individual having two different alleles of a specified gene.

Histones   A class of nuclear proteins involved in maintaining the higher order structure (folding) and function of genomic* DNA.

hnRNA   Heterogeneous nuclear RNA, the immediate product of transcription*, i.e. before the splicing* out of introns* to produce messenger RNA* (mRNA).

Homeobox genes   A family of transcription factors which all possess a highly conserved 180 bp sequence coding for the 'homeobox' DNA-binding domain*.

Homogeneously staining regions   Regions of chromosome visualized on metaphase spreads corresponding to hugely amplified* gene sequences *(see also* **Double minutes**).

Homozygous deletion   Deletion of both alleles at a locus.

Housekeeping genes   Genes that are expressed in most cell types; *cf.* tissue-specific genes, which are expressed only in selected cell types.

Hybridization   The base-pairing of complementary* single strands of nucleic acid that leads to the double-stranded molecule.

Immortalized cells   Cells not restricted by a specific number of cell divisions.

Immunotherapy   Therapies designed to enhance the immune response to infective disease or tumours by vaccination, cytokine administration, adoptive transfer of immune cells or antibody administration.

Initial induced damage   Ionizing-radiation-induced damage after chemical modification (indirect ionization and radical scavenging by thiols or DNA proteins) but before enzymatic repair (i.e. within microseconds of a photon/nucleus interaction).

Inositol   Cyclic sugar molecule with six hydroxyl groups that forms the hydrophilic head group of inositol phospholipids.

Inositol phospholipids (phosphoinositides)   One of a family of lipids containing phosphorylated inositol derivatives. Although minor components of the plasma membrane, they are important in signal transduction in eukaryotic cells.

In situ **hybridization**   The use of labelled single-strand RNA or DNA probes* to detect the presence of complementary* sequences in a cell or tissue.

Integration   Incorporation of foreign (e.g. viral) DNA into host-cell DNA.

Interleukin-2   A protein secreted by activated T lymphocytes that stimulates proliferation of lymphocytes and activates cytotoxic functions of macrophages and lymphocytes.

Introns   Transcribed* sequences spliced* out of mature mRNAs*.

Jumping   A modification of the walking* technique. Clones with large interstitial deletions* are used so that areas of the genome some distance apart can be covered.

Karyotype   The chromosomal composition of a cell.

kb   Kilobases.

Kinase   An enzyme that catalyses the addition of a phosphate group on to a specific residue of a protein or nucleic acid.

Leucine zipper Structural motif seen in many DNA-binding proteins in which two alpha-helices from separate proteins are joined together in a coiled coil, forming a protein dimer.

Ligand A molecule (e.g. growth factor*) that specifically binds to a receptor.

Lineage The developmental ancestry of a particular cell type.

Linkage (genetic or locus linkage) Cosegregation of the two genetic loci*, e.g. coding or non-coding on the same chromosome, at frequencies greater than would be expected at random.

Locus Position on a chromosome.

Lod score A measure of the evidence for linkage. Logarithm to the base 10 of the ratio of the probability of the observed data given linkage divided by the probability of the observed data given no linkage. Lod score of > 3 is statistically taken as evidence for linkage.

Loss of heterozygosity Loss of one allele* at a locus, which can be followed by duplication of information from the remaining locus, leading to homozygosity.

Matrix *see* **Nuclear matrix**

Messengers The term applied to molecules (generally small) that can transmit signals between and within cells.

Methylation Modification of a base by addition of a methyl group. Conversion of cytosine to S-methylcytosine is thought to be associated with transcriptional inactivity of a gene in or near which it occurs.

Missense mutation A mutation where one or more base pairs are replaced by different ones, resulting in a codon that encodes a different amino acid.

mRNA Messenger RNA, the product of DNA transcription* and hnRNA* splicing* that serves as a template for protein translation*.

Multipotent A cell with the capacity to differentiate along one of several cell lineage pathways (*cf.* unipotent, a cell with the capacity to differentiate along only one pathway).

Mutagen A physical or chemical agent that introduces a non-lethal change in the cell's DNA sequence.

Mutation Any alteration in DNA sequence, as a result of point mutation*, deletion*, translocation*, etc.

N-CAM Neural cell adhesion molecule, a plasma membrane glycoprotein expressed on the surface of nerve and glial cells involved in cell–cell adhesion (a member of the immunoglobulin family).

Nested primer A primer used to increase specificity of sequencing or reamplification of a PCR product, located between the two original primers.

Non-coding DNA The 95%+ of the cell's DNA that does not code for amino acid sequences or structural RNAs.

Nonsense mutation An alteration in the DNA sequence resulting in the formation of a stop codon in an open reading frame, causing premature termination of translation.

Northern blot   Standard technique for identifying specific RNA molecules *(see* **Southern blot**, named after the man who first invented this method for DNA analysis).

Nuclear matrix   Scaffold postulated as structural support for chromatin loops and many nuclear enzymes, e.g. topoisomerases.

Nucleoid   Histone*-free DNA prepared from a cell lysed in non-ionic detergent and high salt (2 M NaCl). Nuclear matrix* attachments are preserved.

Nucleosomes   Histone* complexes (octamers) around which the DNA helix is wound as the first level of DNA packing*.

O^6-alkylguanine-DNA alkyltransferase   An enzyme that removes alkyl groups from the O^6 position on guanine following alkylation. It is inactivated by this transfer, i.e. it acts as a suicide inhibitor.

Oligonucleotide   A sequence of several nucleotides.

Open reading frame (ORF)   A sequence of translatable codons* not interrupted by stop codons*, which could therefore code for a polypeptide.

p, q   The short arm (**p**) and long arm (**q**) of a chromosome.

P-glycoprotein   Membrane glycoprotein, coded by MDR, which functions as an active drug efflux pump.

Packing   The process by which centimetres of chromosomal DNA are folded and compressed into a specific higher-order structure in order to fit into the volume of a cell nucleus.

Paracrine   A mechanism of growth stimulation involving the secretion of a growth factor* that interacts with a specific plasma membrane receptor* on a neighbouring cell; *cf.* **endocrine**.

Pedigree   A family tree.

Penetrance   The likelihood that a disease will occur as a result of the presence of a predisposition gene.

Phage (bacteriophage)   A bacterial virus.

Phenotype   The biological expression of a cell or organism's genotype*, e.g. cell morphology, surface receptors expressed.

Phosphorylation/dephosphorylation   The addition/removal of a high-energy phosphate group to/from a specific residue in a protein, resulting in the modulation of some specialized function, e.g. signal pathways*.

Physical map   A map based on actual distances in base pairs between loci*, as opposed to a genetic map, which allocates distances between loci* on the basis of recombination* frequency or, equivalently, genetic linkage* data.

Plasmid   A double-stranded circle of DNA capable of being autonomously replicated in bacteria; useful for DNA cloning*.

Point mutation   Substitution of one base by another.

Polyglutamation   Metabolism of folates or antifolates involving the addition of glutamic acid residues. Leads to retention in the cell and may alter enzyme kinetics.

Polymerase chain reaction (PCR)   An *in vitro* method that uses enzyme synthesis to exponentially amplify specific DNA sequences.

Polymerases   Enzymes that catalyse the addition of deoxyribonucleotides (G, C, A or T DNA) or ribonucleotides (G, C, A or U in RNA) to the 3′ end of a nucleotide chain as part of DNA replication* or RNA transcription*.

Polymorphism   Multiple alternative forms of a protein, e.g. G6PD polymorphism, an RNA or DNA sequence occurring naturally in a population. Used in linkage* analysis.

Postmitotic   Describing a mature cell that can no longer undergo cell division, e.g. neuron.

Primer   Short oligonucleotide that binds to a specific single-stranded target nucleic acid sequence, enabling polymerase to initiate strand synthesis.

Proband   The initially ascertained case in a pedigree or the reference case from which the family is extended.

Probe   A short, specific DNA sequence labelled with 32p or biotin that can be used to detect complementary* sequences in 'test' DNA.

Prokaryotic cells   Cells lacking a cell nucleus, e.g. bacteria.

Promoter/enhancer sequences   DNA sequences that are control points for gene transcription*. Promoter sequences are usually in the vicinity of upstream exon or in the first exon*; enhancer sequences may be many kb* upstream* or downstream* of the gene.

Proofreading ability   The ability of an enzyme to detect and correct mistakes that it has made in DNA synthesis.

Protein kinase   Family of enzymes catalysing the transfer of a high-energy phosphate from ATP to specific residues on proteins, one of the major mechanisms for regulation of protein function.

Proto-oncogene   A cellular gene whose alteration has been shown to be involved in malignant transformation.

Provirus   A viral genome integrated into a host–cell genome.

Ras   One of a large family of GTP-binding proteins that are involved in the transmission of signals from cell-surface receptors to relevant intracellular targets.

Receptor   Protein that binds a specific extracellular signalling molecule (ligand) and initiates a response in the cell. Cell surface receptors, such as the insulin receptor, are located in the plasma membrane, with their ligand-binding site exposed to the external medium. Intracellular receptors, such as steroid hormone receptors, bind ligands that diffuse into the cell across the plasma membrane.

Recessive   Refers to the member of a pair of alleles that fails to be expressed in the phenotype when the dominant member is present. Also refers to the phenotype of an individual that has only the recessive allele.

Recombination   'Crossing over', a mechanism of exchange of genetic material between a pair of homologous chromosomes.

Renaturation of DNA   The ability of denatured* DNA to resume its normal double-strand conformation *(see* **Denaturation**).

Repetitive sequences   Approximately 30% of genomic* DNA consists of repeated, non-coding* nucleotide sequences, i.e. tandem repeats* or satellite DNA, with no known functions.

Replication   Duplication of genomic* DNA during S phase.

Restriction endonuclease   A group of endonucleases each of which cleaves double-stranded DNA at a specific 'recognition site' ('restriction site') determined by the exact DNA sequence. Names indicate the bacterium of origin, e.g. the endonuclease *Eco*RI originates from *E. coli*.

Restriction fragment-length polymorphism (RFLP)   A polymorphism* in the size of restriction fragments due to a sequence difference between alleles*, usually in non-coding* regions.

Restriction fragments   The products of digesting DNA with restriction endonucleases*.

Restriction map   Schema showing the positions of cutting sites of specific restriction enzymes in DNA; often used as a way of characterizing specific genomic* sequences, many kb* in length.

Retroviral transduction   Incorporation of part of a host genome*, e.g. a proto-onco-gene*, into the genome of a newly formed retroviral particle, the presumed origin of the retroviral oncogenes.

Retroviruses   RNA viruses that transcribe* their RNA into DNA using reverse transcriptase* as part of their intracellular life cycle.

Reverse transcriptase   An enzyme, present in retroviruses*, that makes a double-stranded DNA copy from a single-stranded RNA template molecule. The purified enzyme is useful in synthesis of cDNA* *in vitro*.

RNA   Ribonucleic acid.

Second messenger   Cytoplasmic molecules that transmit chemical signals across the cell following growth factor* (first messenger) binding/activation to a cell surface receptor*.

Senescence   Programmed cell ageing.

Sense strand   A sequence of nucleotides coding for a sequence of amino acids, *cf.* **anti-sense**.

Signal pathway   Sequence of chemical interactions that selectively amplify and transmit messages, e.g. growth stimulus, across the cell.

Somatic deletion   Deletion* arising *de novo* in a somatic cell, *cf.* **germ-line deletion**.

Somatic mutation   Mutation* arising *de novo* in a somatic cell, *cf.* **germ-line mutation**.

Southern blot   Standard technique for identifying specific DNA sequences; typically chromosomal DNA is digested with a restriction enzyme and the DNA fragments are separated by gel electrophoresis*. The separated fragments are denatured (converted to single-stranded form) and transferred (blotted) on to special membrane (nitrocellulose filter). A radioactively labelled probe* will hybridize* to a complementary* sequence on the filter, which is detected by autoradiography. In Northern blotting, RNA molecules are separated by electrophoresis without prior digestion.

Splice acceptor site   Intron–exon boundary at the 3' end of the exon.

Splice donor site   Intron–exon boundary at the 5' end of the exon.

Splicing   The process whereby introns* are removed from newly transcribed RNA (heterogeneous nuclear or hnRNA*) to produce messenger RNA* (mRNA).

Src homology 2 (SH2)/Src homology 3 (SH3) domains   Protein domains that mediate protein–protein interactions relevant to intracellular signalling processes. SH2 domains are approximately 100 amino acids in size and form specific interactions with phosphorylated tyrosine residues in an appropriate amino acid sequence context. SH3 domains are smaller, made up of approximately 60 amino acids, that interact with protein motif rich in proline residues.

Start/stop codons   The AUG sequence in mRNA* coding for the amino acid methionine marks the starting point of translation*. UAA, UAG and UGA are stop codons in mRNA which terminate protein translation.

Stem cell   Relatively undifferentiated cell that can continue dividing indefinitely, producing daughter cells that can undergo terminal differentiation into particular cell types.

Supercoiling   Increase or decrease in the number of turns of one strand of DNA about the other in the double helix due to twisting of the DNA about its own axis. DNA packed on a nucleosome is supercoiled and the degree of supercoiling is altered in *vivo* by topoisomerases.

Superfamily   Structurally related genes arising by duplication and divergence of ancestral genes.

Tandem duplication *see* **Tandem repeat**

Tandem repeat   DNA sequences repeated head to tail in genomic DNA. Responsible for homogeneously staining regions (HSRs)

Taq* DNA polymerase**   The heat-stable enzyme used in PCR*, isolated from algae that live in hot springs *(see* **Polymerases)*.

Telomere   Specific DNA sequences located at both ends of a chromosome.

Topoisomerase   Enzymes that relax DNA packing* and open up the helix during replication and transcription*. **Type I**   DNA enzyme that uses a reversible single-strand break to catalyse relaxation of negative supercoiling. **Type II**   Enzyme that uses the energy from ATP to produce negative supercoiling via a reversible double-stranded break.

Transcription   Copying of DNA sequences into complementary* RNA.

Transcription factors   Specific regulatory proteins that control gene expression (transcription*) by recognizing and binding to specific DNA promoter–enhancer* sequences nearby.

Transduction   The transfer of a chemical signal across the cell, e.g. conversion of a growth factor* (first messenger) signal outside the cell into a cytoplasmic signal carried by second messenger molecules.

Transfection   Uptake of foreign DNA by a cell (*in vitro*).

Transfer *see* **Southern blot**

Translation   Protein synthesis based on an mRNA* template.

Translocation   Exchange of chromatin* between chromosomes.

Tumour suppressor gene   Gene whose normal function is involved with the suppression of cell proliferation.

Tyrosine kinase *see* **Protein kinase**

Upstream   Beyond the 5' end of a gene sequence, i.e. preceding the start of coding sequence; *cf.* **downstream**.

Vector   An independently replicated DNA molecule, e.g. a phage or plasmid*, into which a specific DNA sequence can be integrated* and replicated*, e.g. in a bacterial host.

Viral oncogene   A gene in a virus responsible from its oncogenic effects in an animal host.

Walking   A technique of mapping segments of DNA (up to several hundred kb*) through the identification of overlapping DNA fragments in a genomic DNA library*.

Western blot   A technique for identifying specific protein species, analogous to Southern* and Northern blotting*.

Index